0 010 442 017
AF742862
WITHDRAWN
...ORT UNIVERSITY
LIBRARY

98

007

DRAWING AND MATERIALS:
V. E. BOXALL

Recent amendments to BS 308: 1964 should be noted in connection with this book. The following abbreviations are now recommended:

DIA *for diameter* when used in note form

Ø *for diameter* in dimensioning. It precedes the numeral, e.g. Ø 27

R *for radius* when used for dimensioning. It precedes the numeral, e.g. R 15

DRAWING AND MATERIALS

To My Wife and Family

DRAWING AND MATERIALS

A textbook for Engineering Technicians

SI Units

V. E. BOXALL C. Eng., M.I.Mech.E., M.I.E.D.

Formerly Lecturer, Southgate Technical College, London

Artwork and cover design by
GREGORY F. WHITELEY and
M. DE LA MARE

EDWARD ARNOLD

© V. E. BOXALL 1971
First published 1971
by Edward Arnold (Publishers) Ltd
41 Maddox Street, London WIR OAN

ISBN 0 7131 3249 3

All Rights Reserved. No part of this publication may be reproduced, stored in a retrieval system, or transmitted in any form or by any means, electronic, mechanical, photocopying, recording or otherwise, without the prior permission of Edward Arnold (Publishers) Ltd.

LEICESTER POLYTECHNIC
LIBRARY
Acc. No.
Date 20-9-82
Loc./Form
Class. 604.24
BOX

CITY OF LEICESTER POLYTECHNIC
LIBRARY

744.62
B

Printed in Great Britain by
Fletcher & Son Ltd, Norwich

Preface

This book is intended mainly to cover the *Drawing and Materials* syllabus for Part I of the City and Guilds of London Institute's Course 293 for Mechanical Engineering Technicians, and related courses of the regional examining bodies. However, students on other courses, such as those for Electrical Technicians and the new National Certificate courses on Communications, Drawing and Design, will find sections which are relevant to their studies—as will some technical schools and some industrial organisations who are running their own training schemes for draughtsmen and technicians.

Drawing and Materials cover a broad spread of technical knowledge which all practising engineers use daily, at every level of the manufacturing industry. This book sets out the basic details in a form suitable for students; it was written from first hand experience gained in establishing the Drawing and Materials Section at Southgate Technical College with my colleague Leslie S. Delaforce.

In the chapters on Drawing, the text has been kept to a minimum and the fundamentals of geometry and projection have been presented as worked examples in simple programmed stages.

As Britain is now committed to changing completely to the metric system, all drawings will soon be dimensioned in millimetres; this follows the practice already adopted by the main manufacturing organisations for new design. For the most part, metric units have been used in the text, with imperial units inserted only where it was thought they would be helpful to the student during the period of change.

With materials, which form a very extensive subject, a concise account has been given of their history, the methods of extraction and preparation, the forms in which they are available, and their uses. Reference has been made to some of the British Standards covering them.

This information has been presented in a form easily referred to by the student. Although he would not be expected to digest all of it at a reading, it is hoped that it will spark off the desire to obtain a wider and deeper knowledge of materials by reading more specialised works on individual materials and manufactured components, together with the related Standards.

V. E. Boxall

Acknowledgements

I should like to thank all the organisations and individuals who kindly supplied information and materials that helped me in writing this book. I am also grateful for the permission granted by the British Standards Institution, 2 Park Street, London W.1, to include illustrations and tables based on information contained in their publications. Copies of the complete British Standards may be obtained by post from the British Standards Institution, 110-113 Pentonville Road, London N.1.

Contents

		Page
1	Who Requires Drawings?	2
2	The Making of Engineering Drawings	4
3	Metrication	24
4	Geometry	32
5	Practical Tangency	54
6	Orthographic Projection	62
7	Fits and Clearances	80
8	Screw Threads and Fasteners	92
9	Interpenetration and Sheet-Metal Development	108
10	Isometric Drawing, Oblique Drawing and Sketching	144
11	Jigs and Fixtures	156
12	Materials: Selection, Properties and Mechanical Testing	178
13	An Introduction to Plastics	184
14	Iron and Steel	198
15	Copper and Copper Alloys	206
16	Aluminium and Aluminium Alloys	212
17	Other Non-Ferrous Metals	218
18	Bearings and Lubrication	222
19	Insulating Materials	234
20	Machine Drawing	236
	Index	243

Who requires drawings

Chapter 1

Perhaps you remember your first journey on the bus or underground railway when, not being too sure of your way, you asked someone for directions, only to be further confused by a volume of words and much arm waving. Later you found that on every underground station, and in most bus shelters, there are those useful maps, often just line drawings, that make it so simple to see your whereabouts and the direction you should take. You have used a drawing. Information has been conveyed to you by a picture instead of by speech or written words. A foreign visitor would find it much easier to follow the simple map drawing than laconic instructions from busy ticket collectors.

It has been said that a picture is worth ten thousand words. The value of the picture is very apparent in this modern age when, with every piece of equipment purchased—whether a camera, plastic model kit, radio or car—an illustrated pamphlet or book is supplied to give assembly or working instructions.

Drawing is one of our means of communication. We have other general means of communication, such as speech and writing, but these have the disadvantage of being understood only if they are in a familiar language. Drawing is a universal language and can be understood regardless of nationality; it is a precise language.

ENGINEERING DRAWING

Engineering drawing is a system for communication between engineers, and our modern industries are very dependent on it. Mass production methods would be impossible without engineering drawings to convey exactly the ideas of the designers to the production engineers.

The design and drawing office is the heart of any large industrial organisation. It provides the creative ideas and communication, in the form of engineering drawing, that enable the money subscribed by shareholders to be utilised by skilled technicians, craftsmen and machines for production.

In industry, engineering drawing is the only academic subject which is directly involved with every article produced. For every article manufactured there is at least one drawing. It may be a supersonic aircraft, a lawn mower or a common cigarette packet; they all require finely produced, accurate drawings. These drawings must be clear in every detail so that the technicians in factories and laboratories can read the design requirements without ambiguity, and so manufacture the multitude of products our everyday existence demands. See example Figure 1.1

WHO NEEDS ENGINEERING DRAWINGS

Long and detailed lists of those who require engineering drawings could be given but are not necessary here. A study of Figure 1.2 will indicate the areas where engineering drawings are essential for production and industrial communication.

Figure 1.1

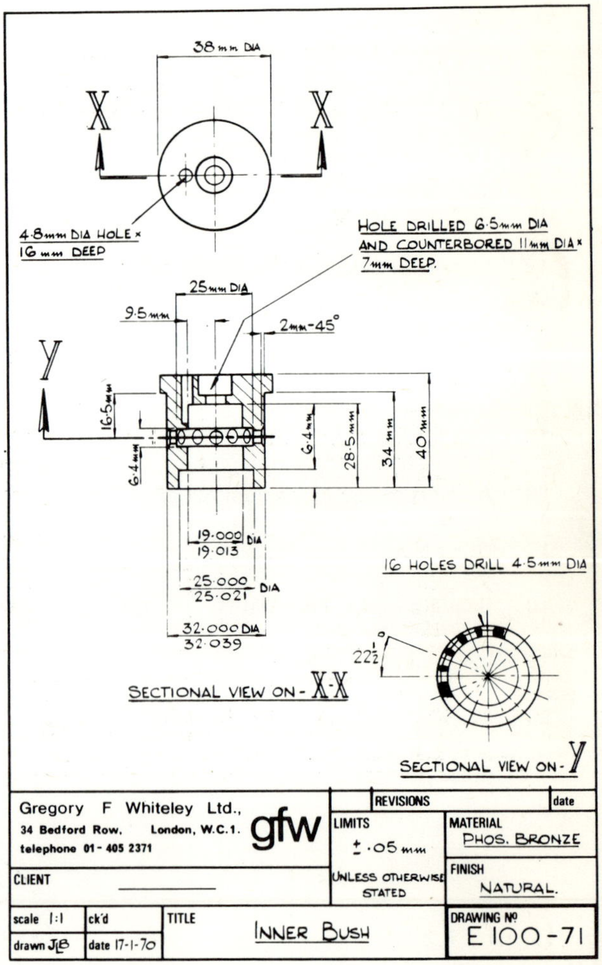

NOTE
The drawing office is the most important department in modern engineering. The drawing office interprets designers' ideas into drawings which are the universal means of communication in engineering.
All buying and production departments are immobile until instructions in the form of drawings are available.

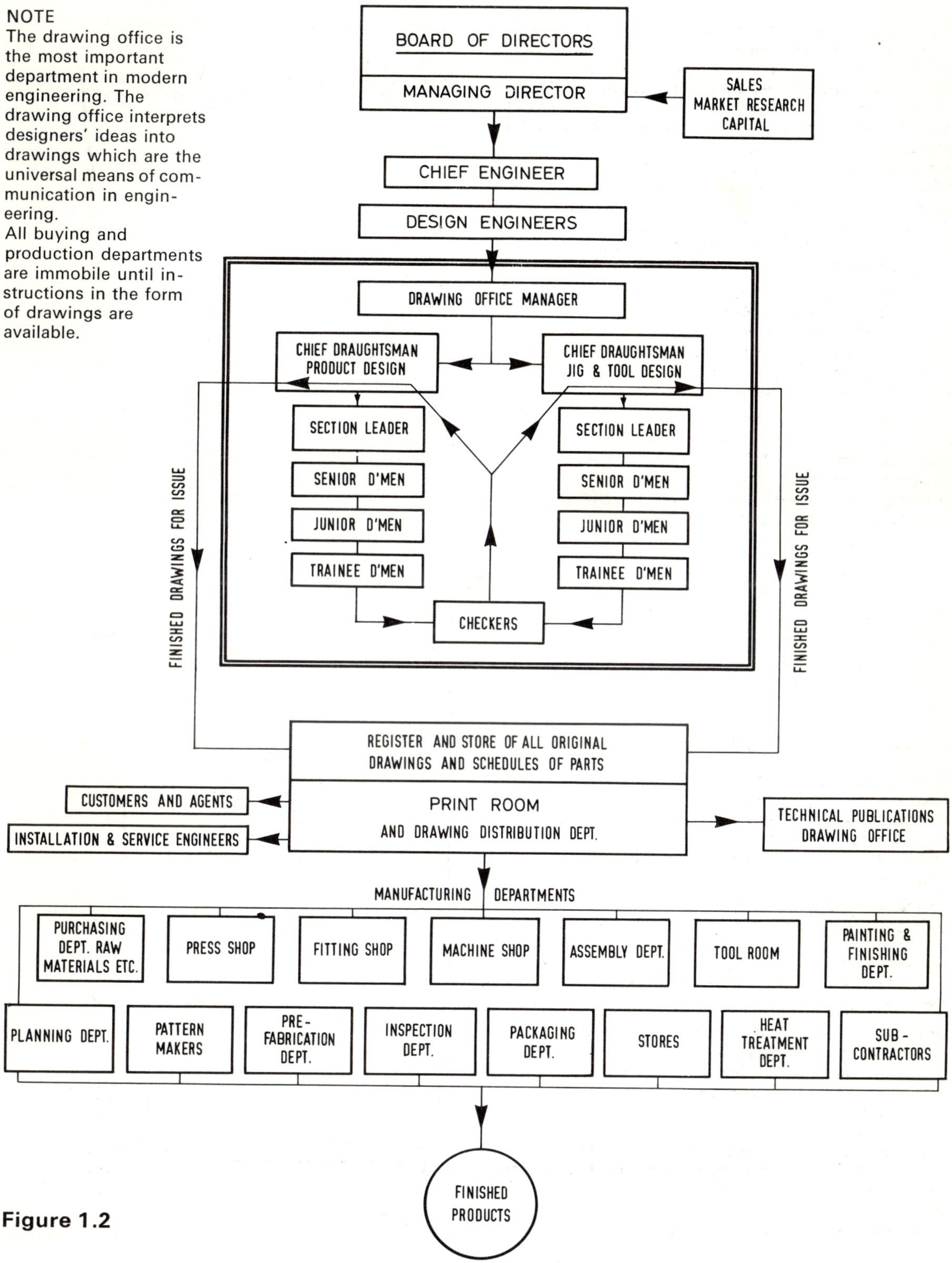

Figure 1.2

Production Route Diagram in an Engineering Works.

To make an engineering drawing one does not need the talent of an artist because there is no need for freehand work or shading to give depth or lighting effects. All lines are drawn to scale and are produced with instruments.

The abilities required to produce professional engineering drawings are:

1. To think in three dimensions.
2. To master the techniques of projection.
3. To produce, at good speed, clean accurate lines of correct thickness.
4. To print rapidly clear block capitals and figures for dimensioning and for written instructions.
5. To understand and apply the conventions recommended in BS 308, which is the British Standard on drawing office practice.
6. To practice, until a skill has been developed, producing in minimum time a drawing of professional appearance.

An engineering drawing is a communication between the designer and the craftsmen carrying out the work. It must be made clearly, precisely and accurately, containing the minimum of drawing effort and giving information once only. It must be complete in all instructions so as to leave production personnel in no doubt of the exact requirements of the designer.

The draughtsman should be proud of his finished work, but should not regard his drawings as works of art; they are communications which must be produced quickly, skilfully and efficiently. It must be remembered that producing excellent engineering drawings does not necessarily make one a designer, but it is impossible to be a good engineering designer without the ability to produce good engineering drawings.

DRAWING BOARDS

Experience has shown that, among the many expensive sophisticated drawing boards and dual-purpose equipment on the market, there is much to recommend for student use the simple block-board drawing board of A2 size (616 mm by 445 mm) or half-imperial size (23″ by 16″), providing it can be raised a few inches at the back and fitted with a polished wooden tee square. This equipment is cheap and easily replaceable, and can be used on any surface of table height without supplying special stools. It is the only practical equipment available for examinations with a large student entry.

Figure 2.2 shows a wooden block glued to a cheap wooden drawing board to make a more satisfactory piece of equipment.

PENCILS

Buy good-quality pencils and keep them for drawing only. The grade best suited to your hand is only found by experience. For cartridge paper, grades of F, H and 2H or 3H should be suitable. Pencils should be sharpened to a chisel point as shown in Figure 2.1, using the sandpaper block to produce and maintain the sharp edge. With this type of point, thin and thick lines can be drawn with the same pencil; it can also provide a characteristic printing of thick/thin type. Some draughtsmen, however, prefer a cone point for printing.

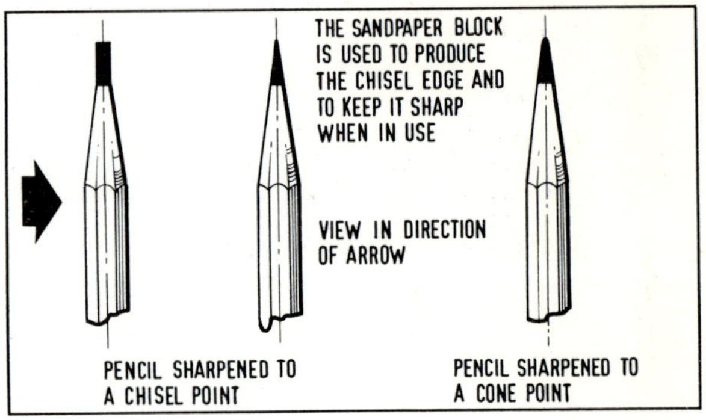

Figure 2.1

DRAWING INSTRUMENTS

It is wise to purchase the best drawing instruments one can afford, and it is often cheapest in the long run to buy a few necessary instruments of professional quality rather than a large box full of shiny cheaper instruments, only to find in a very short time that the latter are inadequate. Buying cheap instruments is like buying cheap fountain-pen sets at Christmas; these sets usually find their way into the dustbin by January, whereas a good pen may last a lifetime.

Compasses

Figure 2.3 shows two small spring bow compasses. Either is satisfactory, the one with the centre adjustment being capable of drawing slightly larger circles.

Two large compasses are also shown, both of which are suitable. The one with the centre adjustment is limited to drawing circles of approximately 125 mm (5″) radius, but the rigid construction allows heavy pressure to be applied. With the other type, heavy pressure will cause spreading of the compass legs, particularly when large circles are being drawn. However, with care and the use of a softer lead, these compasses will produce good-quality work, and circles of more than 125 mm radius can be drawn. They can be fitted with an extension bar to produce the larger radius which is occasionally required.

All compasses must be fitted with shouldered needles, as illustrated, to prevent digging a 'crater' at the centre of a drawing consisting of a number of concentric circles; this provides for greater accuracy.

Set squares

These come in many sizes and, for students producing drawings of A2 or half-imperial size, the illustration gives an indication of the size of set square to purchase. It is also wise to buy those with square edges; the bevel-edged squares are more expensive and are used mainly for ink work.

French curve

Freehand work is not tolerated in engineering drawing, and this piece of equipment allows curves other than circular arcs to be drawn. Experience has proved that it is far more satisfactory than other marketed devices. It is also useful for drawing graphs in science and mathematics problems.

Ruler

To start a drawing course, an inexpensive polished wooden school ruler engraved with centimetre and millimetre divisions and with 1/8″, 1/10″, 1/12″ and 1/16″ divisions is satisfactory. More expensive scales and flexible steel rules can be bought later to suit one's particular requirements.

Eraser

A soft white pencil rubber is necessary for use on cartridge paper. The hard green draughtsman's rubber will destroy the paper surface.

Sandpaper block

This is necessary to produce the chisel edge to one's pencils. A small smooth file is a good substitute.

Drafting tape or clips

To produce accurate engineering drawings, the paper must be firmly attached to the drawing board with adhesive tape or clips. Either method is satisfactory but tape gives a rather more secure fixing.

Figure 2.2

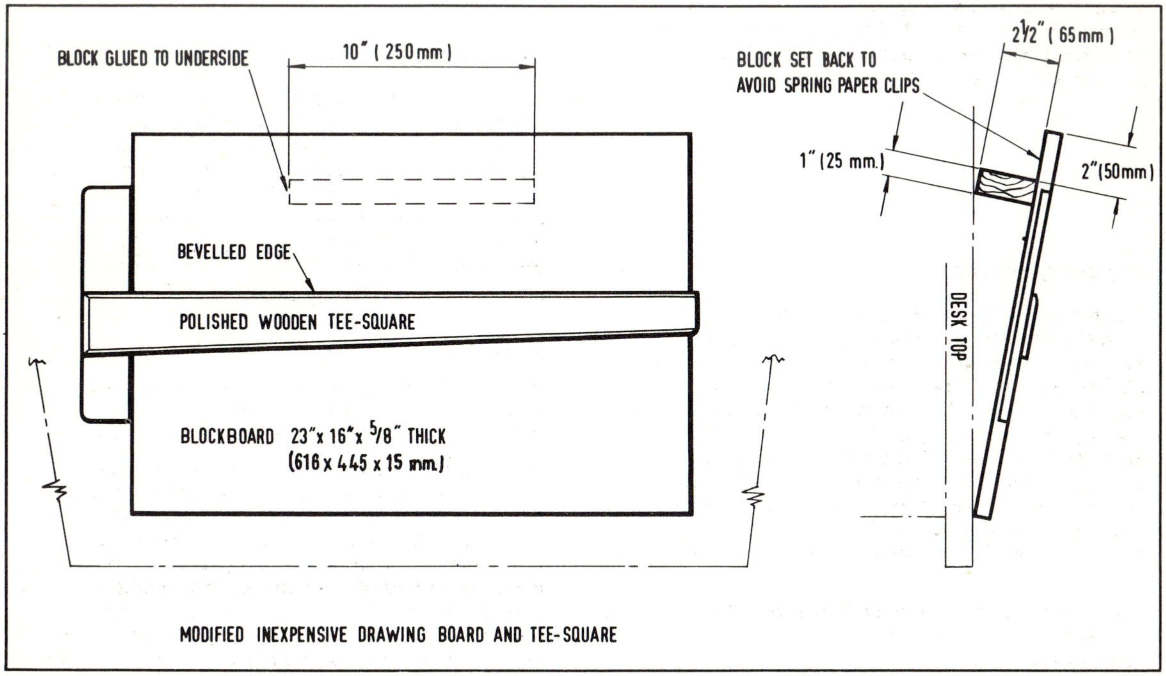

The following is a minimum list of instruments that a student should provide for himself:

One pair of small spring-bow compasses

One pair of large compasses

One 60°–30° set square

One 45° set square

One french curve (No. 15)

Pencils of F, H and 2H or 3H grades of hardness

One polished wood ruler engraved with centimetre and millimetre divisions and with 1/8″, 1/10″, 1/12″ and 1/16″ divisions

One soft white rubber eraser

One sandpaper block or small smooth file

One roll of drafting tape or two spring drafting clips

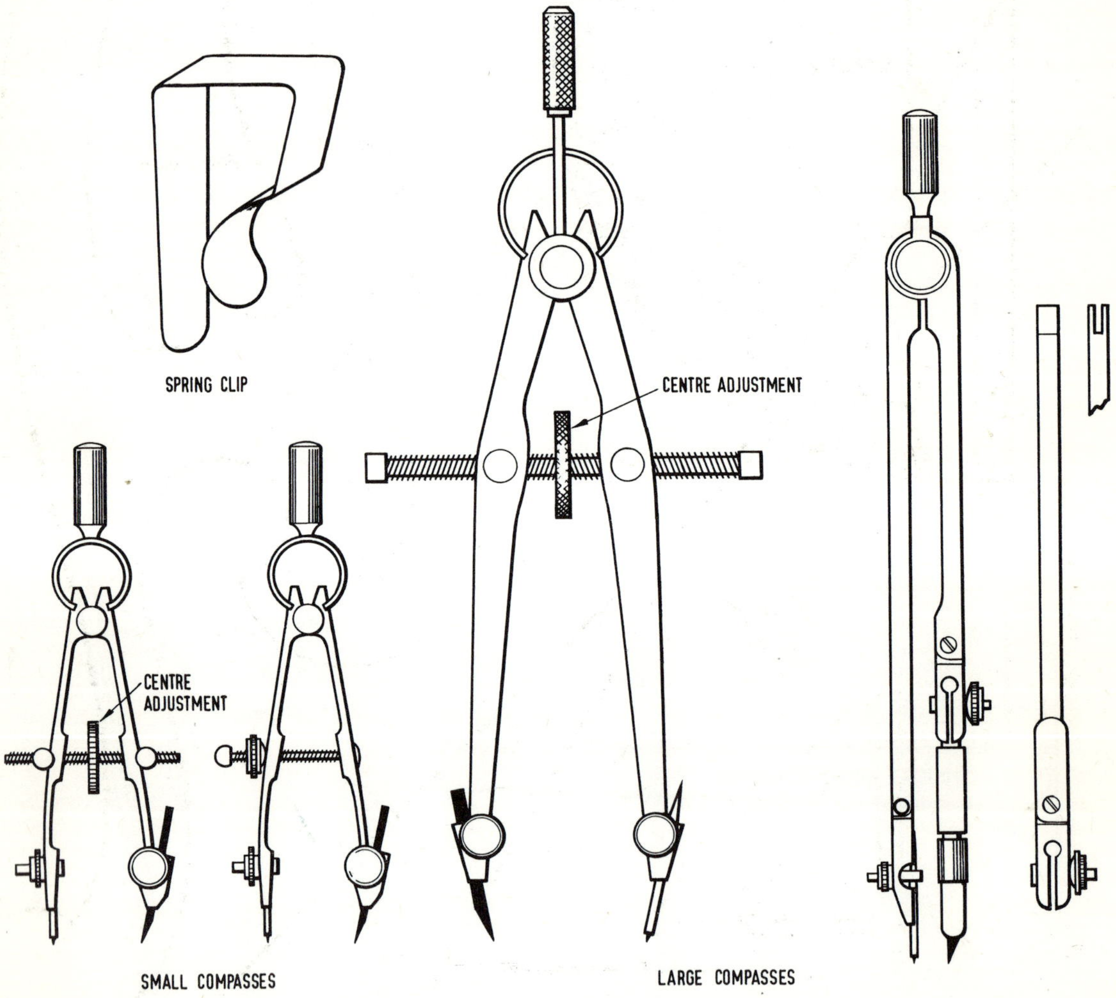

Figure 2.3

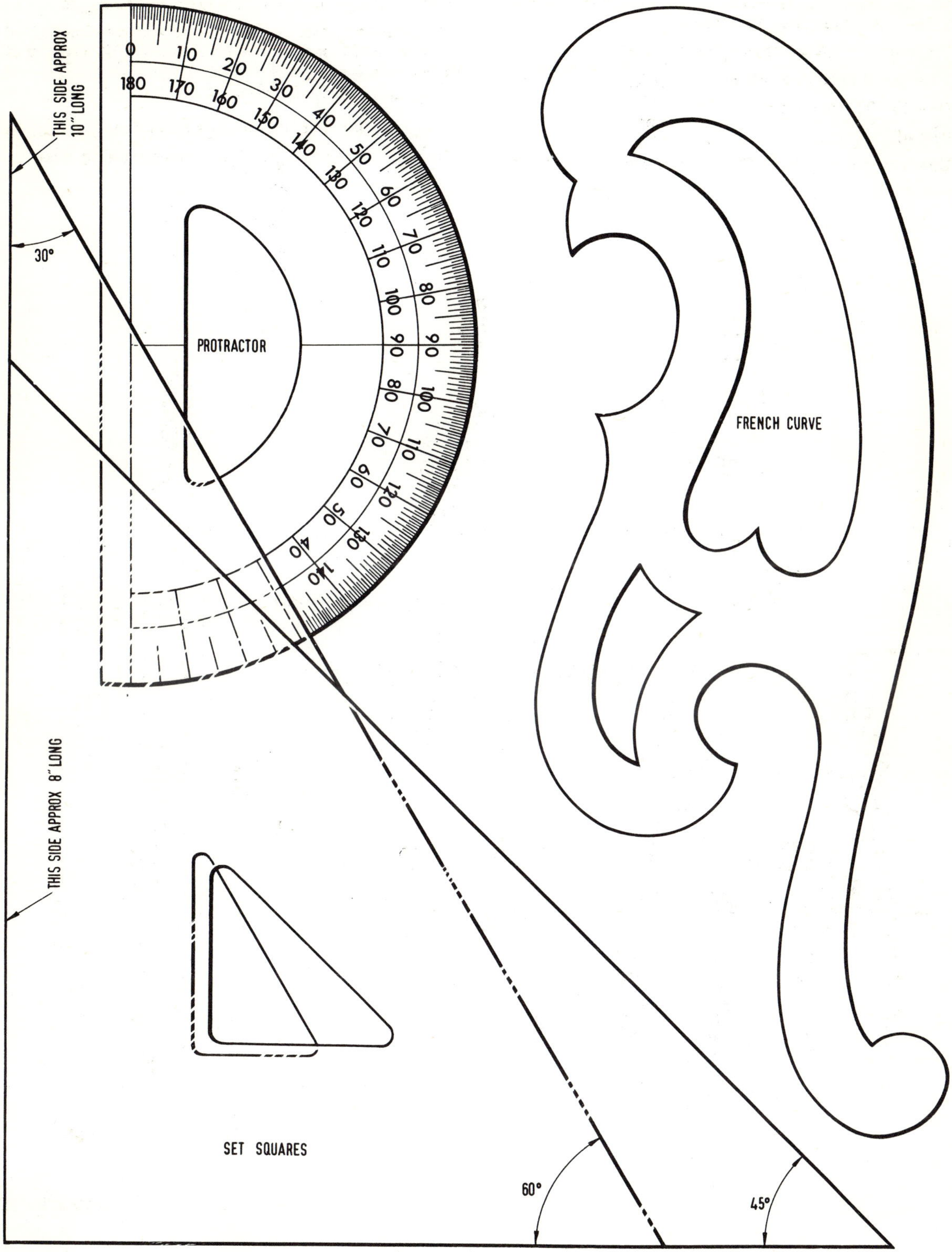

Figure 2.3 ***(cont.)***

The making of engineering drawings

Chapter 2

DRAWING PAPER

It has been found that the best practical results for student work are achieved on inexpensive cartridge with a reasonably smooth surface, of A2 size (594 mm by 420 mm), or half-imperial size (22″ by 15″) if still used. In some colleges paper with a printed border and title block is supplied to students; this not only saves time but adds a professional touch to a series of course drawings.

ORTHOGRAPHIC PROJECTION

This is a system of drawing universally used in engineering to illustrate an object with sufficient clarity and skill to enable any manufacturer to interpret precisely a designer's requirements and to convert them into manufactured parts. It is a method of accurately defining the shape of an object by a system of separated views so arranged that each view is related to the others. The basic views are drawn true to scale and are taken at 90° to each other; when set on paper they are said to be drawn in the principal planes.

There are at present two separate methods of relating the views to each other, namely *first-angle projection* and *third-angle projection*.

Third-angle projection is the system used exclusively in Canada and the United States. First-angle projection is used in Britain and European countries but since World War II many companies, particularly those with world-wide export interests and those influenced by American technology and capital, have changed from first to third angle. This change by industrial groups, coupled with the change to metric units, could be the beginning of acceptance of third-angle projection as the universal system.

HOW DOES FIRST-ANGLE PROJECTION DIFFER FROM THIRD-ANGLE PROJECTION?

The actual drawn views in both methods are exactly the same; the difference lies in the relative position of the views to each other. This can be seen in the final configuration of the lettered sides of the cube laid out in the two systems. For a simple understanding of the two methods make up a cardboard cube and letter the sides as shown in Figure 2.4.

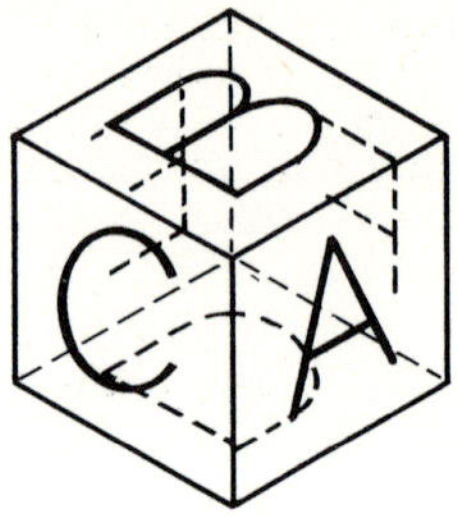

Figure 2.4

Cardboard cube made from white card and lettered

FIRST-ANGLE PROJECTION

1. Place the cube on a piece of paper with face A uppermost, and on the paper directly below draw the view seen.
2. Roll the cube away from you through 90°. Side D will be uppermost. Draw this view on the paper directly below and return the cube to the *starting position with side A uppermost.*

3. Roll the cube towards you, bringing face B uppermost. Draw this view and return the cube to the starting position.

4. Roll the cube to the left, bringing face F uppermost. Draw this view and return the cube to the starting position.
5. Roll the cube to the right, bringing face C uppermost. Draw this view.

6. With face C showing, roll the cube to the right again to bring face E uppermost. Draw this final view.

By rolling the object through a series of 90° turns, a set of related views has been produced illustrating each of the six sides of the cube. An orthographic drawing has been produced in *first-angle projection*, with the views in the six principal planes of projection.

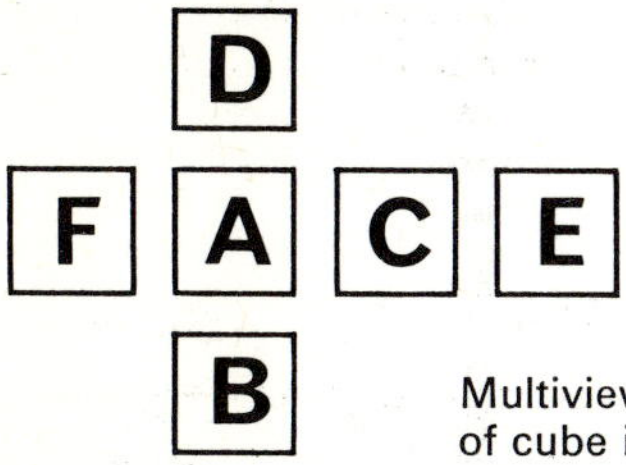

Figure 2.5

Multiview arrangement of sides of cube in first-angle projection

In practice, the views are separated by spaces to allow for dimensioning and instructions. Figure 2.5.

The four views, F,A, C and E are referred to as elevations; B is referred to as the plan and D as the reverse plan.

To illustrate an object it is not necessary to draw all six views; sometimes one view is enough. In other cases, when the object is of complex shape, many views are required, but as long as the object is rolled to bring the required surface into view the relationship of the views is maintained in correct first-angle projection.

THIRD-ANGLE PROJECTION

Take the cardboard cube as illustrated and cut all the edges except the five edges drawn thickly in Figure 2.6.

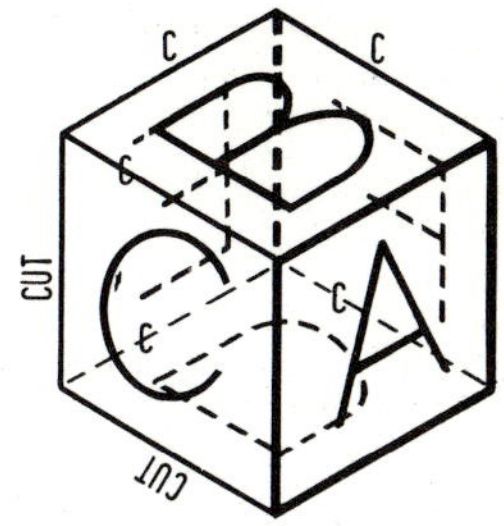

Figure 2.6

When the sides of the cube are opened up, another relationship between the views is shown, producing an orthographic drawing in third-angle projection with views in the six principal planes of projection (Figure 2.7).

Figure 2.7

Multiview arrangement of sides of cube in third-angle projection

The views C, A, F and E are referred to as elevations; B is referred to as the plan and D as the reverse plan. As in first-angle projection, one or many views may be drawn as required to illustrate the subject, the views being separated by spaces to allow for dimensioning and instructions.

Chapter 6 deals more fully with projection, and explains why the two methods are given the titles *first-angle projection* and *third-angle projection*.

The making of engineering drawings — Chapter 2

TYPES OF LINE USED TO PRODUCE ENGINEERING DRAWINGS. (Figure 2.8.)

When a draughtsman makes an engineering drawing he is actually making a photographic negative. For this purpose he draws on good-quality tracing paper, or similar material which is transparent, with dense lines to enable clear photographic prints to be taken for use by those departments requiring drawings.

In all drawing exercises it is essential, even on cartridge paper, to produce drawings having the correct quality of line. The draughtsman uses three basic types of line:

1

Wide or thick line

This type of line is dense and sharp, and has no sign of furriness. It is used for visible outlines, to make the object stand out sharply in relief. It has two other minor applications: one as a wavy line to represent a break when only part of the drawing is required; the other as a thick chain-dotted line to indicate a cutting plane.

2

Narrow or thin line

On an engineering drawing all lines other than the visible outline are dense, sharp and *narrow*, being about one-third the width of a wide line. Thin lines are drawn in various ways to represent various features as indicated in Figure 2.8. It is well to note here that section-lining or hatching is produced by narrow lines usually drawn at 45°, the pitch varying according to the area to be covered.

3

Extra-thin (construction) line

This type of line is used only to lay out and construct work in the process of making a drawing. These lines are so thin that they are not reproduced by the printing process and are usually removed by the draughtsman from the finished drawing. However, particularly in the early stages of student work, construction lines should be clear and sharp and should be left on the exercise for reference purposes. Most examination questions state that they should be clearly shown.

TYPE OF PRINTING RECOMMENDED FOR ENGINEERING DRAWING

For printing, draw faint lines approximately 3 mm apart, and print in alternate spaces, making the characters with downward and horizontal strokes. An aid to printing is to place a set-square edge just below the lower guide lines, and to end all the downward strokes against it. The sample on the right is drawn by hand.

Poor printing and figures can spoil even the best drawing. It is advisable that printing should be practised on every drawing exercise; this will develop a skill in producing with reasonable rapidity a good, clear, simple style without ornament or flourishes.

Printing on engineering drawings is usually in block upright capitals approximately 3 mm high. Sloping printing is permissible, but too great a slope makes the printing difficult to read.

PRINTING ON AN ENGINEERING DRAWING IS
AS IMPORTANT AS THE PICTORIAL VIEWS OF THE
OBJECT, THEREFORE THE CHARACTERS USED MUST
BE OF A SIMPLE AND CLEAR STYLE. BLOCK
CAPITALS ABOUT 3mm HIGH ARE USED FOR
MOST PRINTED INFORMATION, TITLES OF DRAWINGS
BEING PRINTED IN LARGER CHARACTERS. BEFORE
PRINTING, DRAW THIN GUIDE LINES 3mm APART,
PRINT IN ALTERNATE SPACES, THUS LEAVING
APPROXIMATELY 3mm SPACES BETWEEN LINES
OF PRINTING. THE FIRST LETTERS OF EACH LINE
SHOULD LIE DIRECTLY BELOW EACH OTHER, AND
A THIN VERTICAL LINE WILL HELP TO ACHIEVE THIS.

TYPES OF LINE USED FOR
GENERAL ENGINEERING DRAWINGS

EXAMPLE	WHERE USED
Thick	
continuous	Visible outlines to throw drawn object in relief
wavy	Short boundary line at broken section
long chain	Section or viewing plane
Thin	
continuous	Dimension lines. Projection or witness lines. Hatching or section lining. Leader lines for balloon referencing and notes. Outlines for revolved sections.
short dashes	Hidden detail.
long chain	Centre lines. Pitch circles. Path lines indicating travel.
short chain	Developed views. Adjacent parts. Alternative position of moving parts. Feature located in front of sectioned view.
ruled line interrupted with short zigzags	Long break line.

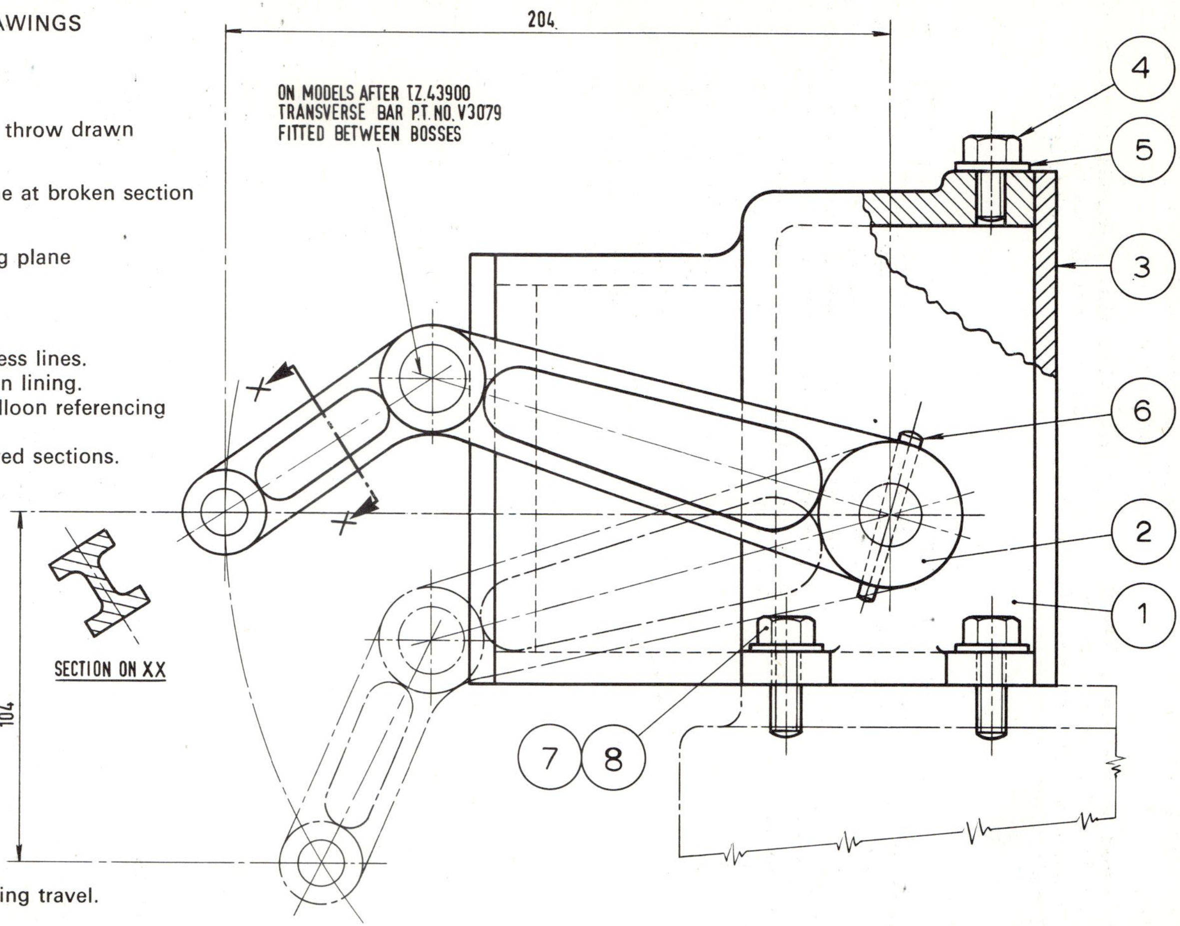

Figure 2.8

Types of line used
for Engineering Drawing.

SECTIONED AND SECTIONAL VIEWS

To illustrate certain features clearly, and especially when preparing drawings of the interior of an assembly, views are drawn as though parts have been cut away and removed. The problem of what views are to be sectioned on an engineering drawing is decided by the experience of the draughtsman. Sectioned views often reduce the amount of drawing necessary to communicate ideas with clarity and dispense with much work on hidden detail. Surfaces which have been sectioned are always hatched with thin lines at 45° to the horizontal unless there are special circumstances such as a clash of direction with the visible outline. The student should familiarise himself with the sectioning conventions recommended in BS 308.

It is important to remember several exceptions to the general rule of sectioning; where certain features are cut longitudinally by the cutting plane these are, for clarity, *not sectioned*. These features are: ribs lying in the plane of the paper; nuts, bolts and washers; shafts and rods; rivets, pins, keys and similar parts. They are shown as solid outside views.

The student often falls into the trap of interpreting the cutting plane too literally instead of applying the accepted conventions of sectioning; on study, these conventions often communicate information effectively, whereas a true representation of the cut surface would only confuse.

Remember that the features noted above as *not to be sectioned* when positioned longitudinally on cutting planes are often introduced as hazards into examination questions.

For examples on sectioning see Figures 2.9 and 2.10.

NUMERALS

The following examples show the type of numeral recommended in BS 308 and consistent with normal practice:

1 2 3 4 5 6 7 8 9 0

DIMENSIONING

Before discussing the dimensioning of engineering drawings, the importance of this subject must be clearly grasped. An engineering drawing consists of two main parts:

1. A pictorial representation to scale of the part to be manufactured.
2. Instructions in the form of dimensions and printed notes.

The student's time is largely occupied in producing pictorial views and learning drawing techniques and conventions. The subject of dimensioning is often overlooked and the dimensions squeezed in as an afterthought. It is forgotten that an excellent detail drawing without dimensions, when given to a machinist in a workshop, is not likely to result in the product that was originally intended.

No matter how carefully drawings are made they are not accurate to engineering standards. The paper on which drawings are printed is usually dimensionally unstable, due to the humidity in the atmosphere and damp printing techniques. Also drawings are often made with instruments which are not capable of the accuracy required in engineering; to draw a line with pencil and scale to an accuracy greater than plus or minus 0·5 mm without going cross-eyed is a near impossibility. Hence pictorial representation is not sufficiently accurate by itself.

On the other hand, if a part is simple in contour, a sketch on the back of an envelope, together with the correct dimensions, will enable an object to be made accurately. Without suggesting that sketching, except for simple ideas on simple objects, is a substitute for good professionally-produced engineering drawings, such an example does illustrate the importance of well-planned, accurate dimensioning.

Every drawing made presents a unique dimensioning problem. How it is dealt with depends on its function and how it relates to the whole design, on the interchangeability of the part being drawn with other production components, on recognised drawing practice, and on experience and common sense in conveying the exact requirements clearly and without ambiguity.

WHAT IS A DIMENSION?

A dimension is a numeral (together with the unit being used) that indicates the size of a particular feature, the numeral usually conveying the main part of the message. Therefore, particular attention must be paid to the form of the numerals. They must be clear, and of a size to leave the craftsman in no doubt as to the required dimensions.

Remember too, that it is becoming common practice to reduce the size of drawings when prints are taken, so a dimension printed with small badly-formed characters becomes unreadable when reduced.

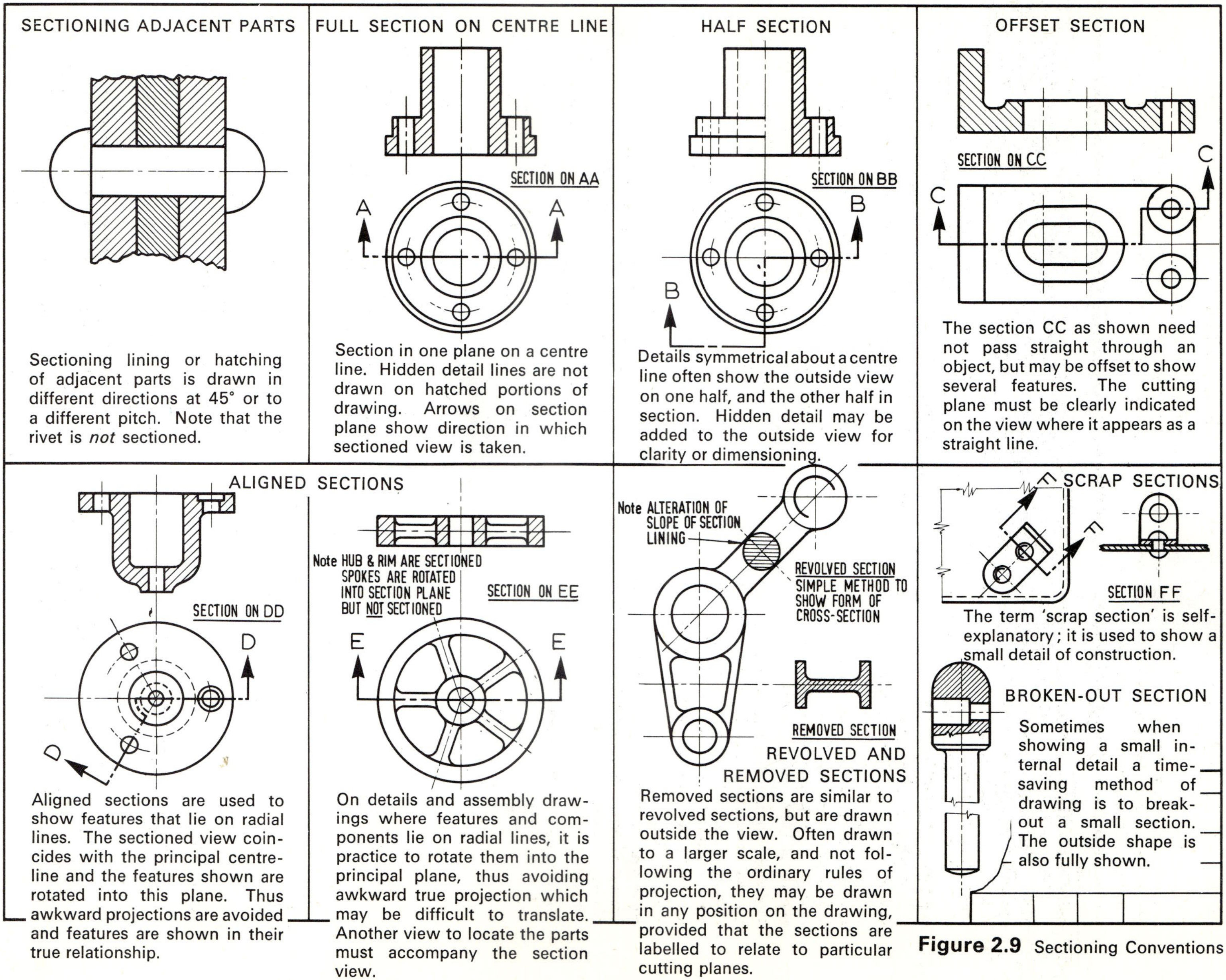

Figure 2.9 Sectioning Conventions

LINEAR DIMENSIONS (INCH UNITS)

Dimensions in feet and inches may be expressed by either the fractional or the decimal method, depending on the policy of the company producing the drawings. Below are recommended examples:

Fractional	*Decimal*
½″	·5″ (*not* 0·5)
1½″	1·5″
12½″	12·5″
19½″	19·5″
2′– 0½″ (*not* 2′–½″)	24·5″
7′– 6½″	90·5″
10′–11½″	131·5″

Decimal points should be bold and given a full letter space, and figures preceding the decimal points should indicate the total distance in inches, i.e.

131·5″ *not* 10′ 11.5″

Fractions should be larger than the whole number and the dividing line between numerator and denominator should be horizontal, except where space is restricted; only then should the line be sloped.

The method of dimensioning when metric units are used is described in Chapter 3.

HOW A DIMENSION IS APPLIED TO A DRAWING

Figures 2.11 to 2.14 show typical examples of how dimensions should be put on engineering drawings. The following should be noted.

1. Extension lines and dimension lines are narrow, sharp and clean.
2. A small gap is left between the extension lines and the features dimensioned.
3. The dimension line ends in arrow heads, which are solid and slim and about 5 mm long.
4. The extension lines end neatly just beyond the dimension lines. Untidy ragged ends must be removed.
5. Dimension lines should be well spaced, and for clarity well clear of the visible outline.
6. It is more convenient, especially in an extensive application of dimensions on a drawing, to put the dimensions above the lines rather than insert them in breaks in the dimension lines. A skeleton framework of the dimension lines is first planned, and the figures then inserted.
7. It is recommended, and usual practice, to place dimensions so that they may be read from either the bottom or the right-hand side of the drawing.
8. In good dimensioning practice all dimensions are positioned outside the visible outline. This is not always possible, e.g. for corner radii or small details in the middle of a large area.

GENERAL PRINCIPLES OF DIMENSIONING

A drawing should carry the minimum dimensions necessary to define the product. Information should appear only once.

All dimensions required for manufacture must be stated. It is not acceptable in modern practice to allow production personnel to deduce or calculate dimensions, or to measure dimensions with a rule. Any discrepancy in dimensioning revealed during production must be reported back to the design office for investigation.

To ensure correct assembly and interchangeability of quantity-produced products, dimensions are referenced to common datum features such as machined mating faces, centre lines of shafts, selected hole centres in multiple-drilled matching parts, a common design datum point or similar selected features. It is interesting to note that the overall control of dimensions of a modern aircraft, on which thousands of design personnel are engaged, is related back to a single common point which is fixed, at the outset of the project, in the centre of the fuselage.

Wherever practicable, standard sizes would be used for drilled and reamed holes, threads, nuts, bolts, pins etc. Commercially supplied stock, such as bright bar, tubes and extruded and rolled sections, should be used for work where the finish and size is acceptable. All toleranced dimensions should be carefully investigated, and the widest limits for correct functioning and assembly should be selected. Attention to these points is necessary to reduce production and inspection costs.

Production processes and methods are not stated on drawings, with the exception of quoting drill sizes where necessary.

All production drawings should contain full information for manufacture, including dimensions, material and required finish.

The student should at this stage consult BS 308 and familiarise himself with the recommendations on dimensioning.

CAST COVER WITH EXTERNAL RIBS

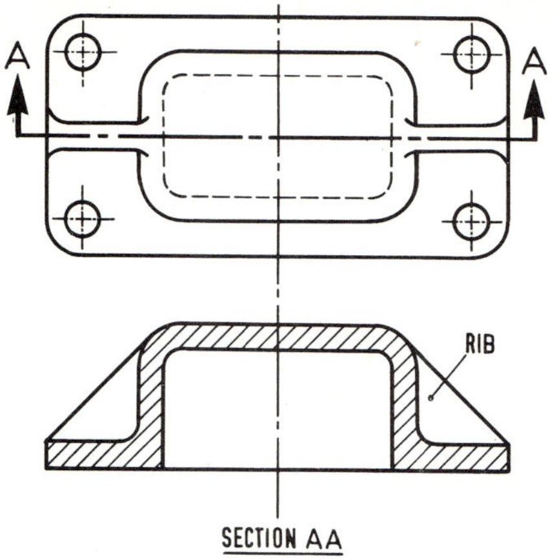

BRACKET WITH EXTERNAL WEB

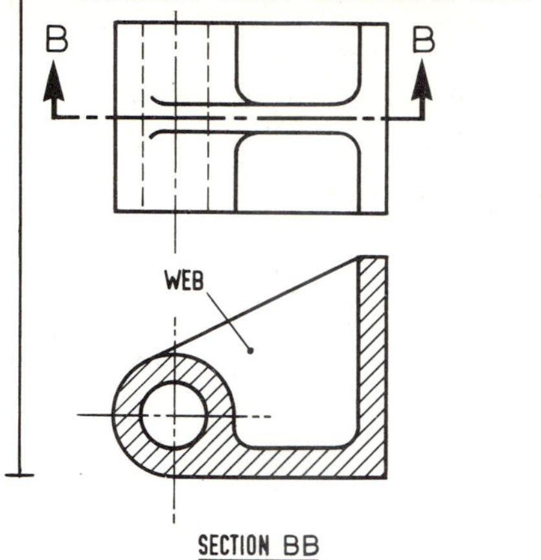

When ribs and webs in sectioned views lie flatwise in the plane of the paper, it is standard drawing practice *not* to section these parts. The view is drawn as though the cutting plane is positioned just in front of the web or rib.

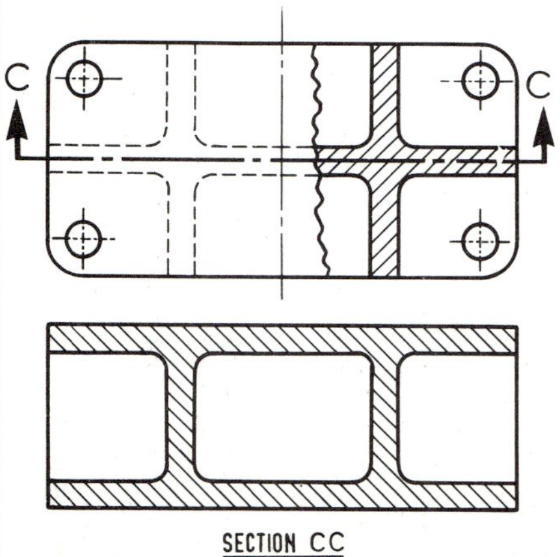

When drawing sectioned views of intersecting web members, section the webs at right angles to the plane of the paper; but *not* the web that lies flatwise in the plane of the paper.

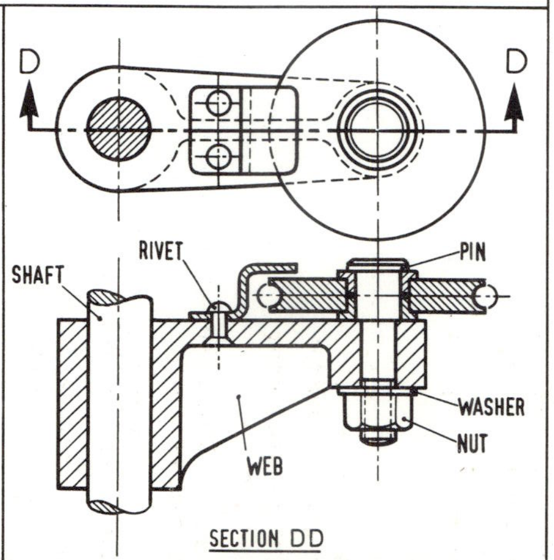

View showing elements which are *not* sectioned when cut longitudinally by the section plane, i.e. shafts, pins, rivets, nuts and webs.

Note those elements which are not sectioned although cut longitudinally by the sectioning plane.

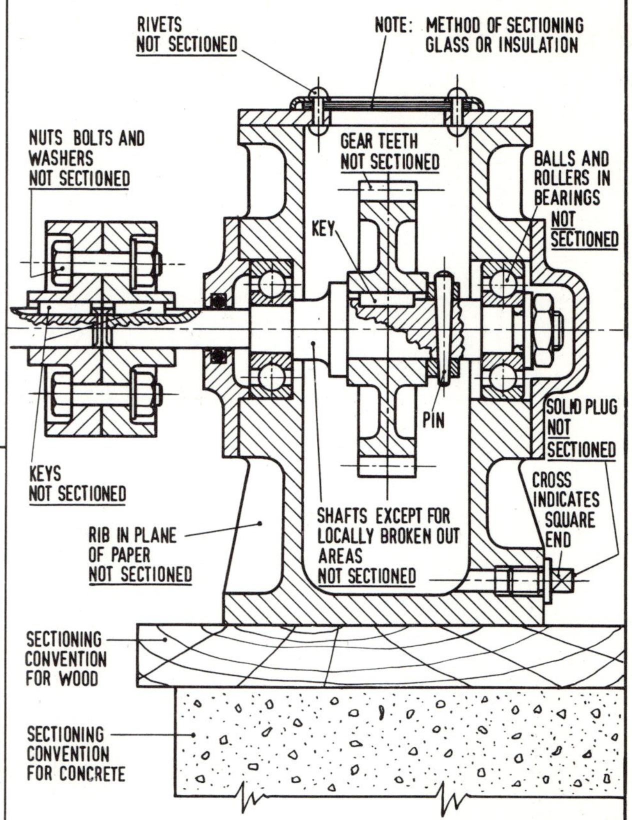

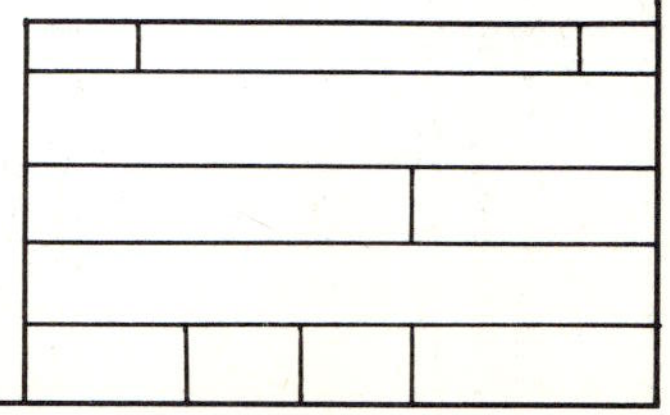

Figure 2.10

Sectioning Conventions

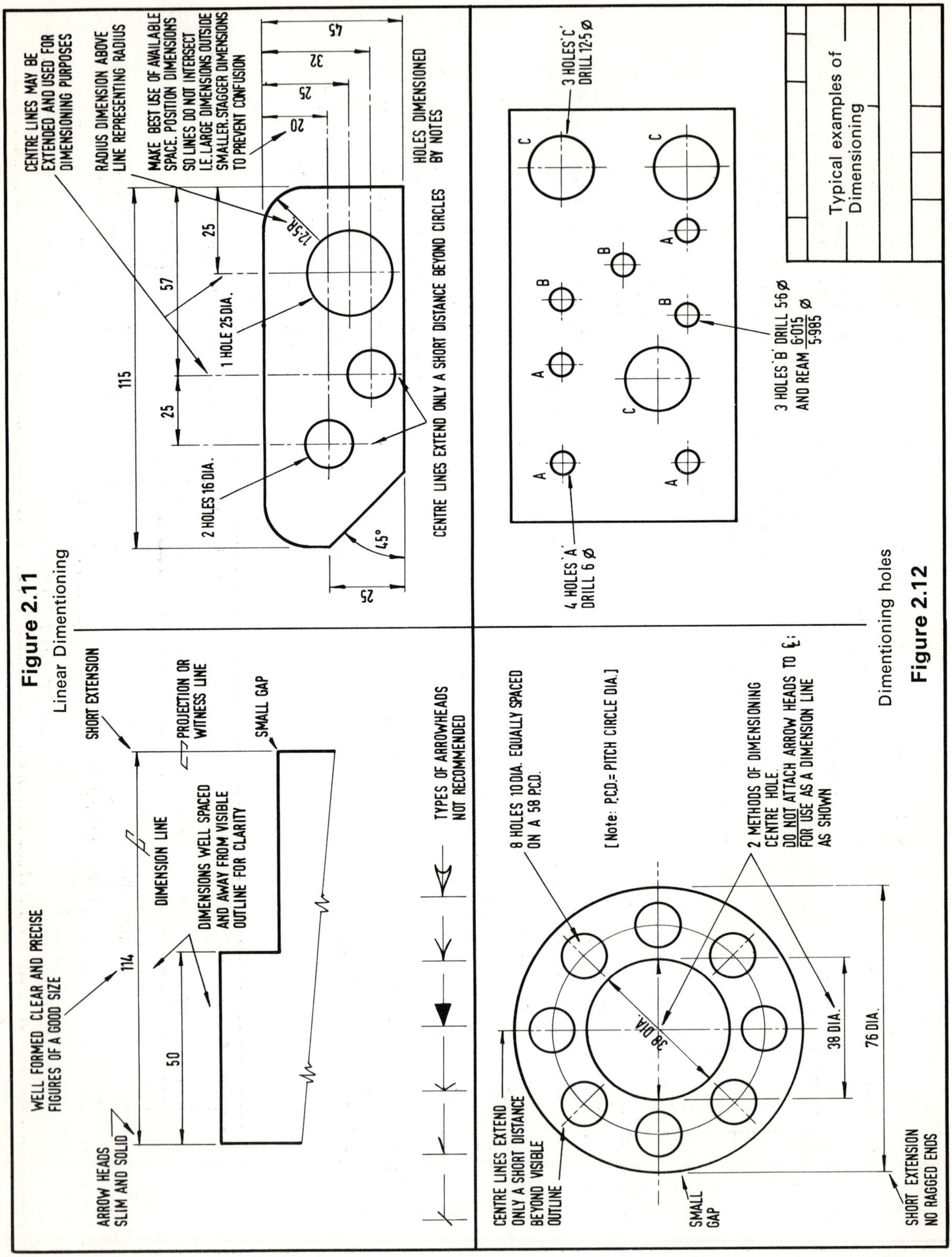

Figure 2.11 Linear Dimentioning

Figure 2.12 Dimentioning holes

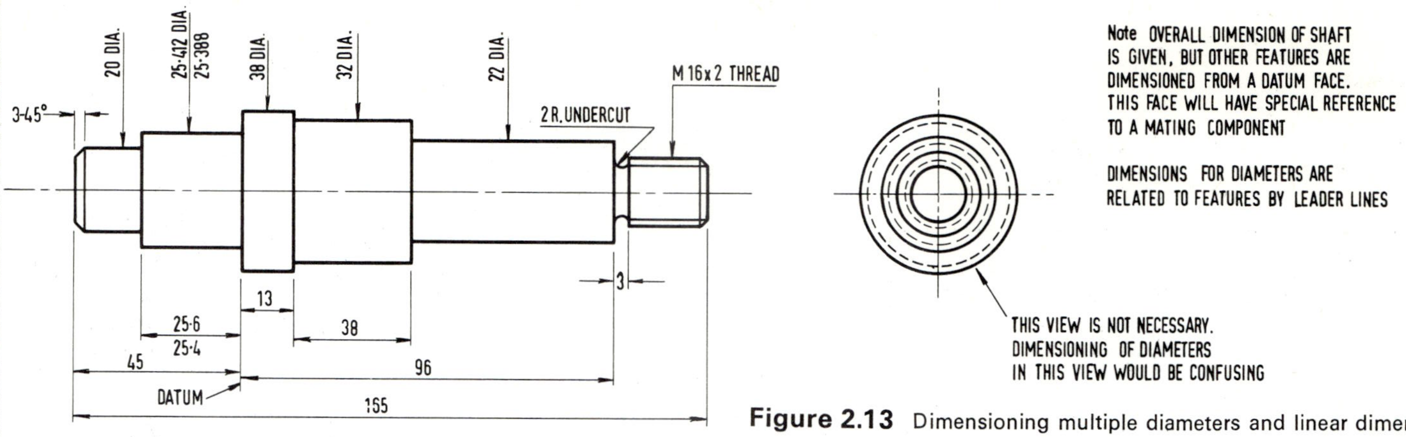

Figure 2.13 Dimensioning multiple diameters and linear dimensions

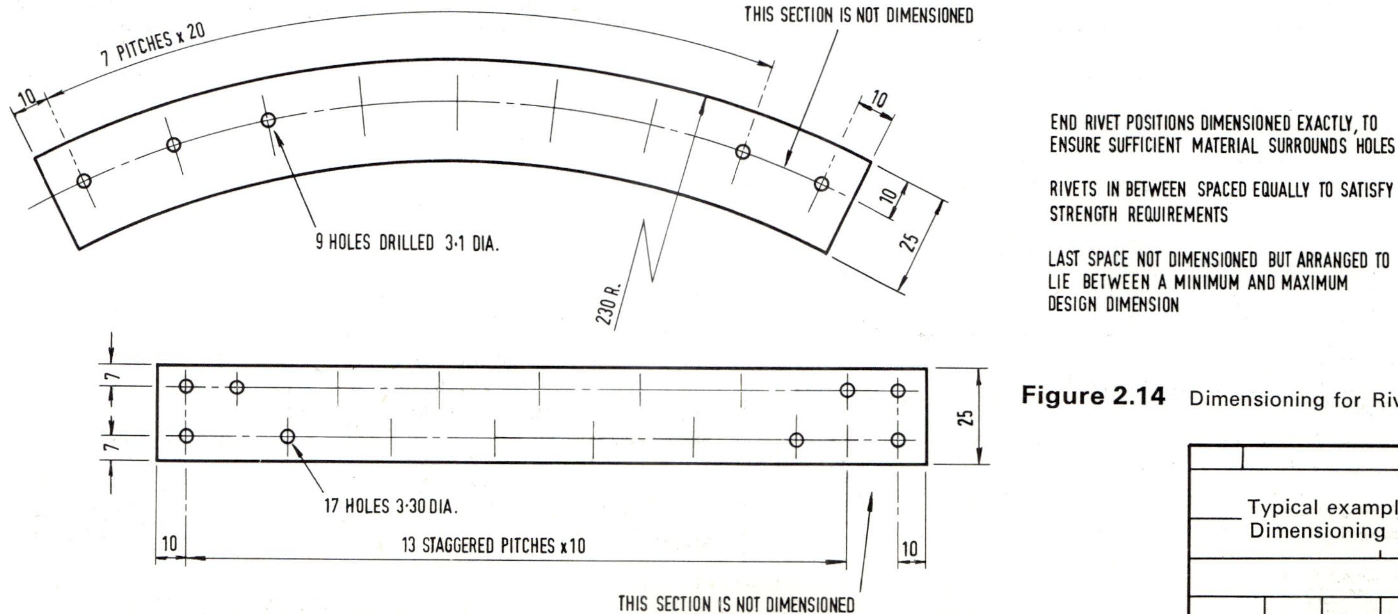

Figure 2.14 Dimensioning for Riveting

Typical examples of Dimensioning

EXERCISES

EXERCISE 2.1

On an A2 or half-imperial sheet of drawing paper draw a border having dimensions of 540 mm × 360 mm, and draw a title block of overall dimensions 100 mm × 60 mm in the bottom right-hand corner.

Using the dimensions given, position the drawings within the border and copy the exercise, using the correct thickness of line. Thick lines should be dense, sharp and about 0·75 mm to 0·1 mm wide. All other lines are thin lines of various types. These should also be dense and sharp, but approximately one-quarter the width of the thick lines.

Do not add dimensions but copy the NOTE adjacent to the title block, carrying out the instructions on printing contained therein. Estimate all undimensioned parts. Complete the title block by printing name, scale, class, date and title in the same manner.

EXERCISE 2.2

Divide an A2 or a half-imperial sheet into four equal parts and draw, in first-angle projection, three views of the objects shown, viewed in the direction of the arrow A, B and C.

All lines to be visible outlines, i.e. sharp, dense and 0·75 to 0·1 mm wide. The views for the first example are shown in faint outline.
All dimensions are in millimetres. Estimate any dimensions not given. Do not dimension the views.

EXERCISE 2.3

Divide an A2 or a half-imperial sheet into four equal parts and draw, in third-angle projection, three views of the objects shown, viewed in the directions of arrows A, B and C.

All lines to be visible outlines, i.e. sharp, dense and 0·75 to 1·0 mm wide.

All dimensions are in millimetres. Estimate any dimensions not given. Do not dimension the views.

EXERCISE 2.4

Divide an A2 or a half-imperial sheet of drawing paper into four equal parts, and draw and dimension the four following components.

Figure 1 shows a rectangular drilled plate which measures 130 mm long and 75 mm wide. Draw full size.

Using the given scale, position and dimension the single large hole (40 mm diameter) from the plate edges. Using the centre-lines of this hole as data, and the given scale, position the six holes of 12 mm diameter and the six of 6 mm diameter.

Fully dimension the drilled plate, dimensioning the twelve small holes from the centre lines of the single hole of 40 mm diameter.

Figure 2 shows a machined bar. Using the following information, draw full size and fully dimension the bar so that all axial faces are dimensioned from the datum shown.

The overall length of the bar is 230 mm and the lengths and sizes of the parts are:

Part	*Length* (mm)	*Diameter* (mm)
A	25	32
B	25	75
C	38	50
D	50	hexagon 22·5 mm A/F
E	38	32
F	38	25
G		thread M20 × 2·5

Figure 3 shows a sheet-metal plate which is to be blanked and pierced with five round holes. The plate, however, is badly dimensioned.

Draw the plate full size and completely redimension the plate to conform to recommendations on dimensioning given in BS 308.

Figure 4 shows two views in third-angle projection of a sheet-metal angle to be riveted into a light metal, highly stressed structure using round-head rivets of 3 mm diameter.

Draw full size the two views as shown, and position and dimension the rivet holes so that accumulation of errors in measuring during production will not result in any rivet being less than 6 mm from any edge of the metal. The pitch between rivets should be not more than 20 mm and not less than 15 mm.

EXERCISE 2.5

Divide an A2 or a half-imperial sheet into four parts as shown and draw the following sections:

Figure 1. Copy Section AA, which is a section across a laminated joint. Finish the drawing with cross-hatching lines, remembering that the outside plates are part of the same bracket.

Figures 2A and 2B show two metal objects, one solid and one webbed, having similar overall dimensions. Draw in orthographic projection the cross-sectional views as indicated by the section planes marked AA.

Figure 3 shows two views in orthographic projection of a pulley. Do not draw the given views but, using the given information, draw an aligned section on the section plane marked BB.

All dimensions are in millimetres.

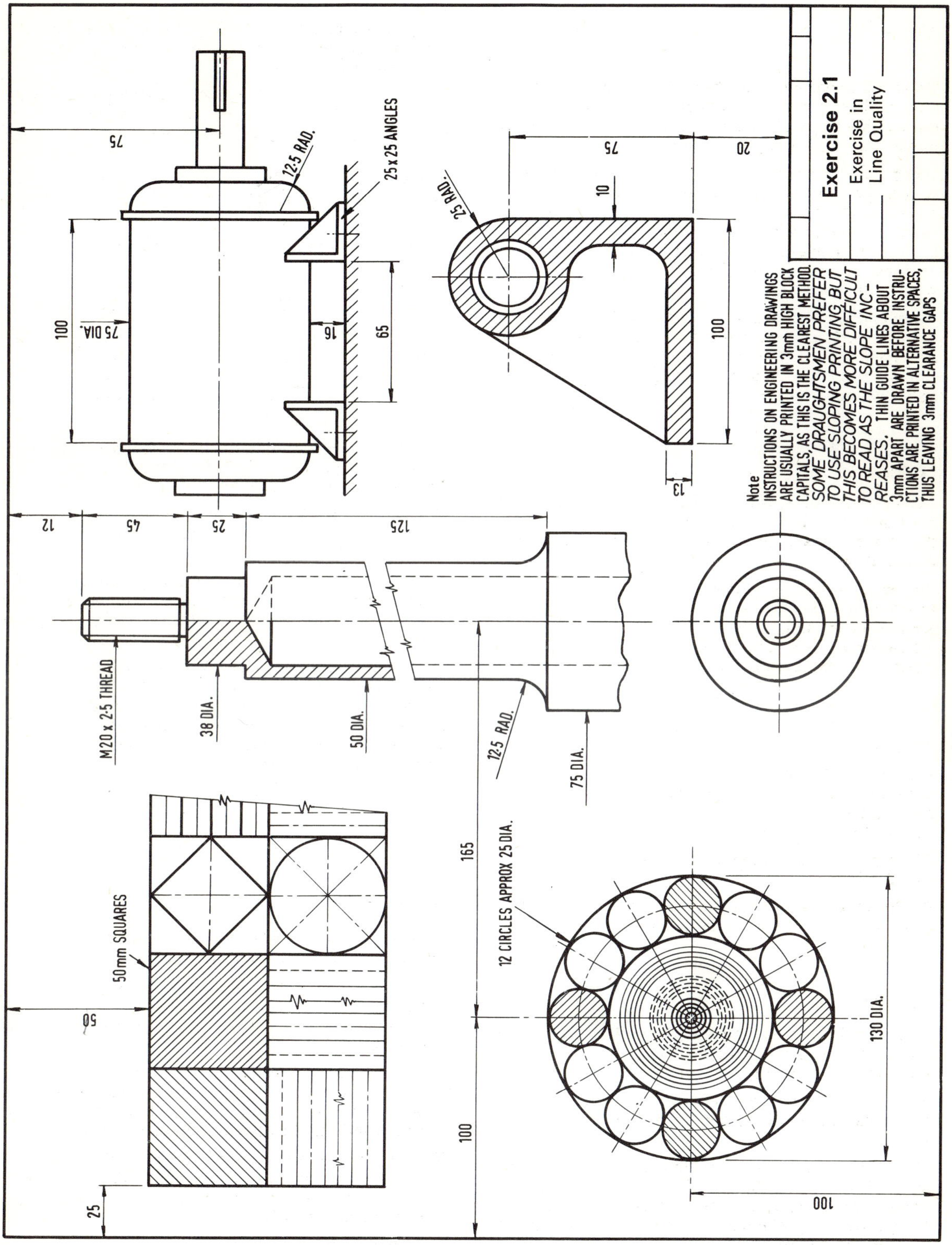
Exercise 2.1
Exercise in Line Quality
Note
INSTRUCTIONS ON ENGINEERING DRAWINGS ARE USUALLY PRINTED IN 3mm HIGH BLOCK CAPITALS, AS THIS IS THE CLEAREST METHOD. SOME DRAUGHTSMEN PREFER TO USE SLOPING PRINTING BUT THIS BECOMES MORE DIFFICULT TO READ AS THE SLOPE INC-REASES. THIN GUIDE LINES ABOUT 3mm APART ARE DRAWN BEFORE INSTRU-CTIONS ARE PRINTED IN ALTERNATIVE SPACES, THUS LEAVING 3mm CLEARANCE GAPS
75
12.5 RAD.
25 x 25 ANGLES
100
75 DIA.
16
65
25 RAD.
75
20
10
100
13
12
45
25
125
M20 x 2.5 THREAD
38 DIA.
50 DIA.
12.5 RAD.
75 DIA.
50mm SQUARES
50
165
12 CIRCLES APPROX 25 DIA.
130 DIA.
100
25
100

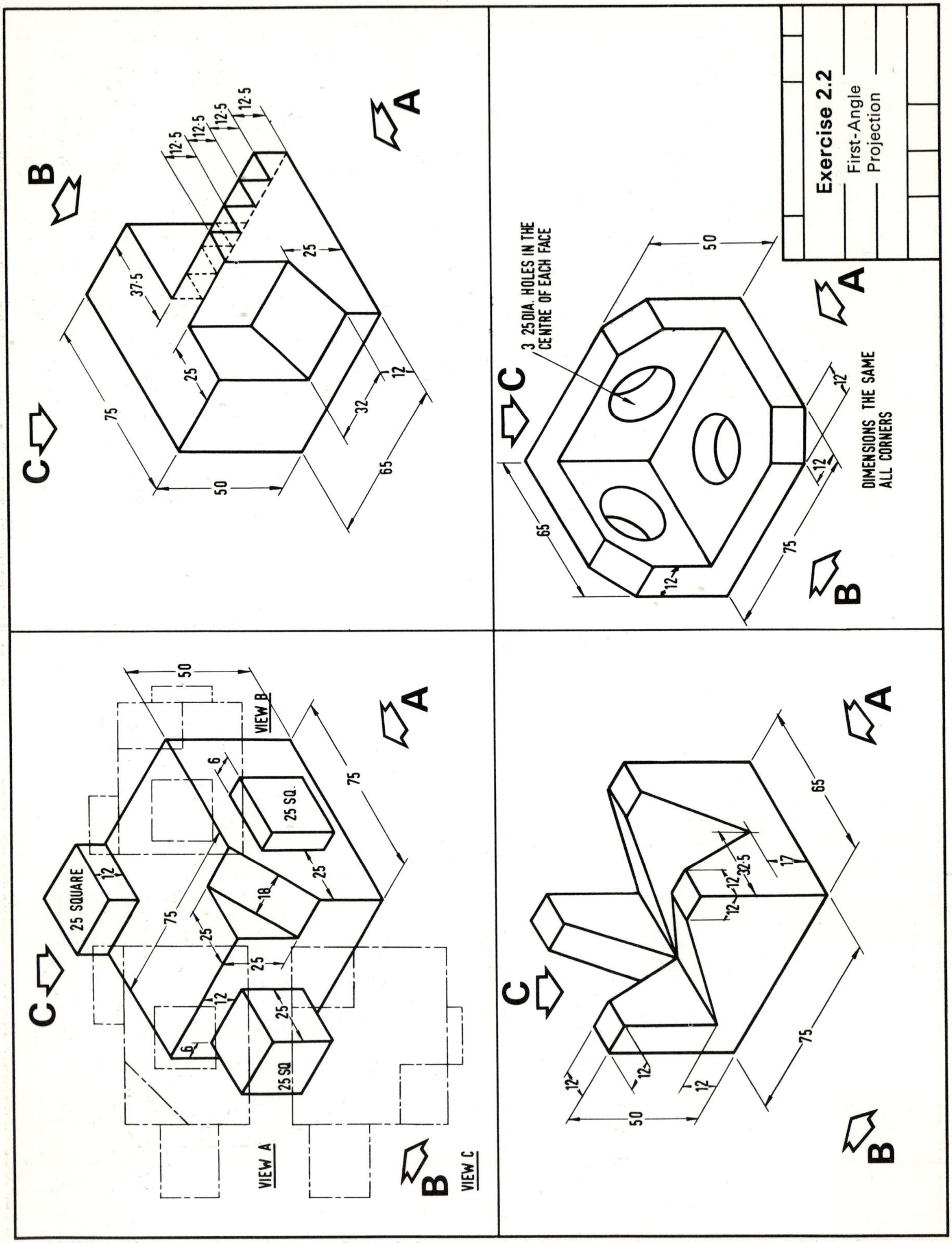
Exercise 2.2
First-Angle Projection
3 25 DIA. HOLES IN THE CENTRE OF EACH FACE
DIMENSIONS THE SAME ALL CORNERS
25 SQUARE
25 SQ.
VIEW A
VIEW B
VIEW C
A
B
C

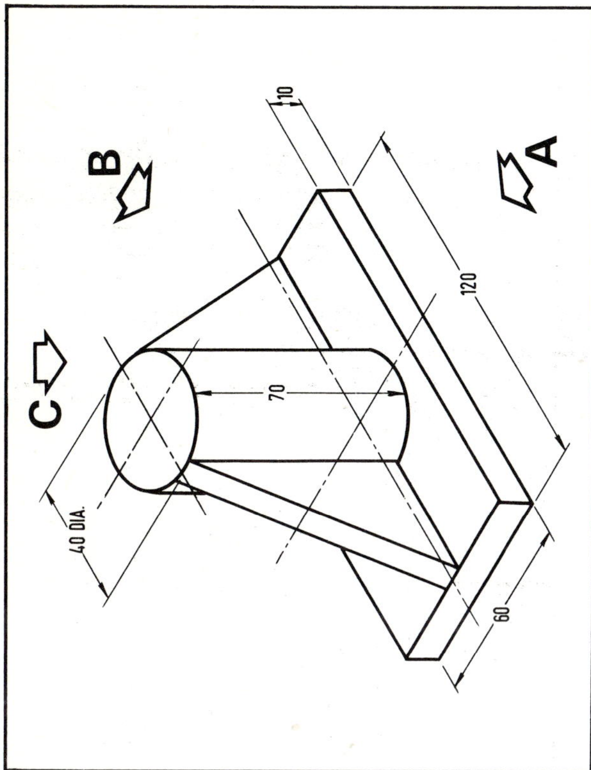
A
B
C
10
120
70
40 DIA.
60

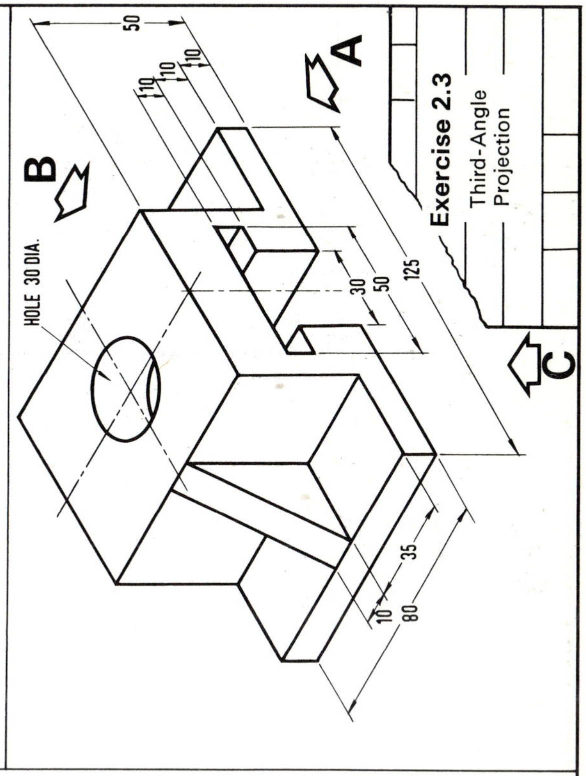
Exercise 2.3
Third-Angle Projection
A
B
C
50
10
10
10
10
HOLE 30 DIA.
30
50
125
35
10
80

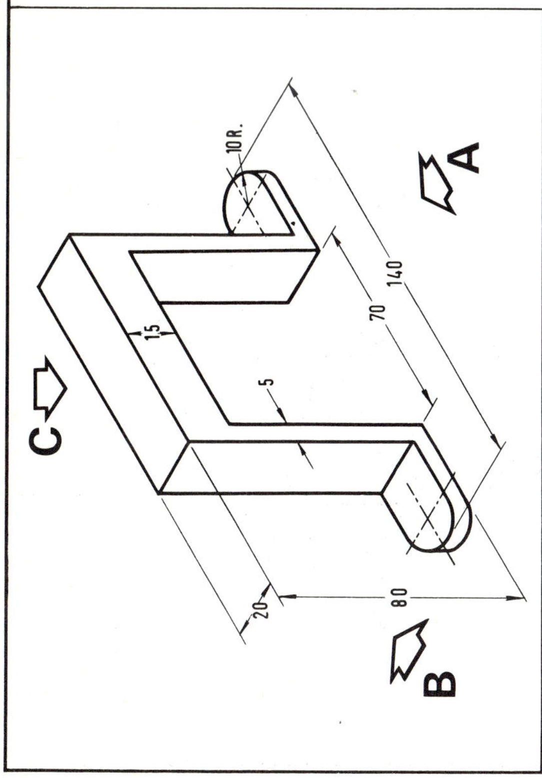
A
B
C
10 R.
140
70
15
5
20
80

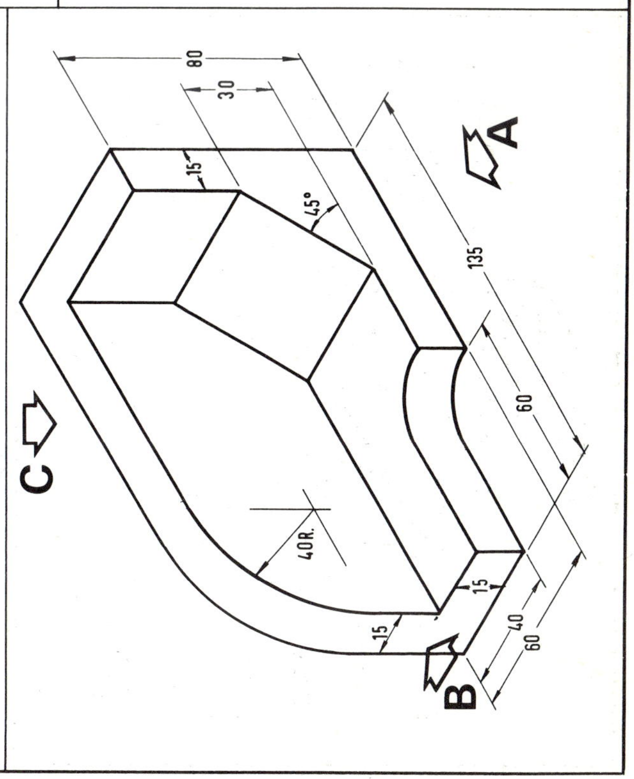
A
B
C
80
30
15
45°
135
60
40 R.
15
15
40
60

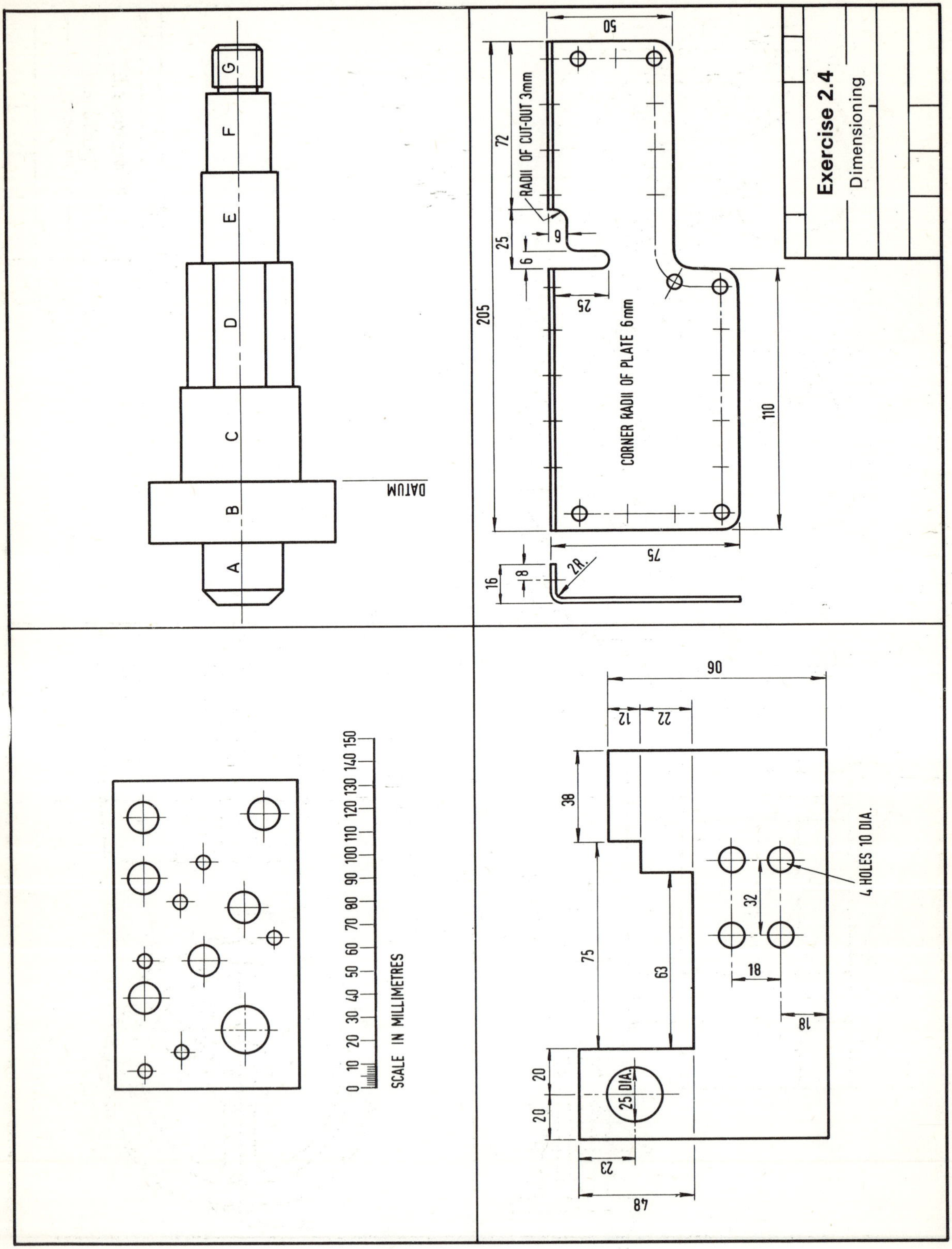

Exercise 2.4
Dimensioning
A
B
C
D
E
F
G
DATUM
RADII OF CUT-OUT 3mm
CORNER RADII OF PLATE 6mm
2R.
50
72
25
6
9
25
205
110
75
16
8
SCALE IN MILLIMETRES
0 10 20 30 40 50 60 70 80 90 100 110 120 130 140 150
90
12
22
38
75
63
4 HOLES 10 DIA.
32
18
18
20
20
25 DIA.
23
48

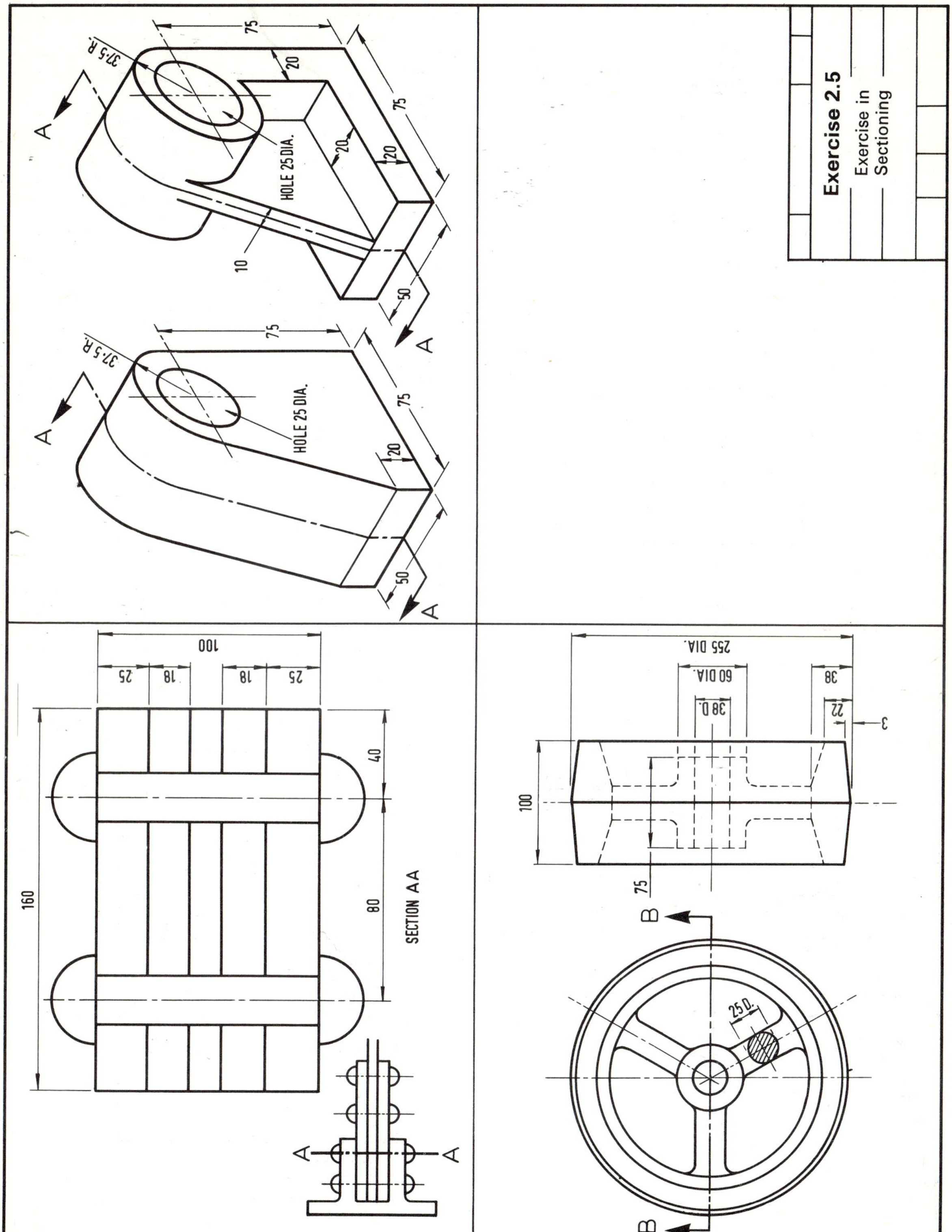
Exercise 2.5
Exercise in Sectioning
HOLE 25 DIA.
37.5 R.
75
20
10
50
A
SECTION AA
100
25
18
160
40
80
255 DIA.
60 DIA.
38 D.
38
22
3
25 D.
B

THE PRESENT SITUATION

A vast amount of international trading is carried on today in goods manufactured in both imperial and metric units, and the difference in the two systems of units can cause many difficulties. For example the loss of a nut has been the common experience of many of us—like the one that disappeared from the bicycle in our early days, and the nut from the spares in the garage that appeared to be the right size but just did not fit. It jammed after a couple of turns, or spoilt the thread on to which it was forced because the thread form was different.

Similar but more serious situations arise many times a day in international industry and commerce. A well-known sixteenth-century quotation could be rewritten, without exaggeration, like this:

For the sake of a nut a component was lost,
For the loss of a component a machine was idle,
For the lack of a machine production was lost,
For the loss of production an order was lost,
For the loss of an order men became idle,
For the loss of employment wages were lost,
For the loss of wages, livelihoods were lost,
All for the sake of a nut.

A situation in the home, where knowledge of engineering is usually very limited, serves to highlight the problem. Take the example of a British manufacturer making a domestic machine which he exports to many countries of the world. It contains a drum which has six studs welded on the base and is fitted to a driving plate with six BSF nuts which are only accessible through punched holes in a structural framework. The nuts, due to vibration, work loose and fall off one by one, and over a period are swept away by the purchaser.

Finally, the machine becomes unserviceable. The services of a technician are sought to mend it, and what happens?

In America, BSF nuts are not standard so, after struggling to fit ANF nuts to BSF studs, the technician tells the housewife she will have to wait several weeks until the correct nuts can be obtained from Britain.
In Italy the same situation arises and the technician tries to fit metric nuts to BSF studs. The Italian housewife is told the same story. Her machine will be out of use for several weeks until nuts of the correct size are available.

A universally adopted set of standards for measurement and manufacture would eliminate many of these difficulties, particularly in the replacement of screwed fasteners, and for the measurement of broken components to facilitate replacement by local manufacture.

IMPERIAL UNITS

Britain is a manufacturing country, and the livelihood of the population depends largely on the flow of the wide variety of manufactured goods exported overseas to provide foreign exchange for the purchase of the food and raw materials Britain needs from abroad. The goods exported, whether ships, aircraft, cars, office machines, domestic equipment, engineering components, radio, shoes, clothing, or any other item on the long list of manufactured goods that leave Britain, have in the main been manufactured to imperial inch units of measurement, with the screwed fasteners also produced in inch units to either Whitworth, BA or Unified form.

The British Standards Institution has for many years been the focal point in activating British industry to manufacture goods to standard quality and sizes, and has produced several thousand Standards which are in most cases specifications in imperial units.

Imperial units have become established by practice from the distant past, and follow no mathematical pattern. There are many romantic stories of how imperial units came into being. How the foot was divided into *twelve* inches. How the yard was divided into *three* feet. How *1760* yards were taken to represent *one* mile which can be subdivided into *eight* furlongs, each of which can be subdivided into *ten* chains each of *twenty-two* yards length. The imperial units for weights and measures, for area and for currency are all equally irrational.

It is a revealing exercise to discover the gaps which exist in our knowledge of imperial units. Do you know that a rod, pole or perch is $5\frac{1}{2}$ yards? When you go into a chemist to buy something prepacked in a bottle, are you sure you understand what fluid ounces are? Imperial units are confusing to the British; to the foreigner used to the decimal system for all measurement, the imperial system is impossible.

In education, students of engineering have for years been confused in their studies by requiring a knowledge of many disciplines, including physics, applied mechanics and electricity, that each have a traditional system of units. There have been different systems of units for force, mass and acceleration. The physicist uses the dyne and poundal as units of force; for mechanical engineering the unit of force is the pound (lbf) and the unit of mass is the slug; for electrical engineering the unit of force in the MKS system is the newton.

GOING METRIC

With such a system of units, which has evolved partly by rule of thumb and agricultural practice, it is not surprising in this age of computers, mechanised counting, tape-controlled machine tools and modular construction that many attempts have been made in Britain to encourage the general adoption of metric weights and measures.

However, it was not until 1962 that the British Standards Institution circulated a statement entitled *Change to the Metric System* to industry in the United Kingdom. This set out the main findings of various investigations into the growing use of the metric system throughout the world.

The Confederation of British Industry made further extensive enquiries and the conclusion was reached that the greater part of British industry favoured the adoption of the metric system. This led to Government support of the changeover which is now taking place in industry and education.

Similar changes are taking place in other countries, e.g. Africa and Australasia, whilst India, Russia, Japan and the countries of Europe are already 'metric'. International co-operation has ensured the adoption of a standard system of metric units known as SI (Système International). See Figure 3.1.

There are various booklets for explaining the SI system of units and derivatives that are used in science and engineering. Of particular interest as basic references in engineering drawing are the following: BS 308: 1964; BS 1991 (Parts 1 to 6): 1961–1967; PD 5686: 1969 (published by BSI).

ENGINEERING DRAWINGS USING METRIC UNITS

1 inch = 25·4 mm exactly

During the period of changeover from inch units to metric units in industry, some existing designs will have the dimensions in inches converted to millimetres. BS 2856: 1957, *Precise conversion of inch and metric sizes on engineering drawings,* clearly sets out how this can be achieved. This standard was issued in 1957 to enable companies who at that time had manufacturing interests in both the United Kingdom and the Continent to inter-convert inch and metric dimensions on engineering drawings.

On new designs it is desirable to exclude inch units completely, and to think and design in metric units only, thus eliminating conversion. In engineering drawings this will largely do away with awkward numerical values containing many figures to the right of the decimal point.

Tables 3.1 and 3.2 give conversion values for inches in decimals to millimetres, and fractions of inches to millimetres; but these will not be necessary for designing when one has become accustomed to thinking in metric units. For the period of acclimatisation remembering a few approximate equivalent dimensions will help in visualising sizes:

0·5 ($\frac{1}{2}$) millimetre is approximately 0·020 inch.
1·5 ($1\frac{1}{2}$) millimetres is approximately 1/16 inch.
25 millimetres is approximately 1 inch.

DIFFERENCES ON ENGINEERING DRAWINGS WHEN METRIC UNITS ARE USED

1. Dimensions will appear as millimetres.
2. Threaded parts will have ISO * metric screw threads.
3. Tolerancing for mating parts and interchangeability will be in millimetres.
4. Hole sizes will be quoted in millimetres.
5. Third-angle projection will be universal.

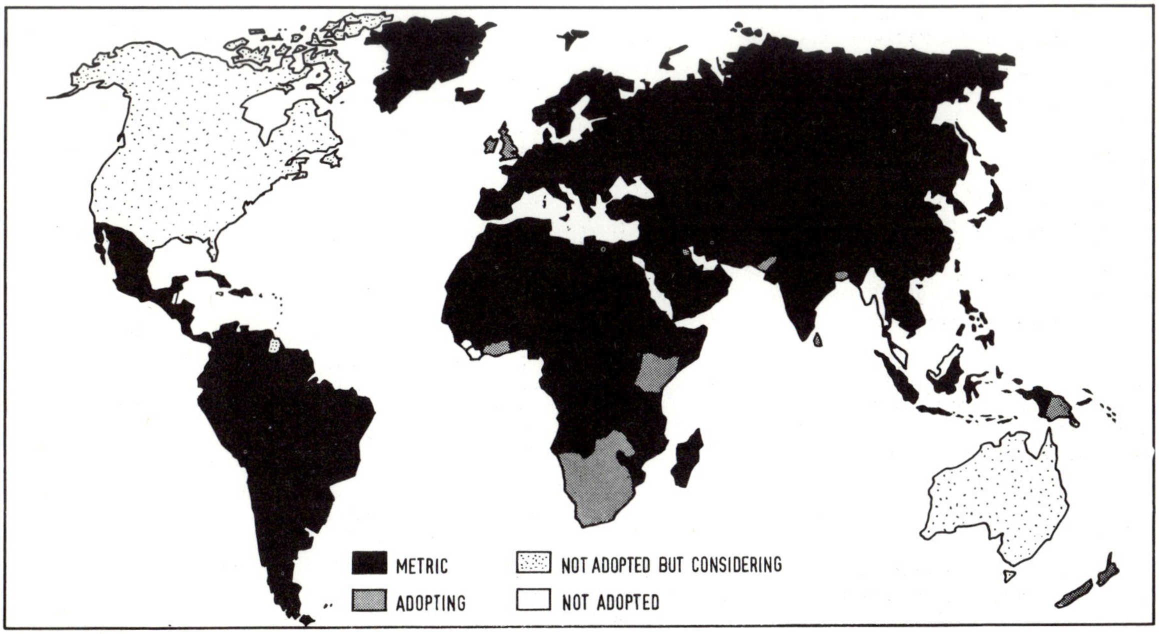

Figure 3.1 Map showing adoption of metric units (1970)

DIMENSIONING

Dimensions on engineering drawings will usually be in millimetres, irrespective of the magnitude of the dimension. There is, however, a difference between the U.K. and continental methods of writing a decimal dimension. The decimal point as used in the U.K. is a full point at half the number height, whereas on the Continent the decimal point is a comma level with the bottom of the numbers and taking a complete number space.

U.K.	*Continental*
12·6	12,6
126·195	126,195**
0·126 1	0,126 1

It can be seen from the second example** of Continental usage that this could be misread in the U.K. as one hundred and twenty-six thousand, one hundred and ninety-five. To avoid confusion with the Continental decimal marker, the British Standards Institution recommends the use of a space instead of a comma to separate groups of three figures, e.g.

12 629 532
1 262 953·2
126 295·32
12 629·532
126·295 32
12·629 532

Note: BS: 308: 1964 gives commas as the decimal marker for metric dimensioning, but this has now been superseded by PD 6424: 1968 which recommends the full point as standard usage.

SCREW THREADS (see Chapter 8)

To replace all the existing threaded connections—whether they are nuts and bolts or screwed components—having BSW, BSF or BA screw threads would be an impossible task, and the only way to make the changeover to acceptable metric threads is to produce all new designs with the new ISO metric screw thread, and discontinue the use of the present standards. The period of changeover may be a little chaotic, but the adoption of international standards will benefit engineers of the future.

TOLERANCING (see Chapter 7)

Table 7.1 gives a summary of limits for holes and shafts; it covers most of the fits required by 90 per cent of the mating parts in engineering. Whereas it has been practice to work in thousandths of an inch (0·001″), it will now be necessary to work in thousandths of a millimetre (0·001 mm).

DRILL HOLES (BS 528: 1959)

The standard for drill sizes used in the past in British industry was a mixture of (1) a series of fractional-sized drills increasing by 1/64th of an inch in diameter, (2) a numbered series, No. 80 to No. 1, covering hole sizes between 0·0135″ and 0·228″ in diameter, and (3) a lettered series, A to Z, covering twenty-six sizes between 0·2340″ and 0·4130″ in diameter.

The numbered and lettered drills catered for sizes between the smaller fractional drills, and were necessary when screw threads had to be tapped, or clearance holes were required for fractional pins, rivets, bolts and screws.

The numbered and lettered series of drills are now obsolete and the new standard BS 328 quotes all drill diameters in millimetres. Table 3.3 is a conversion table giving equivalent metric sizes for the now-obsolete drills.

PROJECTION

The changeover from first-angle to third-angle projection by the major industries and large companies, particularly those with American and Continental interests, is now almost complete. Small industries in some instances still cling to first-angle projection, particularly where their products are for the home market However, third-angle projection is soon likely to be the only method used in industry. Both forms are explained and used in this book.

Figure 3.2 is a detail drawing in third-angle projection using the metric system for dimensioning, tolerancing and screw threads.

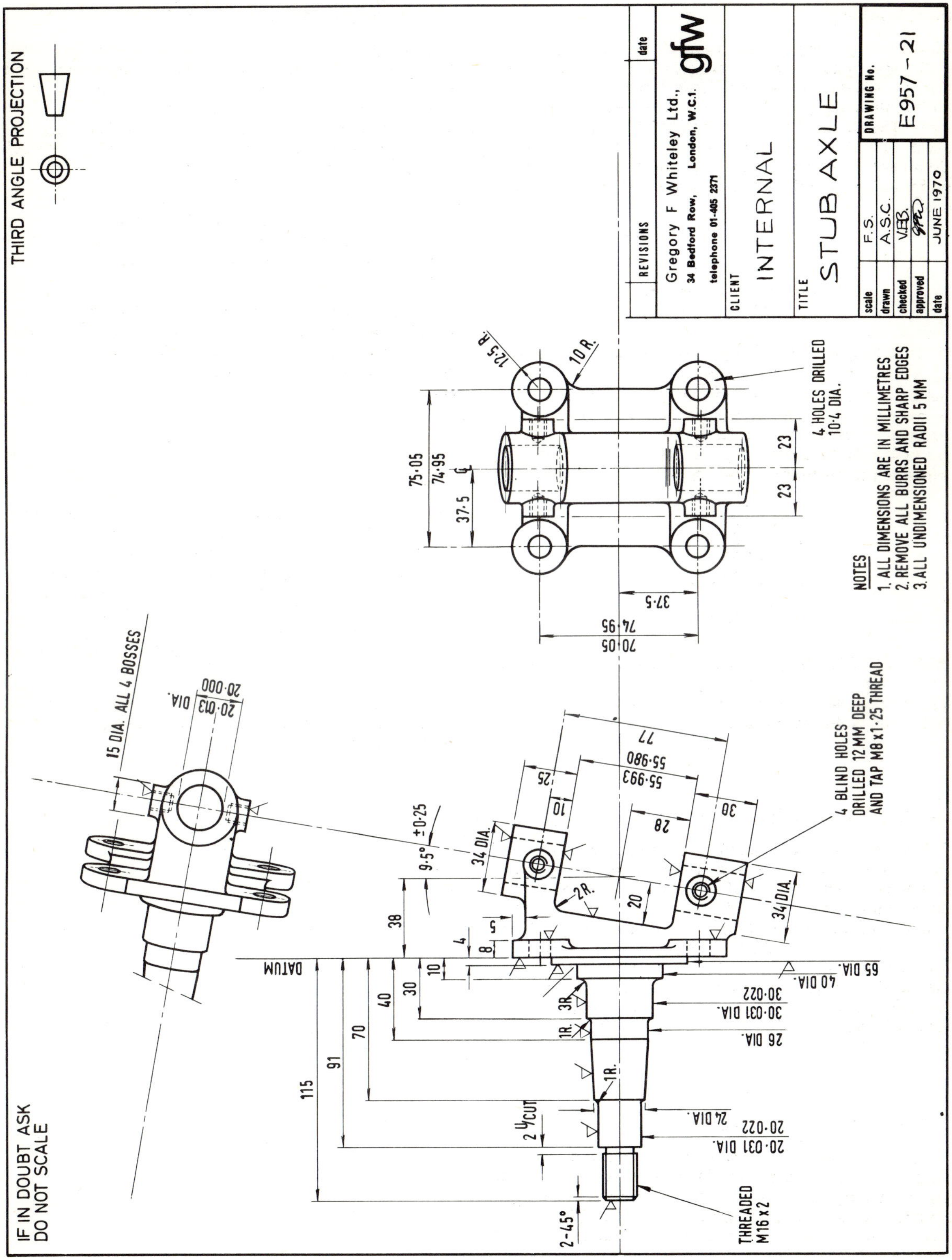
THIRD ANGLE PROJECTION
IF IN DOUBT ASK
DO NOT SCALE
REVISIONS
date
Gregory F Whiteley Ltd.,
34 Bedford Row, London, W.C.1.
telephone 01-405 2371
gfw
CLIENT
INTERNAL
TITLE
STUB AXLE
scale
drawn F.S.
checked A.S.C.
approved V.B.
date JUNE 1970
DRAWING No.
E957 – 21
NOTES
1. ALL DIMENSIONS ARE IN MILLIMETRES
2. REMOVE ALL BURRS AND SHARP EDGES
3. ALL UNDIMENSIONED RADII 5 MM
4 HOLES DRILLED
10·4 DIA.
4 BLIND HOLES
DRILLED 12 MM DEEP
AND TAP M8 x 1·25 THREAD
15 DIA. ALL 4 BOSSES
20·000
20·013 DIA.
THREADED
M16 x 2
DATUM
65 DIA.
40 DIA.
30·022
30·031 DIA.
26 DIA.
24 DIA.
20·022
20·031 DIA.
34 DIA.
9·5° ±0·25
2 ¼ CUT
2-45°
75·05
74·95
37·5
70·05
74·95
55·980
55·993
12·5 R
10 R.
2R.
3R.
1R.

Figure 3.2

TABLE 3.1

Conversion Table inches to millimetres

·001″ to ·499″ in increments of ·001″

inches	0·000	0·001	0·002	0·003	0·004	0·005	0·006	0·007	0·008	0·009
	millimetres									
0·000	—	0·0254	0·0508	0·0762	0·1016	0·1270	0·1524	0·1778	0·2032	0·2286
0·010	0·2540	0·2794	0·3048	0·3302	0·3556	0·3810	0·4064	0·4318	0·4572	0·4826
0·020	0·5080	0·5334	0·5588	0·5842	0·6096	0·6350	0·6604	0·6858	0·7112	0·7366
0·030	0·7620	0·7874	0·8128	0·8382	0·8636	0·8890	0·9144	0·9398	0·9652	0·9906
0·040	1·0160	1·0414	1·0668	1·0922	1·1176	1·1430	1·1684	1·1938	1·2192	1·2446
0·050	1·2700	1·2954	1·3208	1·3462	1·3716	1·3970	1·4224	1·4478	1·4732	1·4986
0·060	1·5240	1·5494	1·5748	1·6002	1·6256	1·6510	1·6764	1·7018	1·7272	1·7526
0·070	1·7780	1·8034	1·8288	1·8542	1·8796	1·9050	1·9304	1·9558	1·9812	2·0066
0·080	2·0320	2·0574	2·0828	2·1082	2·1336	2·1590	2·1844	2·2098	2·2352	2·2606
0·090	2·2860	2·3114	2·3368	2·3622	2·3876	2·4130	2·4384	2·4638	2·4892	2·5146
0·100	2·5400	2·5654	2·5908	2·6162	2·6416	2·6670	2·6924	2·7178	2·7432	2·7686
0·110	2·7940	2·8194	2·8448	2·8702	2·8956	2·9210	2·9464	2·9718	2·9972	3·0226
0·120	3·0480	3·0734	3·0988	3·1242	3·1496	3·1750	3·2004	3·2258	3·2512	3·2766
0·130	3·3020	3·3274	3·3528	3·3782	3·4036	3·4290	3·4544	3·4798	3·5052	3·5306
0·140	3·5560	3·5814	3·6068	3·6322	3·6576	3·6830	3·7084	3·7338	3·7592	3·7846
0·150	3·8100	3·8354	3·8608	3·8862	3·9116	3·9370	3·9624	3·9878	4·0132	4·0386
0·160	4·0640	4·0894	4·1148	4·1402	4·1656	4·1910	4·2164	4·2418	4·2672	4·2926
0·170	4·3180	4·3434	4·3688	4·3942	4·4196	4·4450	4·4704	4·4958	4·5212	4·5466
0·180	4·5720	4·5974	4·6228	4·6482	4·6736	4·6990	4·7244	4·7498	4·7752	4·8006
0·190	4·8260	4·8514	4·8768	4·9022	4·9276	4·9530	4·9784	5·0038	5·0292	5·0546
0·200	5·0800	5·1054	5·1308	5·1562	5·1816	5·2070	5·2324	5·2578	5·2832	5·3086
0·210	5·3340	5·3594	5·3848	5·4102	5·4356	5·4610	5·4864	5·5118	5·5372	5·5626
0·220	5·5880	5·6134	5·6388	5·6642	5·6896	5·7150	5·7404	5·7658	5·7912	5·8166
0·230	5·8420	5·8674	5·8928	5·9182	5·9436	5·9690	5·9944	6·0198	6·0452	6·0706
0·240	6·0960	6·1214	6·1468	6·1722	6·1976	6·2230	6·2484	6·2738	6·2992	6·3246
0·250	6·3500	6·3754	6·4008	6·4262	6·4516	6·4770	6·5024	6·5278	6·5532	6·5786
0·260	6·6040	6·6294	6·6548	6·6802	6·7056	6·7310	6·7564	6·7818	6·8072	6·8326
0·270	6·8580	6·8834	6·9088	6·9342	6·9596	6·9850	7·0104	7·0358	7·0612	7·0866
0·280	7·1120	7·1374	7·1628	7·1882	7·2136	7·2390	7·2644	7·2898	7·3152	7·3406
0·290	7·3660	7·3914	7·4168	7·4422	7·4676	7·4930	7·5184	7·5438	7·5692	7·5946
0·300	7·6200	7·6454	7·6708	7·6962	7·7216	7·7470	7·7724	7·7978	7·8232	7·8486
0·310	7·8740	7·8994	7·9248	7·9502	7·9756	8·0010	8·0264	8·0518	8·0772	8·1026
0·320	8·1280	8·1534	8·1788	8·2042	8·2296	8·2550	8·2804	8·3058	8·3312	8·3566
0·330	8·3820	8·4074	8·4328	8·4582	8·4836	8·5090	8·5344	8·5598	8·5852	8·6106
0·340	8·6360	8·6614	8·6868	8·7122	8·7376	8·7630	8·7884	8·8138	8·8392	8·8646
0·350	8·8900	8·9154	8·9408	8·9662	9·9916	9·0170	9·0424	9·0678	9·0932	9·1186
0·360	9·1440	9·1694	9·1948	9·2202	9·2456	9·2710	9·2964	9·3218	9·3472	9·3726
0·370	9·3980	9·4234	9·4488	9·4742	9·4996	9·5250	9·5504	9·5758	9·6012	9·6266
0·380	9·6520	9·6774	9·7028	9·7282	9·7536	9·7790	9·8044	9·8298	9·8552	9·8806
0·390	9·9060	9·9314	9·9568	9·9822	10·0076	10·0330	10·0584	10·0838	10·1092	10·1346
0·400	10·1600	10·1854	10·2108	10·2362	10·2616	10·2870	10·3124	10·3378	10·3632	10·3886
0·410	10·4140	10·4394	10·4648	10·4902	10·5156	10·5410	10·5664	10·5918	10·6172	10·6426
0·420	10·6680	10·6934	10·7188	10·7442	10·7696	10·7950	10·8204	10·8458	10·8712	10·8966
0·430	10·9220	10·9474	10·9728	10·9982	11·0236	11·0490	11·0744	11·0998	11·1252	11·1506
0·440	11·1760	11·2014	11·2268	11·2522	11·2776	11·3030	11·3284	11·3538	11·3792	11·4046
0·450	11·4300	11·4554	11·4808	11·5062	11·5316	11·5570	11·5824	11·6078	11·6332	11·6586
0·460	11·6840	11·7094	11·7348	11·7602	11·7856	11·8110	11·8364	11·8618	11·8872	11·9126
0·470	11·9380	11·9634	11·9888	12·0142	12·0396	12·0650	12·0904	12·1158	12·1412	12·1666
0·480	12·1920	12·2174	12·2428	12·2682	12·2936	12·3190	12·3444	12·3698	12·3952	12·4206
0·490	12·4460	12·4714	12·4968	12·5222	12·5476	12·5730	12·5984	12·6238	12·6492	12·6746

TABLE 3.1 ***(cont.)***

Conversion Table inches to millimetres

·500″ to 1·000″ in increments of ·001″

inches	0·000	0·001	0·002	0·003	0·004	0·005	0·006	0·007	0·008	0·009
	millimetres									
0·500	12·7000	12·7254	12·7508	12·7762	12·8016	12·8270	12·8524	12·8778	12·9032	12·9286
0·510	12·9540	12·9794	13·0048	13·0302	13·0556	13·0810	13·1064	13·1318	13·1572	13·1826
0·520	13·2080	13·2334	13·2588	13·2842	13·3096	13·3350	13·3604	13·3858	13·4112	13·4366
0·530	13·4620	13·4874	13·5128	13·5382	13·5636	13·5890	13·6144	13·6398	13·6652	13·6906
0·540	13·7160	13·7414	13·7668	13·7922	13·8176	13·8430	13·8684	13·8938	13·9192	13·9446
0·550	13·9700	13·9954	14·0208	14·0462	14·0716	14·0970	14·1224	14·1478	14·1732	14·1986
0·560	14·2240	14·2494	14·2748	14·3002	14·3256	14·3510	14·3764	14·4018	14·4272	14·4526
0·570	14·4780	14·5034	14·5288	14·5542	14·5796	14·6050	14·6304	14·6558	14·6812	14·7066
0·580	14·7320	14·7574	14·7828	14·8082	14·8836	14·8590	14·8844	14·9098	14·9352	14·9606
0·590	14·9860	15·0114	15·0368	15·0622	15·0876	15·1130	15·1384	15·1638	15·1892	15·2146
0·600	15·2400	15·2654	15·2908	15·3162	15·3416	15·3670	15·3924	15·4178	15·4432	15·4686
0·610	15·4940	15·5194	15·5448	15·5702	15·5956	15·6210	15·6464	15·6718	15·6972	15·7226
0·620	15·7480	15·7734	15·7988	15·8242	15·8496	15·8750	15·9004	15·9258	15·9512	15·9766
0·630	16·0020	16·0274	16·0528	16·0782	16·1036	16·1290	16·1544	16·1798	16·2052	16·2306
0·640	16·2560	16·2814	16·3068	16·3322	16·3576	16·3830	16·4084	16·4338	16·4592	16·4846
0·650	16·5100	16·5354	16·5608	16·5862	16·6116	16·6370	16·6624	16·6878	16·7132	16·7386
0·660	16·7640	16·7894	16·8148	16·8402	16·8656	16·8910	16 9164	16·9418	16·9672	16·9926
0·670	17·0180	17·0434	17·0688	17·0942	17·1196	17·1450	17·1704	17·1958	17·2212	17·2466
0·680	17·2720	17·2974	17·3228	17·3482	17·3736	17·3990	17·4244	17·4498	17·4752	17·5006
0·690	17·5260	17·5514	17·5768	17·6022	17·6276	17·6530	17·6784	17·7038	17·7292	17·7546
0·700	17·7800	17·8054	17·8308	17·8562	17·8816	17·9070	17·9324	17·9578	17·9832	18·0086
0·710	18·0340	18·0594	18·0848	18·1102	18·1356	18·1610	18·1864	18·2118	18·2372	18·2626
0·720	18·2880	18·3134	18·3388	18·3642	18·3896	18·4150	18·4404	18·4658	18·4912	18·5166
0·730	18·5420	18·5674	18·5928	18·6182	18·6436	18·6690	18·6944	18·7198	18·7452	18·7706
0·740	18·7960	18·8214	18·8468	18·8722	18·8976	18·9230	18·9494	18·9738	18·9992	19·0246
0·750	19·0500	19·0754	19·1008	19·1262	19·1516	19·1770	19·2024	19·2278	19·2532	19·2786
0·760	19·3040	19·3294	19·3548	19·3802	19·4056	19·4310	19·4564	19·4818	19·5072	19·5326
0·770	19·5580	19·5834	19·6088	19·6342	19·6596	19·6850	19·7104	19·7358	19·7612	19·7866
0·780	19·8120	19·8374	19·8628	19·8882	19·9136	19·9390	19·9644	19·9898	20·0152	20·0406
0·790	20·0660	20·0914	20·1168	20·1422	20·1676	20·1930	20·2184	20·2438	20·2692	20·2946
0·800	20·3200	20·3454	20·3708	20·3962	20·4216	20·4470	20·4724	20·4978	20·5232	20·5486
0·810	20·5740	20·5994	20·6248	20·6502	20·6756	20·7010	20·7264	20·7518	20·7772	20·8026
0·820	20·8280	20·8534	20·8788	20·9042	20·9296	20·9550	20·9804	21·0058	21·0312	21·0566
0·830	21·0820	21·1074	21·1328	21·1582	21·1836	21·2090	21·2344	21·2598	21·2852	21·3106
0·840	21·3360	21·3614	21·3868	21·4122	21·4376	21·4630	21·4884	21·5138	21·5392	21·5646
0·850	21·5900	21·6154	21·6408	21·6662	21·6916	21·7170	21·7424	21·7678	21·7932	21·8186
0·860	21·8440	21·8694	21·8948	21·9202	21·9456	21·9710	21·9964	22·0218	22·0472	22·0726
0·870	22·0980	22·1234	22·1488	22·1742	22·1996	22·2250	22·2504	22·2758	22·3012	22·3266
0·880	22·3520	22·3774	22·4028	22·4282	22·4536	22·4790	22·5044	22·5298	22·5552	22·5806
0·890	22·6060	22·6314	22·6568	22·6822	22·7076	22·7330	22·7584	22·7838	22·8092	22·8346
0·900	22·8600	22·8854	22·9108	22·9362	22·9616	22·9870	23·0124	23·0378	23·0632	23·0886
0·910	23·1140	23·1394	23·1648	23·1902	23·2156	23·2410	23·2664	23·2918	23·3172	23·3426
0·920	23·3680	23·3934	23·4188	23·4442	23·4696	23·4950	23·5204	23·5458	23·5712	23·5966
0·930	23·6220	23·6474	23·6728	23·6982	23·7236	23·7490	23·7744	23·7998	23·8252	23·8506
0·940	23·8760	23·9014	23·9268	23·9522	23·9776	24·0030	24·0284	24·0538	24·0792	24·1046
0·950	24·1300	24·1554	24·1808	24·2062	24·2316	24·2570	24·2824	24·3078	24·3332	24·3586
0·960	24·3840	24·4094	24·4348	24·4602	24·4856	24·5110	24·5364	24·5618	24·5872	24·6126
0·970	24·6380	24·6634	24·6888	24·7142	24·7396	24·7650	24·7904	24·8158	24·8412	24·8666
0·980	24·8920	24·9174	24·9428	24·9682	24·9936	25·0190	25·0444	25·0698	25·0952	25·1206
0·990	25·1460	25·1714	25·1968	25·2222	25·2476	25·2730	25·2984	25·3238	25·3492	25·3746
1·000	25·4000	—	—	—	—	—	—	—	—	—

TABLE 3.2

Conversion of fractions of an inch to decimals of an inch and millimetres

Decimals of inch	*Fraction of inch*	*Millimetres*	*Decimals of inch*	*Fraction of inch*	*Millimetres*
0·015 625	1/64	0·396 875	0·515 625	33/64	13·096 875
0·031 250	1/32	0·793 750	0·531 250	17/32	13·493 750
0·046 875	3/64	1·190 625	0·546 875	35/64	13·890 625
0·062 500	1/16	1·587 500	0·562 500	9/16	14·287 500
0·078 125	5/64	1·984 375	0·578 125	37/64	14·684 375
0·093 750	3/32	2·381 250	0·593 750	19/32	15·081 250
0·109 375	7/64	2·778 125	0·609 375	39/64	15·478 125
0·125 000	1/8	3·175 000	0·625 000	5/8	15·875 000
0·140 625	9/64	3·571 875	0·640 625	41/64	16·271 875
0·156 250	5/32	3·968 750	0·656 250	21/32	16·668 750
0·171 875	11/64	4·365 625	0·671 875	43/64	17·065 625
0·187 500	3/16	4·762 500	0·687 500	11/16	17·462 500
0·203 125	13/64	5·159 375	0·703 125	45/64	17·859 375
0·218 750	7/32	5·556 250	0·718 750	23/64	18·256 250
0·234 375	15/64	5·953 125	0·734 375	47/64	18·653 125
0·250 000	1/4	6·350 000	0·750 000	3/4	19·050 000
0·265 625	17/64	6·746 875	0·765 625	49/64	19·446 875
0·281 250	9/32	7·143 750	0·781 250	25/32	19·843 750
0·296 875	19/64	7·540 625	0·796 875	51/64	20·240 625
0·312 500	5/16	7·937 500	0·812 500	13/16	20·637 500
0·328 125	21/64	8·334 375	0·828 125	53/64	21·034 375
0·343 750	11/32	8·731 250	0·843 750	27/32	21·431 250
0·359 375	23/64	9·128 125	0·859 375	55/64	21·828 125
0·375 000	3/8	9·525 000	0·875 000	7/8	22·225 000
0·390 625	25/64	9·921 875	0·890 625	57/64	22·621 875
0·406 250	13/32	10·318 750	0·906 250	29/32	23·018 750
0·421 875	27/64	10·715 625	0·921 875	59/64	23·415 625
0·437 500	7/16	11·112 500	0·937 500	15/16	23·812 500
0·453 125	29/64	11·509 375	0·953 125	61/64	24·209 375
0·468 750	15/32	11·906 250	0·968 750	31/32	24·606 250
0·484 375	31/64	12·303 125	0·984 375	63/64	25·003 125
0·500 000	1/2	12·700 000	1·000 000	1	25·400 000

TABLE 3.3

Equivalent millimetre drill sizes for obsolete number and letter drills

Number drill	*Decimal inch equivalent*	*Nearest mm drill*	*Decimal inch equivalent*
80	0·013 5	0·35	0·013 8
79	0·014 5	0·38	0·015 0
78	0·016 0	0·40	0·015 7
77	0·018 0	0·45	0·017 7
76	0·020 0	0·50	0·019 7
75	0·021 0	0·52	0·020 5
74	0·022 5	0·58	0·022 8
73	0·024 0	0·60	0·023 6
72	0·025 0	0·65	0·025 6
71	0·026 0	0·65	0·025 6
70	0·028 0	0·70	0·027 6
69	0·029 2	0·75	0·029 5
68	0·031 0	0·80	0·031 5
67	0·032 0	0·82	0·032 3
66	0·033 0	0·85	0·033 5
65	0·035 0	0·90	0·035 4
64	0·036 0	0·92	0·036 2
63	0·037 0	0·95	0·037 4
62	0·038 0	0·98	0·038 6
61	0·039 0	1·00	0·039 4
60	0·040 0	1·00	0·039 4
59	0·041 0	1·05	0·041 3
58	0·042 0	1·05	0·041 3
57	0·043 0	1·10	0·043 3
56	0·046 5	1·20	0·047 2
55	0·052 0	1·30	0·051 2
54	0·055 0	1·40	0·055 1
53	0·059 5	1·50	0·059 0
52	0·063 5	1·60	0·063 0
51	0·067 0	1·70	0·066 9
50	0·070 0	1·80	0·070 9
49	0·073 0	1·85	0·072 8
48	0·076 0	1·95	0·076 8
47	0·078 5	2·00	0·078 7
46	0·081 0	2·05	0·080 7
45	0·082 0	2·10	0·082 7
44	0·086 0	2·20	0·086 6
43	0·089 0	2·25	0·088 6
42	0·093 5	2·40	0·094 5
41	0·096 0	2·45	0·096 5
40	0·098 0	2·50	0·098 4
39	0·099 5	2·55	0·100 4
38	0·101 5	2·60	0·102 4
37	0·104 0	2·65	0·104 3
36	0·106 5	2·70	0·106 3
35	0·110 0	2·80	0·110 2
34	0·111 0	2·80	0·110 2
33	0·113 0	2·85	0·112 2
32	0·116 0	2·95	0·116 1
31	0·120 0	3·00	0·118 1
30	0·128 5	3·30	0·129 9
29	0·136 0	3·50	0·137 8
28	0·140 5	3·60	0·141 7
27	0·144 0	3·70	0·145 7
26	0·147 0	3·70	0·145 7
25	0·149 5	3·80	0·149 6
24	0·152 0	3·90	0·153 5
23	0·154 0	3·90	0·153 5
22	0·157 0	4·00	0·157 5
21	0·159 0	4·00	0·157 5
20	0·161 0	4·10	0·161 4
19	0·166 0	4·20	0·165 4
18	0·169 5	4·30	0·169 3
17	0·173 0	4·40	0·173 2
16	0·177 0	4·50	0·177 2
15	0·180 0	4·60	0·181 1
14	0·182 0	4·60	0·181 1
13	0·185 0	4·70	0·185 0
12	0·189 3	4·80	0·189 0
11	0·191 0	4·90	0·192 9
10	0·193 5	4·90	0·192 9
9	0·196 0	5·00	0·196 8
8	0·199 0	5·10	0·200 8
7	0·210 0	5·10	0·200 8
6	0·204 0	5·20	0·204 7
5	0·205 5	5·20	0·204 7
4	0·209 0	5·30	0·208 7
3	0·213 0	5·40	0·212 6
2	0·221 0	5·60	0·220 5
1	0·228 0	5·80	0·228 3
Letter drill			
A	0·234 0	5·90	0·232 3
B	0·238 0	6·00	0·236 2
C	0·242 0	6·10	0·240 2
D	0·246 0	6·20	0·244 1
E	0·250 0	6·40	0·252 0
F	0·257 0	6·50	0·255 9
G	0·261 0	6·60	0·259 8
H	0·266 0	6·80	0·267 7
I	0·272 0	6·90	0·271 7
J	0·277 0	7·00	0·275 6
K	0·281 0	7·10	0·279 5
L	0·290 0	7·40	0·291 3
M	0·295 0	7·50	0·295 3
N	0·302 0	7·70	0·303 1
O	0·316 0	8·00	0·315 0
P	0·323 0	8·20	0·322 8
Q	0·332 0	8·40	0·330 7
R	0·339 0	8·60	0·338 6
S	0·348 0	8·80	0·346 5
T	0·358 0	9·10	0·358 3
U	0·368 0	9·30	0·366 1
V	0·377 0	9·60	0·378 0
W	0·386 0	9·80	0·385 8
X	0·397 0	10·10	0·397 6
Y	0·404 0	10·30	0·405 5
Z	0·413 0	10·50	0·413 4

Geometry

Chapter 4

Geometry is an extensive subject, and this chapter covers only basic geometric constructions used in engineering drawing. Some of the constructions are draughtsman's approximations, which are satisfactory for ordinary drawings where extreme accuracy is not required and time saving is a more important factor. For large full-scale layout and loft work, templates, dies and sheet metal work, more accurate constructions are often necessary.

GEOMETRIC CONSTRUCTIONS

Figure 4.1

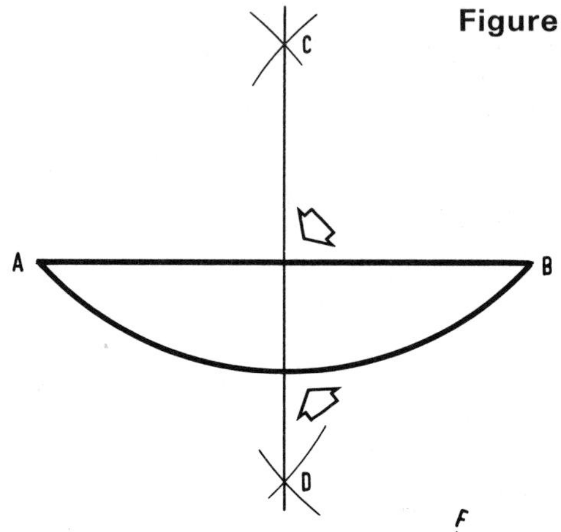

4.1 TO BISECT A GIVEN LINE (OR ARC) AB

1. Set compasses to a distance greater than half AB.
2. With centre A, draw arcs above and below line.
3. With centre B, draw arcs intersecting previous arcs, establishing points C and D.
4. Join CD.
5. The point where CD crosses the line (or arc) is the mid-position.

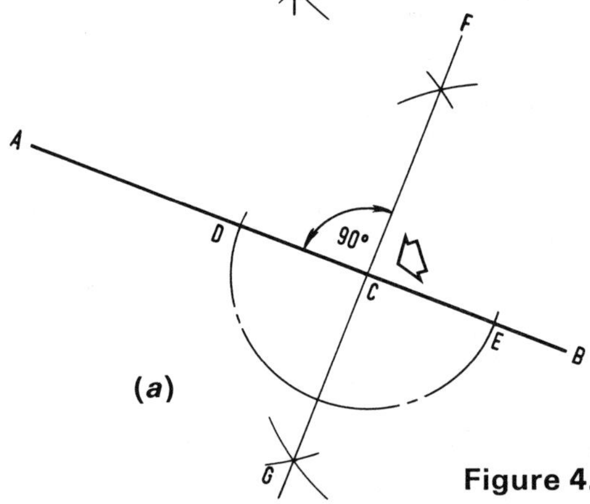

(*a*)

Figure 4.2

4.2 TO ERECT A LINE PERPENDICULAR TO A GIVEN LINE AB AT A GIVEN POINT C

Geometric method (*a*)

1. Open compasses to any suitable radius and, with C as centre, draw an arc DE cutting the given line AB.
2. Bisect the distance between D and E as in 4.1.
3. The bisector line FG is perpendicular to AB at point C.

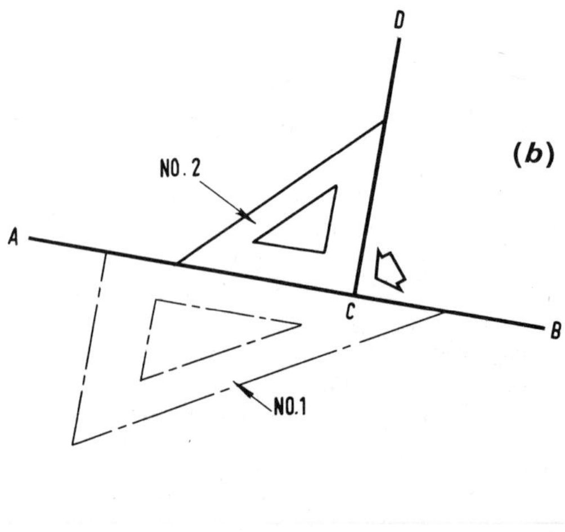

(*b*)

Draughtsman's method (*b*)

1. Carefully line up the side of No. 1 set square with the given line AB.
2. Place No. 2 set square with the 90° angle against No. 1 set square at point C.
3. Draw the required perpendicular line CD.

Figure 4.3

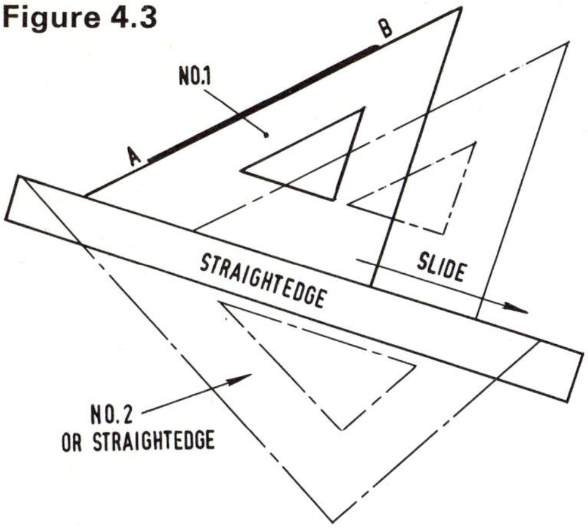

4.3 TO DRAW PARALLEL STRAIGHT LINES TO A GIVEN LINE AB

Draughtsman's method

1. Line up the edge of No. 1 set square carefully against the given line AB.
2. Place the edge of No. 2 set square against No. 1 set square.
3. Slide No. 1 set square to the required positions and draw parallel lines, making sure that No. 2 set square does not move in the process.

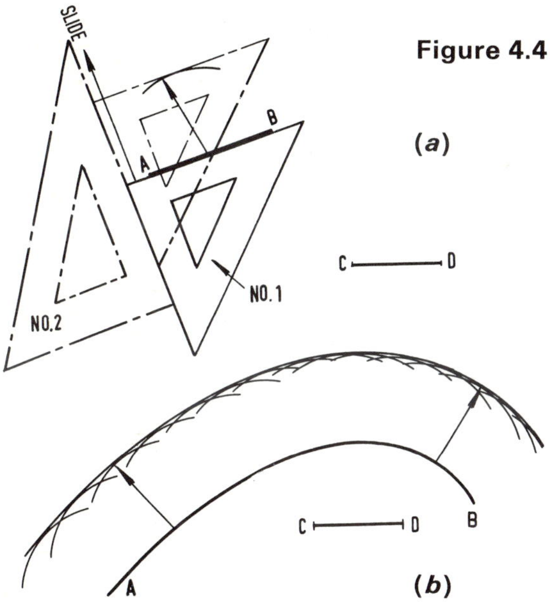

Figure 4.4

(*a*)

(*b*)

4.4 TO DRAW A LINE PARALLEL TO AN EXISTING LINE AT A GIVEN DISTANCE

Straight line (*a*)

1. Let AB be the existing line and CD the given distance.
2. Set the compasses to the given distance CD and, using any point on AB as centre, draw an arc.
3. Using method 4.3 draw a parallel line tangent to the arc.

Curved line (*b*)

1. Let AB be the existing line and CD the given distance.
2. Set the compasses to the given distance CD and, using any points on AB as centres, draw a series of arcs.
3. Using a french curve, draw a smooth line tangent to the arcs.

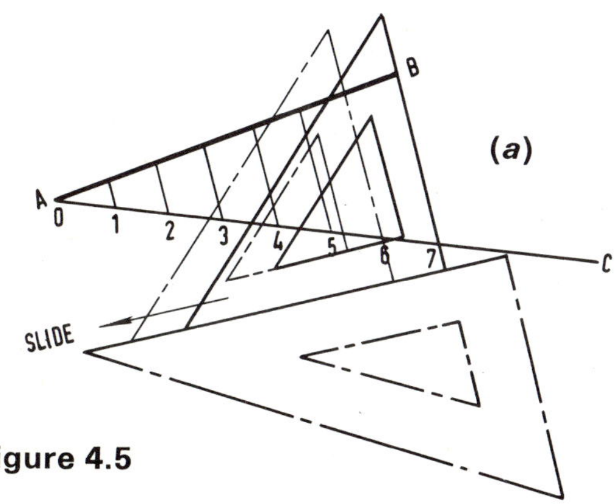

(*a*)

Figure 4.5

4.5 TO DIVIDE A GIVEN LINE AB INTO A NUMBER OF EQUAL PARTS (FOR EXAMPLE, 7 PARTS)

Geometric method (*a*)

1. Draw a line AC at any convenient angle to the given line AB.
2. Using compasses or dividers, set off 7 equal divisions along AC, so that the last division point, 7, falls approximately below the end of the given line.
3. Join point 7 to B.
4. Using method 4.3, draw parallel lines from division points on AC to touch AB, thus dividing it into 7 equal parts.

4.5 (CONTINUED) TO DIVIDE A GIVEN LINE AB INTO A NUMBER OF EQUAL PARTS (FOR EXAMPLE, 7 PARTS)

Draughtsman's method (b)

1. From one end of the given line AB draw a perpendicular BC.
2. Place a ruler so that 7 equal divisions on the ruler lie between the end of the given line, point A, and the perpendicular BC.
3. Mark the positions of the divisions and drop perpendiculars onto AB, thus dividing line AB into 7 equal parts.

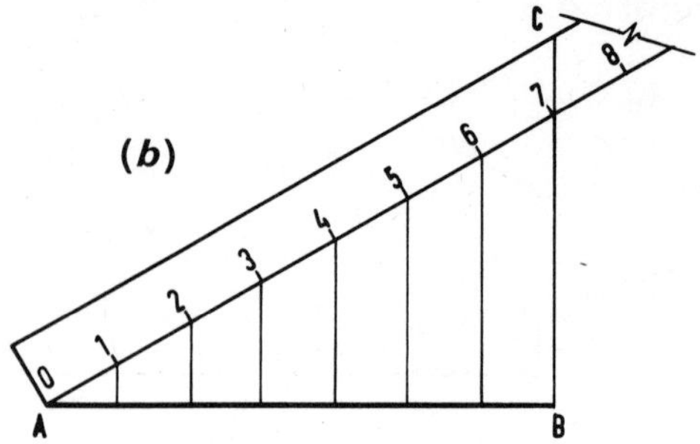

(*b*)

4.6 TO BISECT AN ANGLE

1. Angle BAC is the given angle.
2. With centre A and a convenient radius, draw an arc cutting the lines containing the angle at points D and E.
3. With centres D and E, and compasses set at a greater distance than half DE, draw arcs to intersect at point O.
4. Join OA. The line OA bisects the angle BAC.

By this method, angle BAC can be sub-divided into 4, 8, 16 parts etc., but not an intermediate number of parts.

Figure 4.6

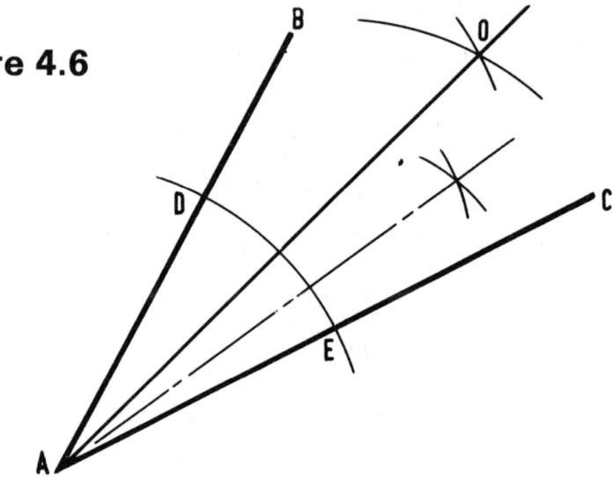

4.7 TO CONSTRUCT ANGLES INCREASING BY 15° FROM 0° TO 180°

1. Draw the base line AB.
2. Select any point O as centre along line AB, and, with compass set at a convenient radius OC, draw a semicircle CD.
3. Using radius OC, step off distances carefully around the semicircle, starting at C. The semicircle will divide equally into 3 parts.
4. Join O through points E and F, thus dividing the semicircle into 3 angles, each 60°.

Figure 4.7

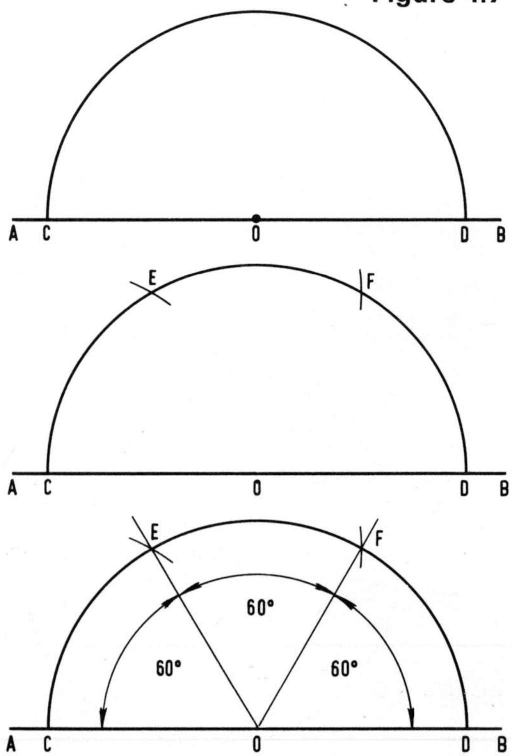

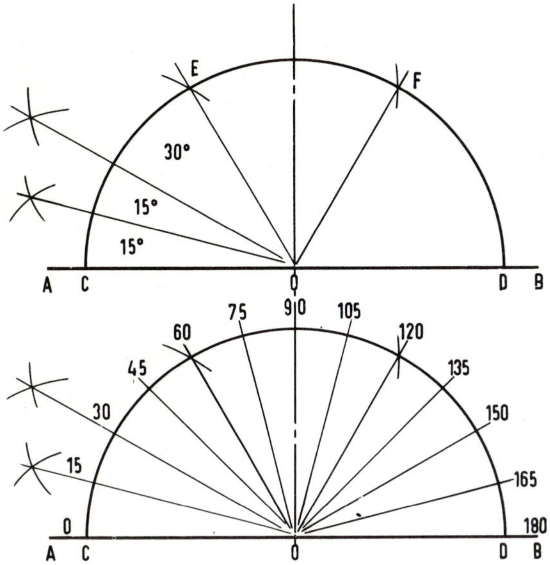

Figure 4.7 *(cont.)*

5. Bisect each angle, making 6 angles of 30°.

6. Bisect each 30° angle, making 12 angles of 15°.

7. Any angle from 0° to 180° increasing by 15° can be measured off.

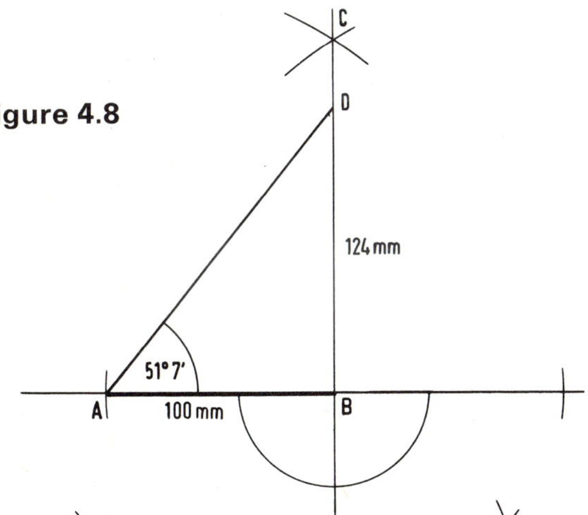

Figure 4.8

4.8 TO DRAW AN ACCURATE ANGLE USING TRIGONOMETRIC TANGENT TABLES

1. The angle required is 51° 7′.
2. Draw a base line and measure off AB 100 mm long as accurately as possible.
3. Using method 4.2, erect a perpendicular BC.
4. From tangent tables read off the value for 51° 7′, and multiply by the base length of 100 mm.
   From tables, tan 51° 7′ = 1·2400
   1·2400 × base length 100 mm = 124 mm
5. Establish point D along the perpendicular by accurately measuring 124 mm from point B.
6. Join AD.
7. Angle DAB is 51° 7′ by construction.

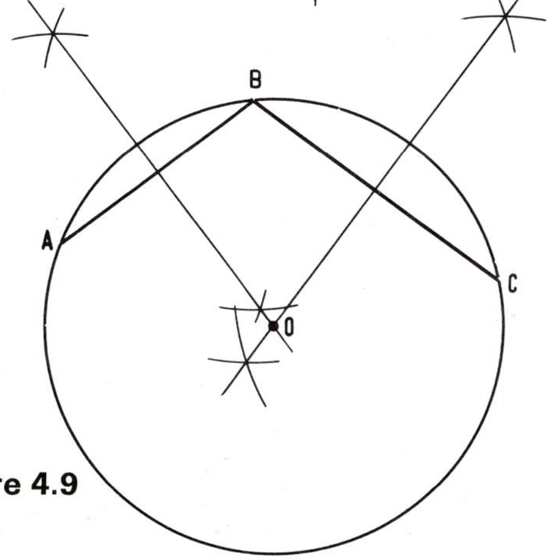

Figure 4.9

4.9 TO LOCATE THE CENTRE OF A GIVEN CIRCLE

1. Draw chords AB and BC as shown.
2. Perpendicularly bisect the chords and let the bisector lines intersect at O.
3. O is the centre of the given circle.

4.10 TO DRAW A CIRCLE OR ARC TO PASS THROUGH THREE GIVEN POINTS

1. Join the given points A, B and C.
2. Perpendicularly bisect AB and BC and let the bisectors intersect at O.
3. With centre O, set the compasses to distance OA and draw a circle which will pass through the 3 given points.

Figure 4.10

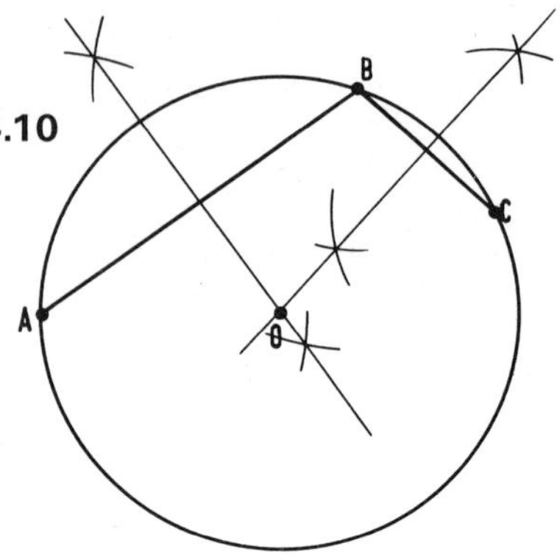

4.11 TO DRAW AN ARC TANGENT TO TWO LINES MEETING AT 90°

1. ABC is the given 90° corner.
2. Set compasses to the required corner radius and, with centre B, draw an arc cutting AB at D and BC at E.
3. With the same radius, and centres D and E, draw arcs intersecting at O.
4. With the same radius, and centre O, draw the arc between points D and E.

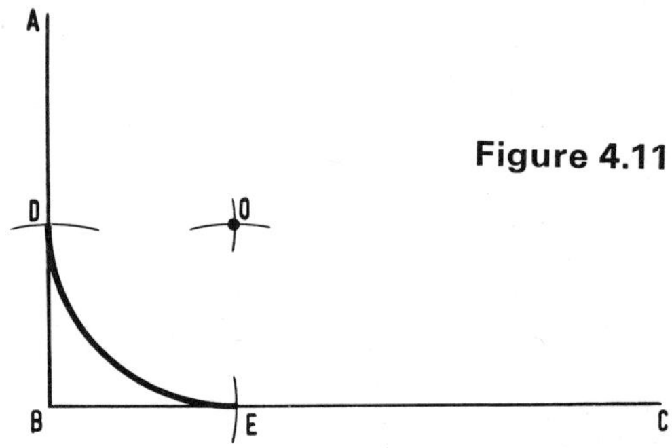

Figure 4.11

4.12 TO DRAW AN ARC TANGENT TO TWO LINES MEETING AT ANY ANGLE

1. Let ABC be the given corner.
2. Set the compasses to the required corner radius and, with any points on AB and BC as centre, draw arcs.
3. Draw lines parallel to AB and BC and tangent to the arcs, intersecting at point O.
4. From point O draw perpendiculars to AB and BC, intersecting AB and BC at D and E respectively.
5. With centre O and radius OD, draw the blending arc between D and E.

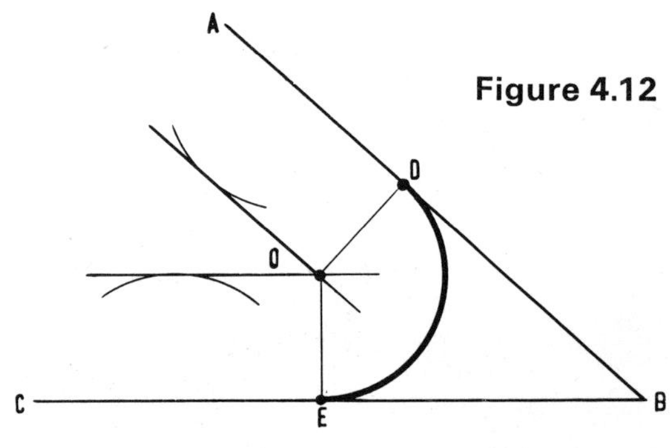

Figure 4.12

4.13 TO DRAW A LINE TANGENT TO A CIRCLE THROUGH A GIVEN POINT ON THE CIRCLE

Geometric method (*a*)

1. The given point A is on a circle with centre O.
2. With centre A and compasses set at radius AO of the circle, draw an arc cutting the circle at B.
3. With the same radius and centre B, draw an arc intersecting the previously drawn arc at C.
4. Bisect arc BC using method 4.1.
5. The perpendicular bisector will be tangential to the circle and pass through the given point A.

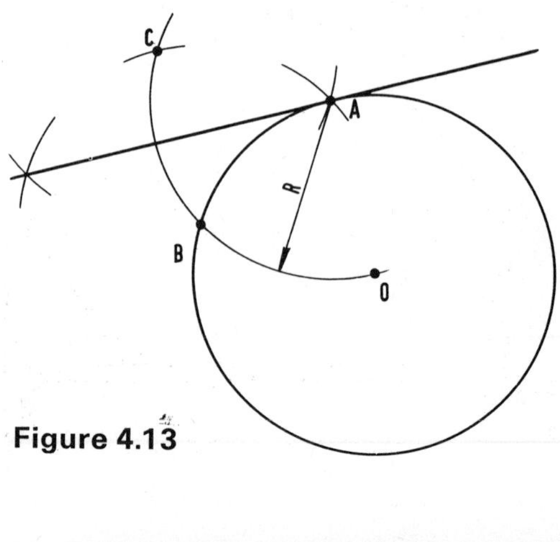

(*a*)

Figure 4.13

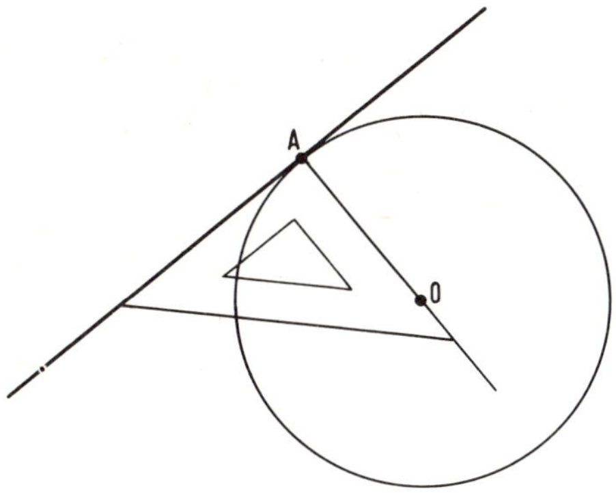

(*b*)

Figure 4.13 *(cont.)*

Draughtsman's method (*b*)

1. The given point A is on a circle with centre O.
2. Place the right-angled corner of a set square precisely at the given point A, with an adjacent side passing through centre O.
3. The other adjacent side of the right angle of the set square is tangential to the circle at point A.
4. Draw a line along this side and extend as necessary.

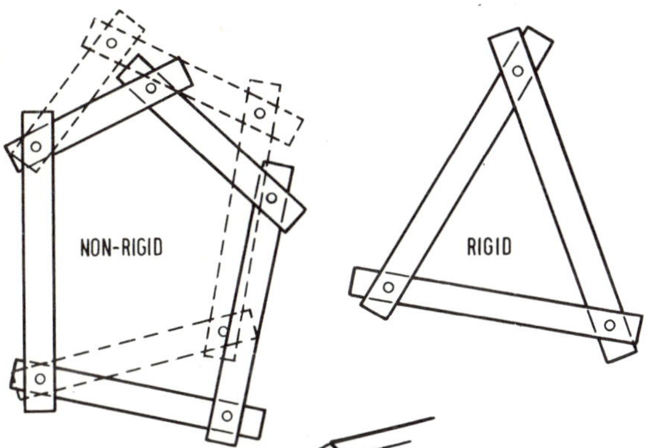

TRIANGLES

The triangle is a three-sided plane figure in which the three internal angles add up to 180°. It is most important in engineering because it can be defined precisely by linear dimensioning. This can be demonstrated by loosely jointing three pieces of wood together with a single nail at each joint. It will be seen that this gives a rigid structure in the plane of the figure.

Four or more pieces jointed in the same way can be pushed into a multitude of shapes; hence it is an unstable structure that can have many interpretations. Because of the inherent rigidity of the triangle and the preciseness with which it can be defined by simple geometrical and trigometrical methods, it is constantly used in engineering.

It can be used for fixing the centres of gear shafts in a multi-gear-drive gear box. It is essential for the layout of sheet-metal developments, and can be everywhere seen in structural engineering, particularly in bridge, crane and roof-truss construction.

It is said that the Egyptian builders used a triangle with sides in the proportions of 3, 4 and 5 to produce a right-angled triangle for laying out the precise squareness of the Pyramids.

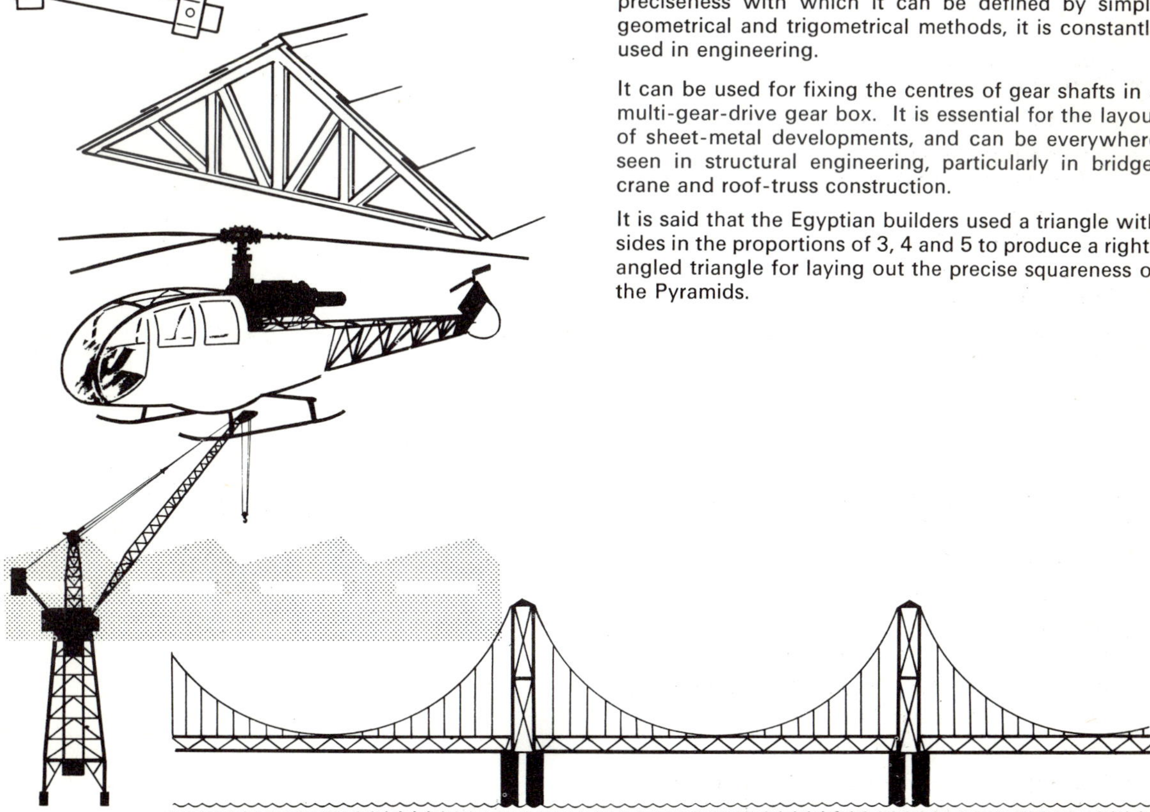

TRIANGLES: GENERAL TERMS

A triangle is a plane figure contained by three straight lines. The most horizontal line is usually termed the *base* and the angles at each end of it the *base angles*. The other two lines are known as *sides*, and the angle at the top is known as the *vertical angle*. The *height* or *altitude* is the perpendicular distance from the base to the *vertex*, or upper angular point. In every triangle the three angles added together equal 180°.

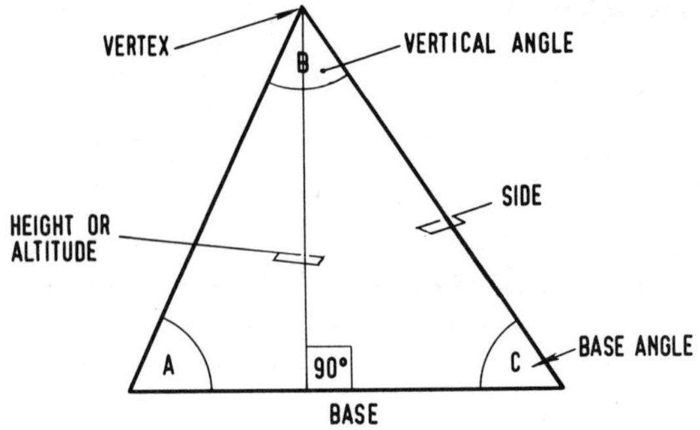

TYPES OF TRIANGLE

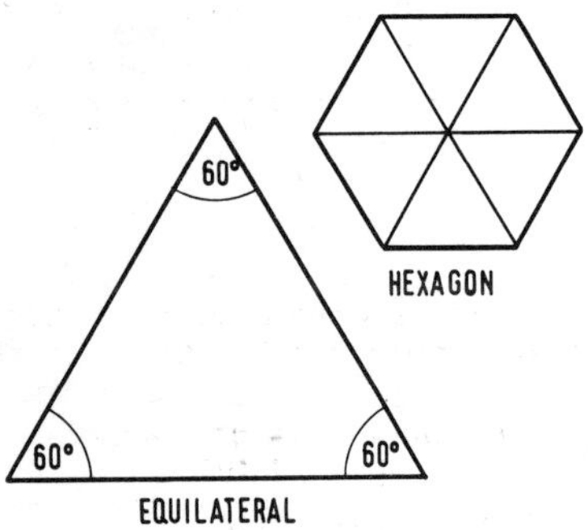

Equilateral

Has all three sides equal in length, and all angles 60°. Note that a hexagon is a pile of equilateral triangles.

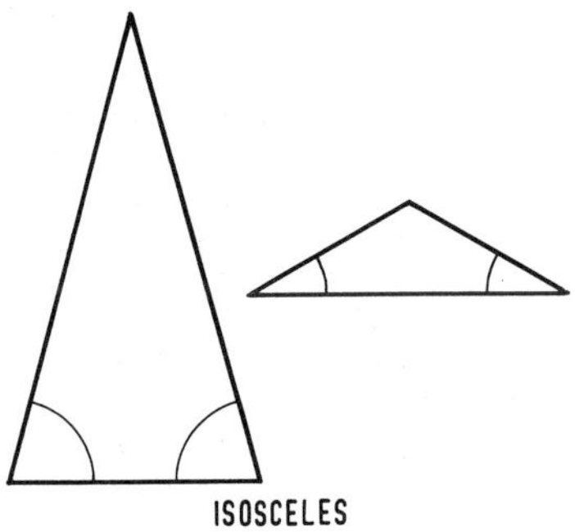

Isosceles

Has two sides equal and two angles equal.

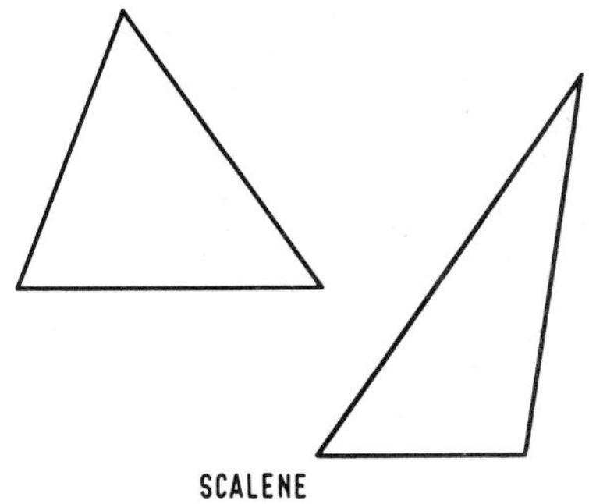

Scalene

All sides and angles are unequal.

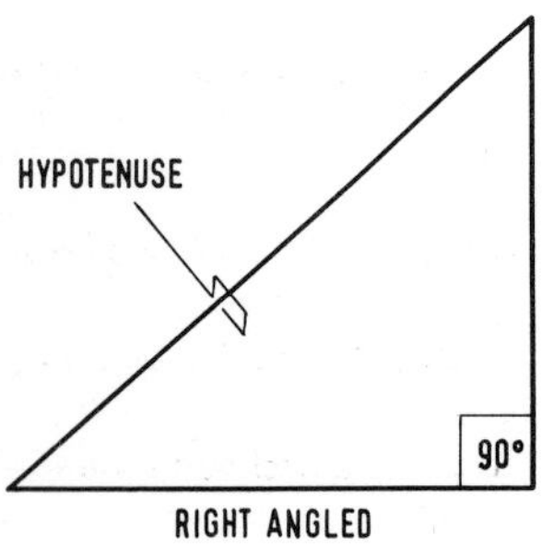

Right-angled

Has one of its angles 90°. The side opposite the right angle is called the hypotenuse. The right-angled triangle is of special importance because it is the basis of all trigonometric ratios. The theorem of Pythagoras, an ancient Greek mathematician, states that the area of a square drawn on the hypotenuse is equal to the sum of the areas of squares drawn on the other two sides.

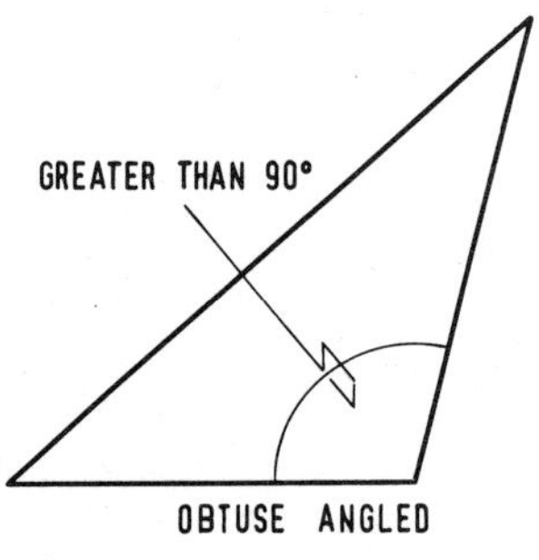

Obtuse-angled

Has one of its angles greater than 90°.

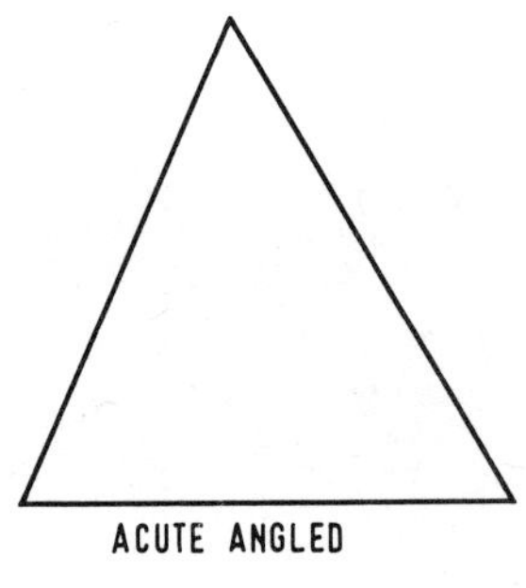

Acute-angled

Has each of its three angles less than 90°.

Figure 4.14

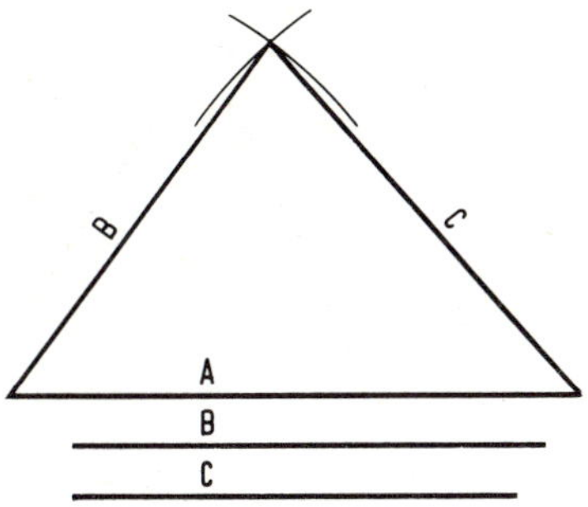

4.14 TO CONSTRUCT A TRIANGLE GIVEN THE LENGTHS OF THE THREE SIDES

1. Draw line A.
2. With one end of A as centre, and compasses set to length B, draw an arc.
3. With the other end of A as centre, and compasses set to length C, draw an intersecting arc.
4. Join the ends of A to the intersection of the arcs.

Figure 4.15

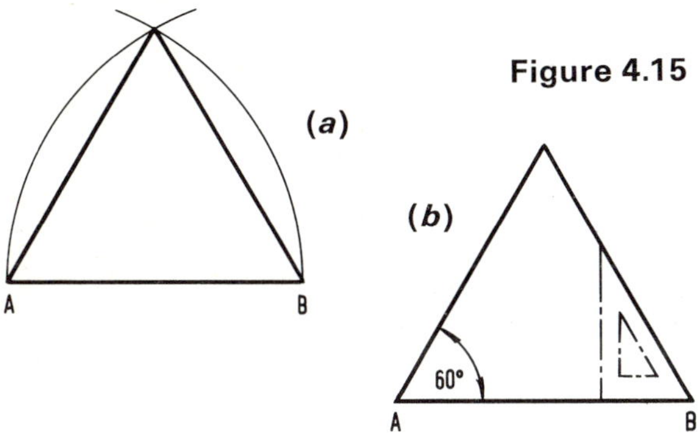

4.15 TO CONSTRUCT AN EQUILATERAL TRIANGLE ON A GIVEN BASE

Geometric method (*a*)

1. Draw the given base AB.
2. With centres A and B, and compasses set to length AB, draw intersecting arcs.
3. Join the ends of AB to the intersection of the arcs.

Draughtsman's method (*b*)

1. Draw the base.
2. With a 30°–60° set square, draw lines at 60° from the ends of the base to intersect at the vertex.

Figure 4.16

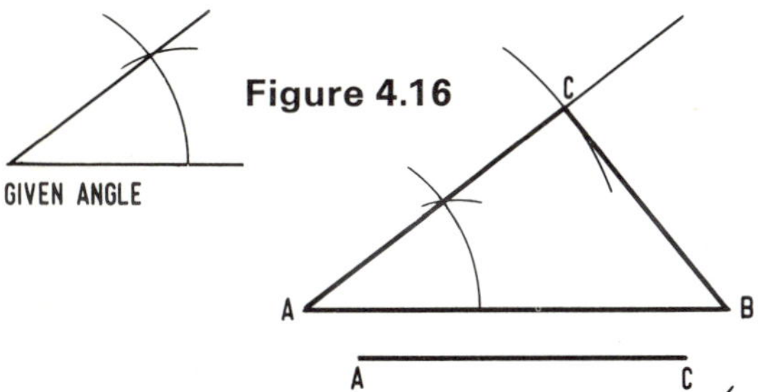

4.16 TO CONSTRUCT A TRIANGLE GIVEN TWO SIDES AB AND AC AND THE ANGLE BETWEEN THEM

1. Draw the given side AB as a base line.
2. At A construct the given angle, using a protractor if necessary.
3. Along the second line enclosing the angle mark off the length AC, using compasses.
4. Join BC.

Figure 4.17

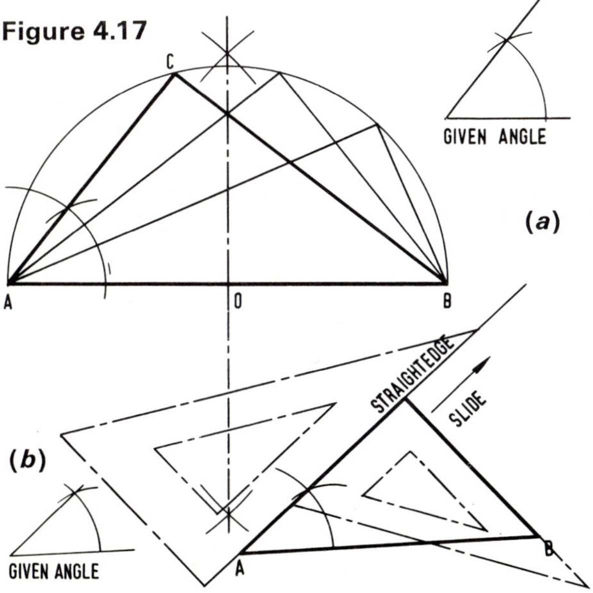

4.17 TO DRAW A RIGHT-ANGLED TRIANGLE GIVEN THE HYPOTENUSE AB AND ONE ANGLE OTHER THAN 90°

Geometric method (*a*)

1. Draw the given line AB and find centre O by perpendicularly bisecting.
2. With centre O and radius OA, draw a semicircle between A and B.
3. At A construct the given angle and continue the line to point C on the semicircle.
4. Join C to B.

Note: Any triangle drawn with the diameter of a circle as base and its vertex on the circumference is a right-angled triangle.

Draughtsman's method (*b*)

1. Draw the given line AB.
2. At A construct the given angle.
3. Place a straight-edge on the line enclosing the angle and slide the right angle of the set square along the straight-edge to coincide with point B.
4. Draw the closing line of the triangle.

4.18 TO SHOW THAT THE SQUARE ON THE HYPOTENUSE OF A RIGHT-ANGLED TRIANGLE IS EQUAL TO THE SUM OF THE SQUARES ON THE OTHER TWO SIDES

1. Draw a right-angled triangle having the right-angled sides 3″ and 4″ long. It will be found that the hypotenuse is 5″ long.
2. Construct squares on each side and divide these into 1″ squares.
3. Count the squares. It will be seen that the number or squares on the hypotenuse, and hence the area, is equal to the sum of the other two squares.

Figure 4.18

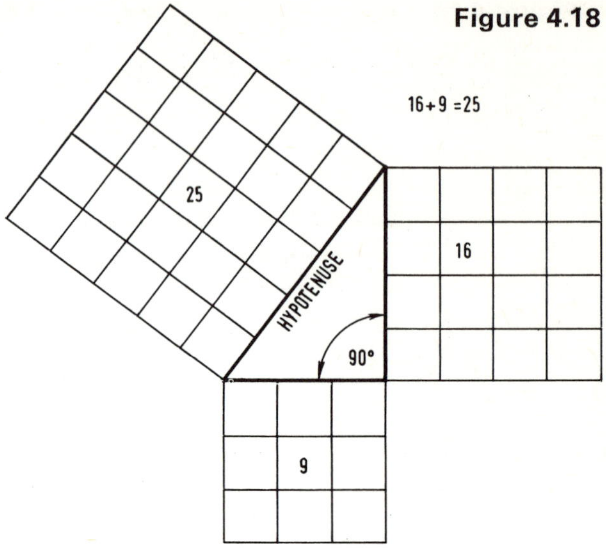

4.19 TO DRAW A CIRCLE PASSING THROUGH THE THREE VERTICES OF A TRIANGLE ABC (CIRCUMSCRIBING CIRCLE)

1. Draw any triangle ABC.
2. Perpendicularly bisect each side.
3. The bisectors will intersect at a common point O which is central between points A, B and C.
4. With centre O and radius OA, draw the circumscribing circle.

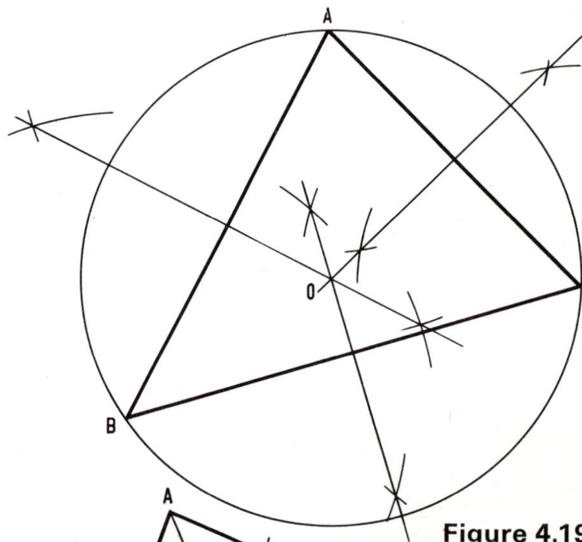

Figure 4.19

4.20 TO DRAW A CIRCLE INSIDE A TRIANGLE SO THAT EACH SIDE OF THE TRIANGLE IS TANGENTIAL TO THE CIRCLE (INSCRIBED CIRCLE)

1. Draw any triangle ABC.
2. Bisect each angle.
3. The bisectors will be found to pass through a common point O which is the centre of the required circle.
4. With centre O, and compasses set to the perpendicular distance to any side, draw the inscribing circle.

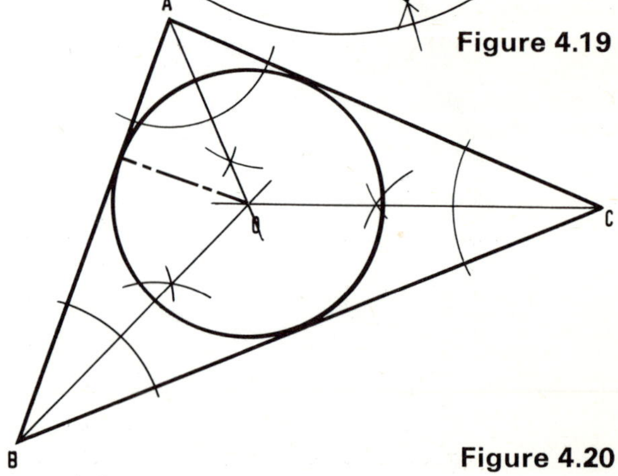

Figure 4.20

THE ELLIPSE

The ellipse is a curve that is obtained when a cylinder or cone is cut by a sectioning plane not at right angles to the axis.

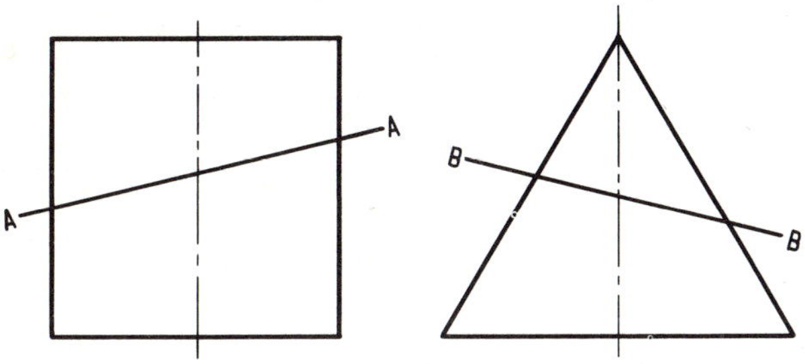

CYLINDER CONE

In engineering drawing the ellipse appears on drawings mainly as the result of the projection of views. It appears in pictorial form to give the craftsman a visual impression of the object under construction, and not to supply dimensional information.

Sometimes a true ellipse form may be required on aircraft structures or submarine hulls for design purposes, and in these particulars the draughtsman must be able to produce the exact shape properly dimensioned.

Hence the draughtsman must have methods which are suitable for accuracy when required, and also methods by which drawings may be prepared in the minimum of time to give close pictorial approximations.

As the pictorial application of an ellipse is needed far more often than the true dimensionally-defined shape, it is wise to memorise *one only* of the numerous methods that are given in textbooks, and to keep to this method. Section 4.21 shows a simple two-arc construction produced by compasses.

4.21 TO DRAW AN ELLIPSE BY APPROXIMATE TWO-ARC CONSTRUCTION

Figure 4.21

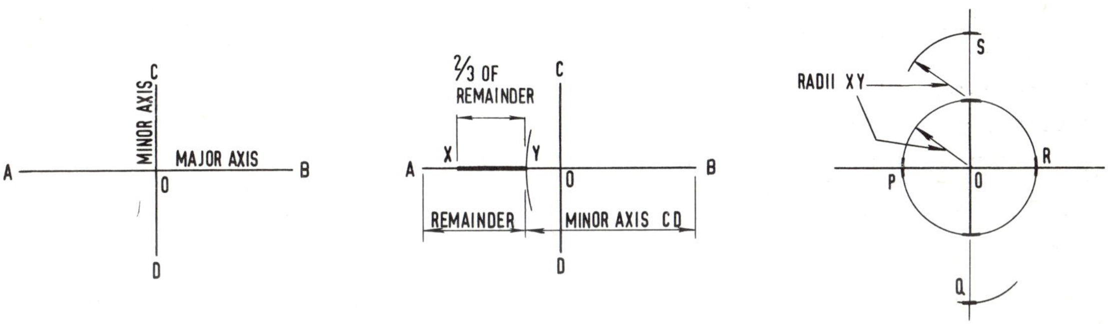

(a) Set out major and minor axes AB and CD.

(b) Mark off the length of the minor axis along the major axis, and select 2/3 of the remainder XY as shown, using method 4.5 if necessary.

(c) With distance XY in the compasses, step off from the centre as follows:

1. Once either side of the main axis.
2. Twice either side of the minor axis (produce the minor axis if necessary to establish centres P, Q, R and S).

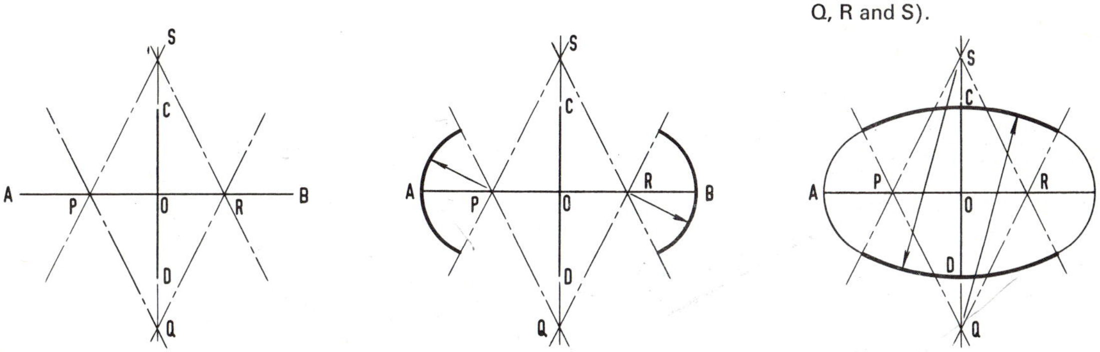

(d) Join centres P, Q, R and S in diamond pattern to set up limits for drawing circular arcs.

(e) With centre P, and radius to the end of the major axis (PA), draw an arc between the diamond construction lines. Repeat with centre R.

(f) With centre Q and radius QC, draw an arc blending with the previously drawn arcs. Repeat with centre S, completing the ellipse.

4.22 TO DRAW AN ELLIPSE USING THE CONCENTRIC-CIRCLE METHOD

1. Draw major and minor axes AB and CD, and use these as diameters to draw circles.
2. Divide the circles into a number or equal parts with radial lines, using a 60°–30° set square.
3. Where the radial lines cut the inner and outer circles, draw horizontal and vertical lines respectively. The intersections of the horizontal and vertical lines are points on the ellipse.
4. Using a french curve, draw a fair curve through the plotted parts.

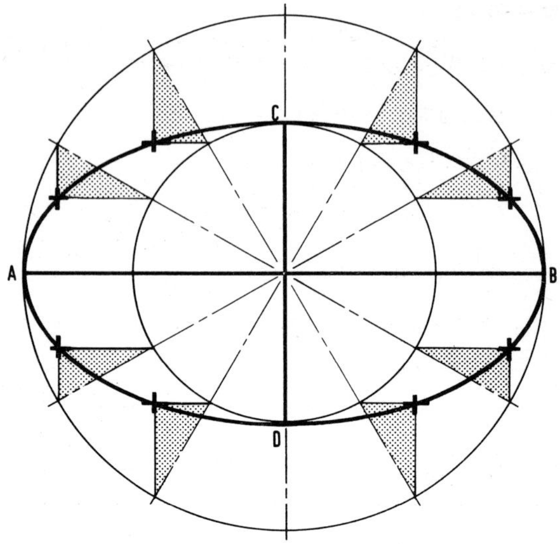

Figure 4.22

4.23 TRUE ELLIPSE DIMENSIONED FOR MANUFACTURING PURPOSES

For an accurately-dimensioned true ellipse for constructional purposes, horizontal and vertical co-ordinates from x and y axes must be given.
The mathematical formula for an ellipse is:

$$\frac{x^2}{a^2} + \frac{y^2}{b^2} = 1$$

Figure 4.23 shows the meaning of the symbols in the above equation.

If the formula for an ellipse is

$$\frac{x^2}{a^2} + \frac{y^2}{b^2} = 1$$

by rearrangement:

$$y = \sqrt{\frac{b^2}{a^2}(a^2 - x^2)}$$

Thus, if the major and minor axes are known, values of y can be calculated for given values of x. Accurate dimensions for a set of points can then be calculated, and the information placed on a drawing as a TABLE OF CO-ORDINATES FOR x AND y.

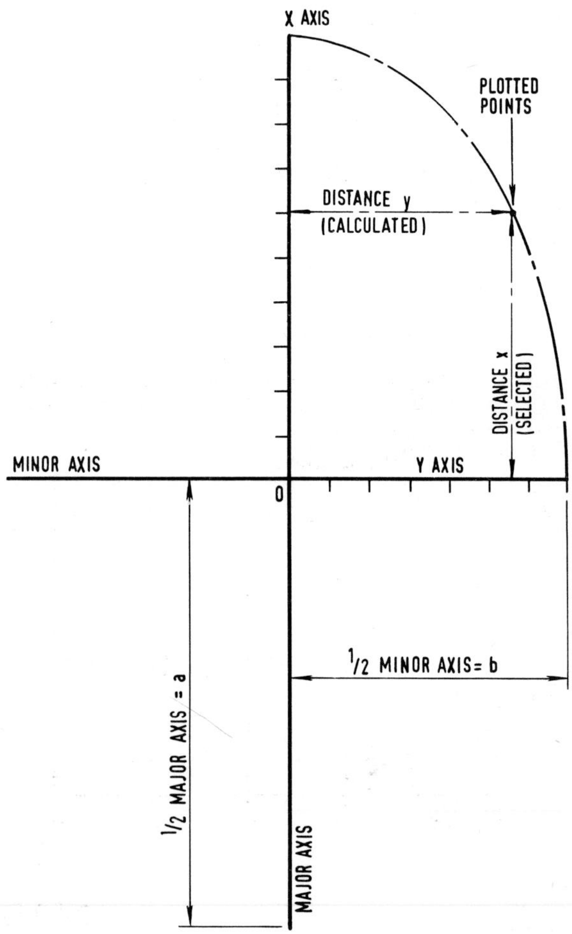

Figure 4.23

EXAMPLE

Draw a fully dimensioned true ellipse having a major axis of 200 mm and a minor axis of 80 mm.

The formula for the ellipse is $\frac{x^2}{a^2} + \frac{y^2}{b^2} = 1$

where $a = 100$ mm ($\frac{1}{2}$ major axis)
$b = 40$ mm ($\frac{1}{2}$ minor axis)
$x = 10$ mm, 20 mm, 30 mm etc. up to 100 mm

y is calculated from the formula $y = \sqrt{\frac{b^2}{a^2}(a^2 - x^2)}$

Draw up a table of values for x and y co-ordinates.

TABLE OF CO-ORDINATES FOR ELLIPSE (IN MILLIMETRES)											
Given values of x	0	10	20	30	40	50	60	70	80	90	100
Calculated values of y	40	39·80	39·19	38·16	36·66	34·64	32·00	28·57	24·00	17·44	0

Specimen calculations

Calculation of y when $x = 10$ mm

$$y = \sqrt{\frac{b^2}{a^2}(a^2 - x^2)}$$
$$y = \sqrt{\frac{40^2}{100^2}(100^2 - 10^2)}$$
$$y = \sqrt{\frac{1600}{10\,000}(10\,000 - 100)}$$
$$y = \sqrt{\frac{1600 \times 9900}{10\,000}}$$
$$y = \sqrt{1584}$$
$$y = \mathit{39{\cdot}80} \text{ mm}$$

Calculation of y when $x = 40$ mm

$$y = \sqrt{\frac{b^2}{a^2}(a^2 - x^2)}$$
$$y = \sqrt{\frac{40^2}{100^2}(100^2 - 40^2)}$$
$$y = \sqrt{\frac{1600}{10\,000}(10\,000 - 1600)}$$
$$y = \sqrt{\frac{1600 \times 8400}{10\,000}}$$
$$y = \sqrt{1344}$$
$$y = \mathit{36{\cdot}66} \text{ mm}$$

The above table of results gives x and y co-ordinates for one quarter of the ellipse. The full ellipse is obtained by symmetrical plotting.

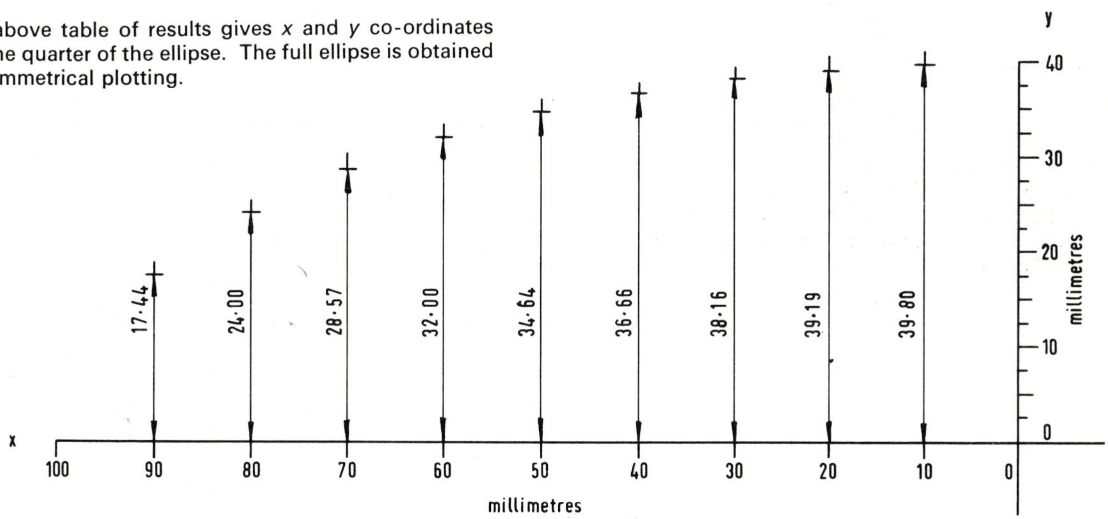

PLOT OF POINTS FROM TABLE

THE HEXAGON

The hexagon is a regular six-sided figure which is regularly occurring in engineering drawing because it is the shape of most nuts and bolts. Hexagonal bar is one of the standard stock sections available for the manufacture of machine details which have to be screwed tight with standard spanners.

Figure 4.24

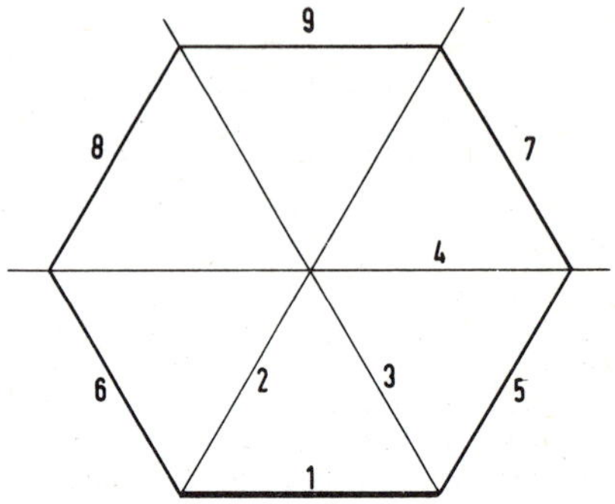

Nos SHOW SEQUENCE OF DRAWING LINES

4.24 TO DRAW A HEXAGON HAVING SIDES OF A GIVEN LENGTH

1. Draw one side, making it the required length.
2. Using a 60°–30° set square, draw a pile of 6 equilateral triangles on this side, as shown.

Figure 4.25

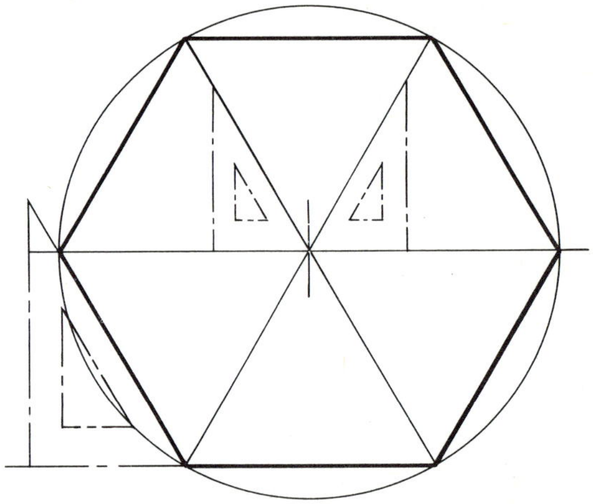

4.25 TO DRAW A HEXAGON OF GIVEN DIMENSIONS ACROSS THE CORNERS

1. Draw a circle of diameter equal to the distance across the corners.
2. With a 60°–30° set square, divide the circle into 6 equal parts.
3. Using the 60°–30° set square, draw chords in the six parts to produce the hexagon.

Figure 4.26

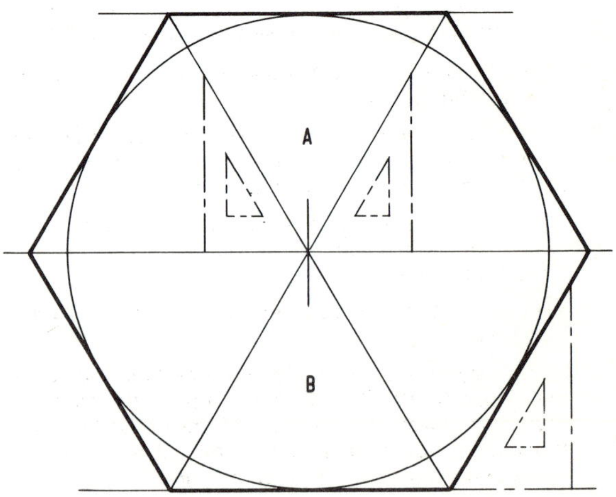

4.26 TO DRAW A HEXAGON OF GIVEN DIMENSIONS ACROSS THE FLATS

1. Draw a circle of diameter equal to the distance across the flats.
2. Draw horizontal construction lines tangential to the top and bottom of the circle.
3. With a 60°–30° set square, draw 2 lines at 60° to the horizontal passing through the centre of the circle, continuing the lines to cut the top and bottom tangential lines and form two equilateral triangles A and B.
4. Complete the pile of equilateral triangles with a 60°–30° set square as shown in Figure 4.24.

THE REGULAR POLYGON

A polygon is a regular many-sided figure which is occasionally needed in engineering drawing. A number of geometric methods of constructing polygons can be found in standard text books. It is difficult to remember them all, so it is best to learn a simple principle which applies to any polygon and can be stored in the memory for use when the need arises. Note two points:

1. As the complete polygon is formed round 360°, each side forms at the centre an angle which must be equal to 360° divided by the number of sides.
2. The external angle is also 360° divided by the number or sides.

   Imagine walking round the polygon. At each corner, in order to continue along the next side, the direction must be changed by an angle. To complete the full circuit of 360° this change of direction must be made as many times as there are sides in the polygon. This angle is known as the external angle and is equal to 360° divided by the number of sides.

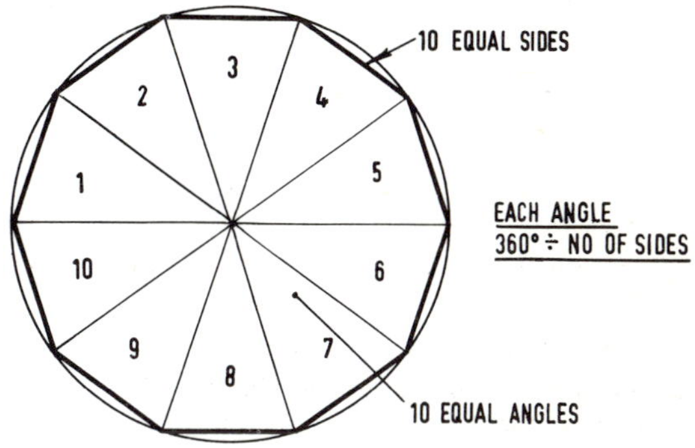

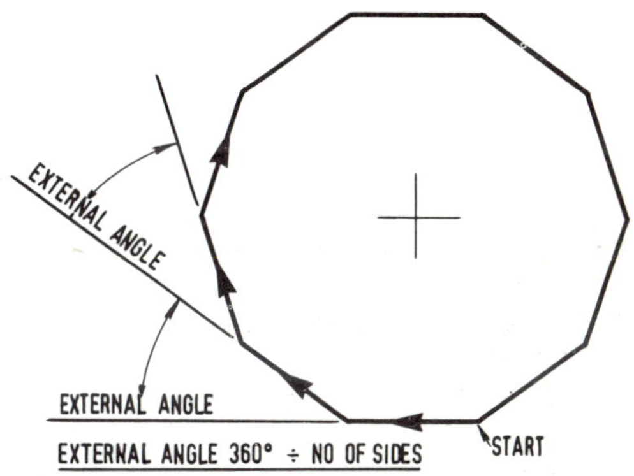

4.27 TO DRAW A REGULAR POLYGON HAVING A GIVEN NUMBER OF SIDES AND EACH SIDE A GIVEN LENGTH

Problem: To draw a regular polygon of 13 sides, each side being 25 mm in length.

1. With centre O, draw a circle of convenient radius and lay off as accurately as possible an angle equal to 360° divided by 13 (27·7° approximately).
2. Take the chordal distance AB in the dividers and, with slight adjustment if necessary, step round the circle 13 times exactly.
3. Join centre O to the 13 points with radial construction lines.
4. Accurately bisect one angle and draw a line parallel to the bisector line, half the side length of the polygon (12·5 mm) away. With radius OP draw a circle. Join the intersection points of the radial lines and the circle to construct the required polygon.

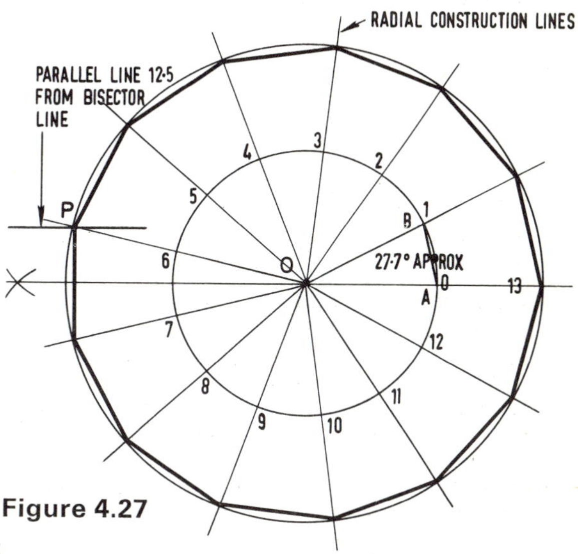

Figure 4.27

GEOMETRIC CURVES (LOCI)

There are many geometric curves which can be drawn for exercise purposes, but few of these occur in engineering applications. Those described here have a direct application to everyday engineering, and the student would be well advised to commit the constructions to memory.

ARCHIMEDEAN SPIRAL

Two applications of this curve are the groove on a gramophone record and the operating scroll of the self-centring lathe chuck. The curve is the locus (or path) of a point moving outwards at a uniform rate as it rotates at a constant angular velocity around a fixed point.

Figure 4.28

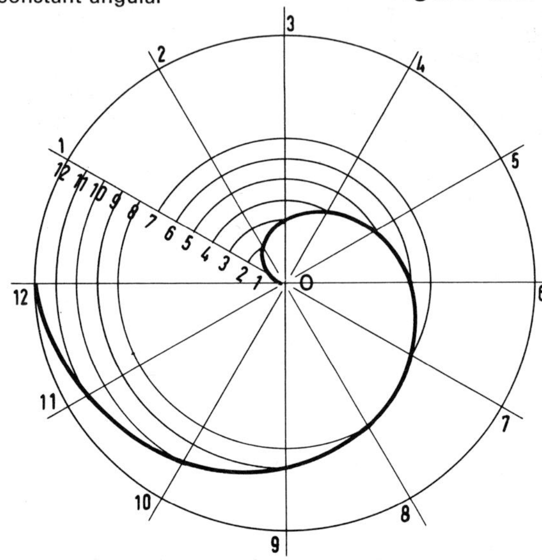

ARCHIMEDIAN SPIRAL

4.28 TO DRAW AN ARCHIMEDEAN SPIRAL HAVING A SIDEWAYS DISPLACEMENT OF 36 mm IN ONE COMPLETE REVOLUTION

1. With centre O, draw a circle of 36 mm radius.
2. Divide the circle into 12 equal angles of 30° each and number them consecutively from 1 to 12.
3. Divide radius No. 1 into twelve equal parts, numbering the divisions from 1 to 12.
4. Using compasses, transfer the points on radius No. 1 to the corresponding radial lines.
5. Join the plotted points with a smooth curve, starting at the centre and moving outward to the circumference.

THE INVOLUTE OF A CIRCLE

The involute can best be described as the path traced out by a point on a taut string as it is unwound from a cylinder. The profile of modern machine-cut gears is based on the involute curve.

4.29 TO DRAW AN INVOLUTE OF A CIRCLE

1. Draw any circle.
2. Draw a straight line equal in length to the circumference of the circle.
3. Divide the circle into a number of equal angles (say 12) and divide the straight line into the same number of equal parts.

   Note: that one straight division is equal in length to one circular arc division.
4. Number the divisions on both the circle and the straight line from 1 to 12.
5. Draw lines tangent to the points on the circle.
6. With compasses, transfer distances 01, 02, 03 etc. from the straight line to the corresponding tangent lines.
7. Draw a smooth curve through the plotted points.

Figure 4.29

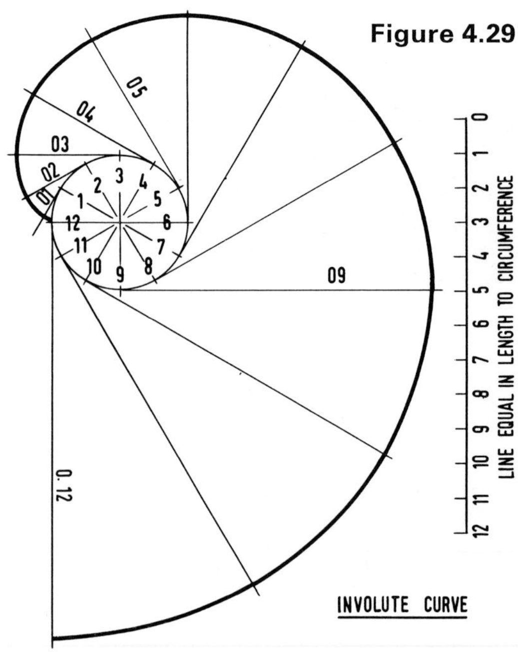

INVOLUTE CURVE

THE CYCLOID

The cycloid is the path traced out by a point on a circle as it rolls along a flat surface. The profile of rack teeth on cast gears is based on the cycloid curve.

4.30 TO DRAW A CYCLOID

1. Draw a circle with centre C and radius R. Tangent to it at its lowest point draw a horizontal straight line AB equal in length to the circumference of the circle. (In one revolution point P on the circle will roll from point A and arrive at point B, tracing out a cycloid in between.)
2. Divide the circle and the line AB into the same number of equal parts (say 12) and number the points from 1 to 12.
3. At the points on the straight line AB erect vertical ordinates.
4. Through the points on the circle draw horizontal projection lines. The centre line is the path line of the centre of the circle, and for each consecutive one-twelfth of a revolution the circle centre will coincide with the vertical ordinates at C1, C2, C3 etc.
5. With centres C1, C2, C3 etc., and compasses set at distance R, plot the position of point P on the corresponding horizontal projection lines.
6. Join the plotted points with a smooth curve between A and B.

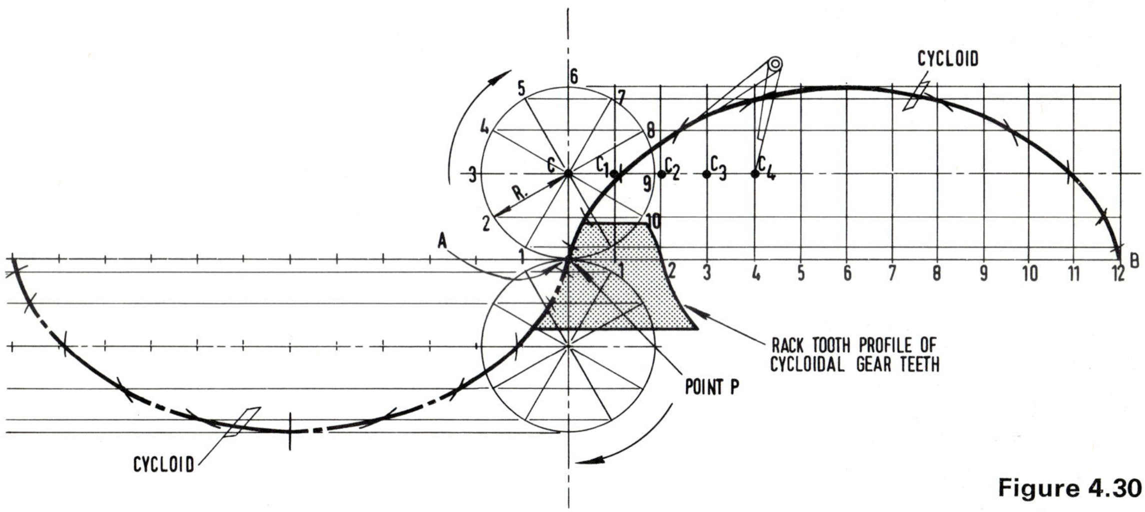

Figure 4.30

THE EPICYCLOID

The *epicycloid* is a curve traced out by a point on a circle as it rolls on the *outside* of the circumference of another circle.

THE HYPOCYCLOID

The *hypocycloid* is a curve traced out by a point on a circle as it rolls on the inside of another circle.

4.31 TO CONSTRUCT AN EPICYCLOID

Use same procedure for constructing a hypocycloid.

1. A rolling circle of centre O is in contact at point P with a fixed arc AB, struck from centre C. *Note* that as the circle rolls point P will generate an epicycloid curve.
2. Divide the rolling circle into a number of equal parts (say 8) and number each point.
3. Lay off round the fixed circumference from point P an equal number of divisions, each one equal in length to the rolling circle divisions, and number as shown.
4. With centre C and radius CO, strike an arc DO to represent the path of the rolling circle, centre O.
5. With radial lines, join centre C to the points on the fixed circumference and continue them to cut arc DO, thus positioning the centre of the rolling circle at each one-eighth of a revolution—O1, O2, O3 etc.
6. With centre C, strike concentric arcs through the points on the rolling circle to establish the height position of point P above the fixed circle at every one-eighth of a revolution.
7. With centres O1, O2, O3 etc., and radius OP in the compasses, plot points on the corresponding numbered concentric arcs drawn in (6) above.
8. Join the plotted points with a smooth curve starting from P.

Figure 4.31 shows an epicycloid curve blending with a hypocycloid curve. The immediate portion either side of the blending point is the profile of cycloidal cast gears, and also of the gears, blanked out of sheet material, that are used in the watch and clock-making industry.

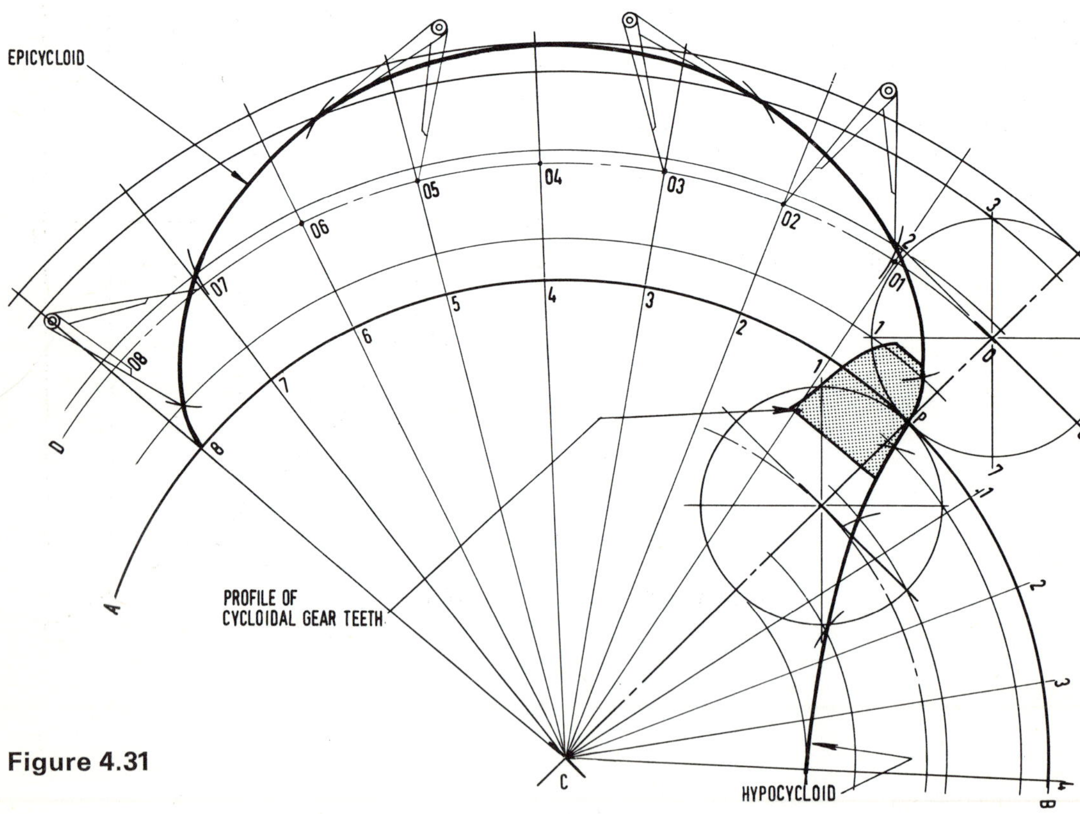

Figure 4.31

THE PARABOLA

This is a conic section; it is that curve which is obtained by a cutting plane parallel to the sloping side of a cone.

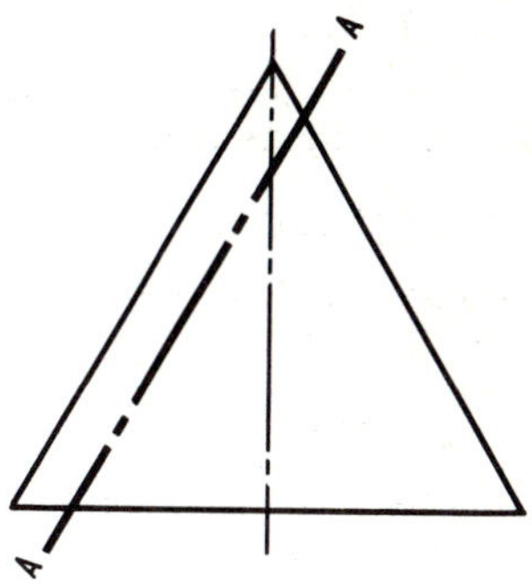

It is also the curve that is generated by a point moving along a path, keeping equidistant from a point called a *focus* and a straight line called a *directrix.*

Reflectors generated from the parabola have the ability to transmit parallel rays of heat, light and energy from a point source of transmission situated at the focus.

Conversely a parabolic reflector can condense weak parallel rays of energy from a great distance and concentrate them at the focal point, where they can be detected and measured.

Parabolic reflectors are to be found in searchlights, headlights, spotlights, electric heating appliances, reflecting telescopes and radio and radar aerials.

4.32 TO DRAW A PARABOLA GIVEN A DIRECTRIX AND A FOCUS

1. AB is the given directrix, and F is the focus.
2. Draw line MN through the focus point F, and at right angles to the directrix AB, cutting AB at O.
3. Bisect the distance between the directrix AB and focus F to establish the vertex V of the parabola.
4. From point V mark off a convenient number of points, numbering them 1, 2, 3 etc., and erect ordinates at these points parallel to the directrix.
5. With distance O1 in the compass and F as centre, strike an arc cutting ordinate No. 1. With distance O2 in the compass and F as centre, strike an arc cutting ordinate No. 2. These are points on the parabola. Repeat with distance O3, O4 etc., cutting corresponding ordinates.
6. Draw a smooth curve through the plotted points.

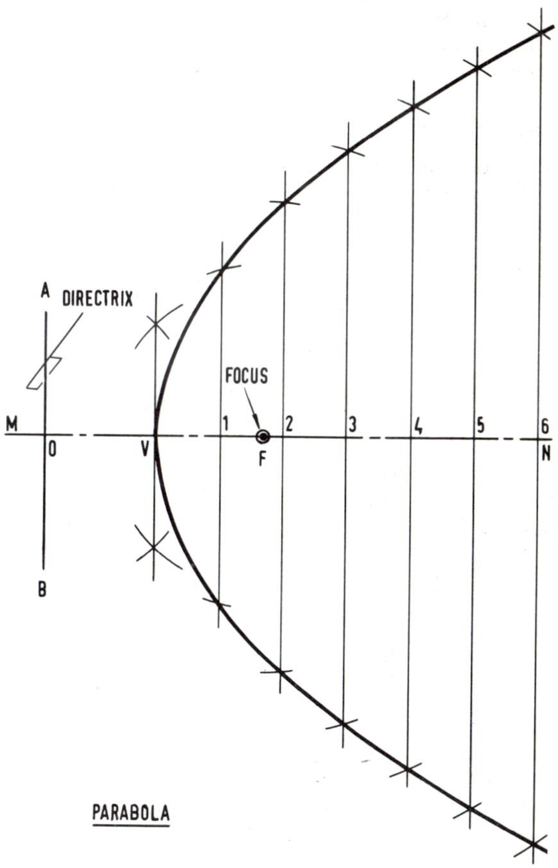

Figure 4.32

4.33 TO DRAW A PARABOLA BY THE RECTANGLE METHOD

1. Given the length AB and the width MN of the required parabola, draw a rectangle as shown.
2. Divide the length AB into a number of equal parts and number them in the direction shown. Draw lines from the numbered points converging to point O.
3. Divide the half-width ON into the same number of equal parts, and number the points outward from the vertex O. From these points draw lines parallel to the axis OX.
4. The intersection points of the corresponding numbered lines are points on the parabola.
5. Draw a smooth curve through the plotted points.

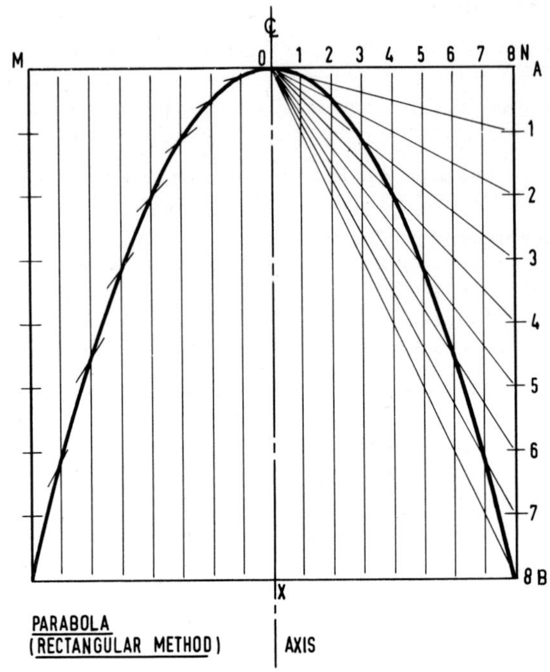

Figure 4.33

Geometry

Chapter 4

EXERCISES

EXERCISE 4.1

Divide an A2 or a half-imperial sheet of paper into twelve parts with two horizontal lines and three vertical lines as shown on page 51, and draw a title block in the bottom right-hand corner square. Draw the following geometric exercises, one in each square, clearly showing all construction.

(a) Draw a straight line 114 mm long at 20° to the horizontal. At a point 53 mm from one end construct a perpendicular.

(b) Draw a straight line at 15° to the horizontal and on it mark off a distance of 80 mm. Using the method of two squares, draw seven more lines 80 mm long parallel to the first line at 10 mm spacing. The ends of all the lines must be on perpendiculars drawn from the first line.

(c) Draw a horizontal line 117 mm long and divide it geometrically into thirteen equal parts.

(d) Construct an angle of 52° 26′ using tangent tables (use 10 cm as a base line), and then geometrically bisect it into four equal angles.

(e) With the origin at the bottom left-hand corner of the next rectangular space, draw horizontal and vertical *x* and *y* axes. Plot points whose (*x*, *y*) co-ordinates are (18,65), (40,88), (93,70) measured in millimetres. By geometric construction find the centre of a circle which will pass through the three plotted points, and state its (*x*, *y*) co-ordinates. Measure and state the diameter of the circle in millimetres.

(f) Draw a circle so that the three sides of a triangle are tangential to its circumference. The length of the sides of the triangle are 79 mm, 84 mm and 103 mm. State the diameter of the circle in millimetres.

(g) What is the diameter of a circle which will just pass through the points of a triangle having sides 70 mm, 77 mm and 85 mm?

(h) Construct an equilateral triangle having an area of 25 cm^2. Measure and state length of side in millimetres.

(j) Draw a regular hexagon having sides 53 mm long.

(k) Construct a regular heptagon (7-sided polygon) having sides 46 mm long.

(l) Draw a regular hexagon measuring 83 mm A/F (across flats).

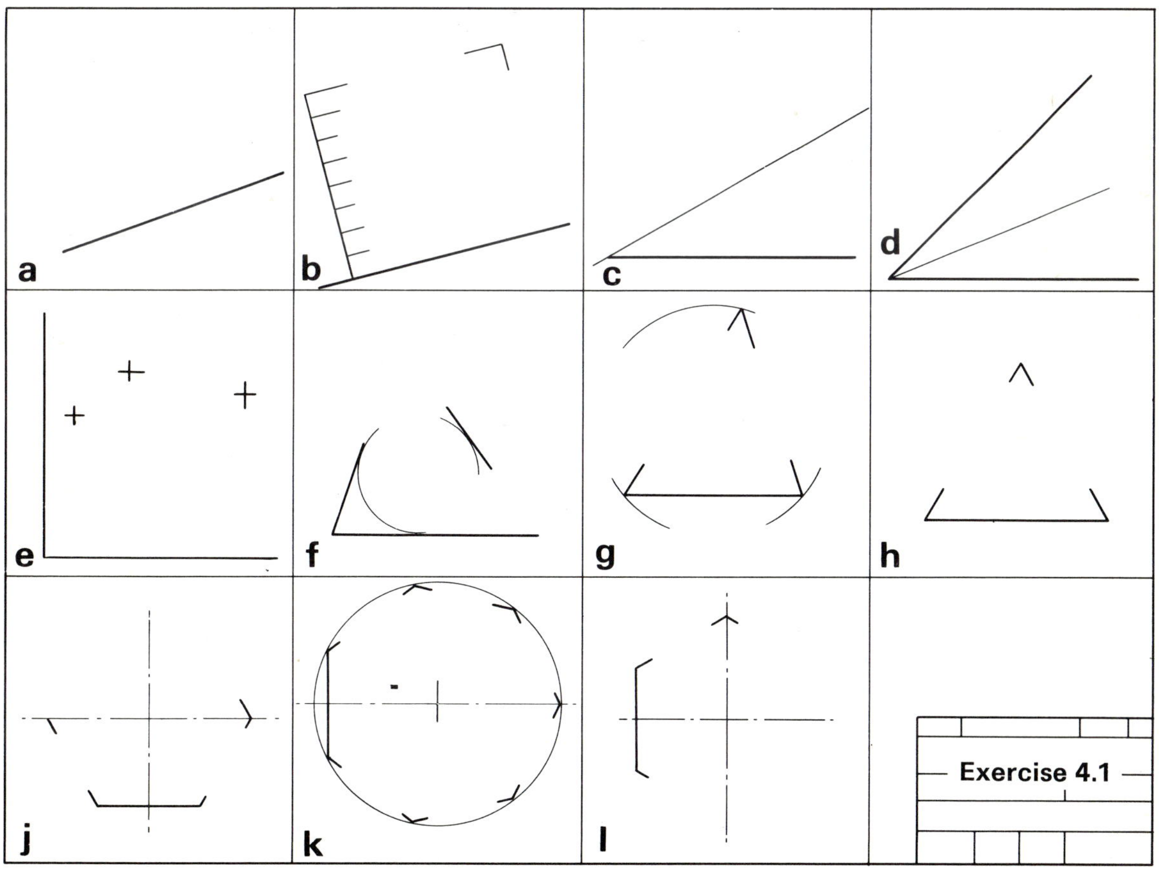

EXERCISE 4.2

Divide an A2 or a half-imperial sheet of drawing-paper into three parts as shown, and construct the following ellipses:

(a) Using a two-arc construction, draw an approximate ellipse having a major axis of length 124 mm and a minor axis of length 80 mm.

(b) Using the method of concentric circles, draw an ellipse having major and minor axes 124 mm and 80 mm long respectively.

(c) The formula for the ellipse is $\frac{x^2}{a^2} + \frac{y^2}{b^2} = 1$. Using this formula, draw up a table of (x, y) co-ordinates to plot a quarter of an ellipse having major and minor axes of 400 mm and 240 mm respectively. Plot the position of the calculated points and draw a smooth curve through them with a french curve or spline.

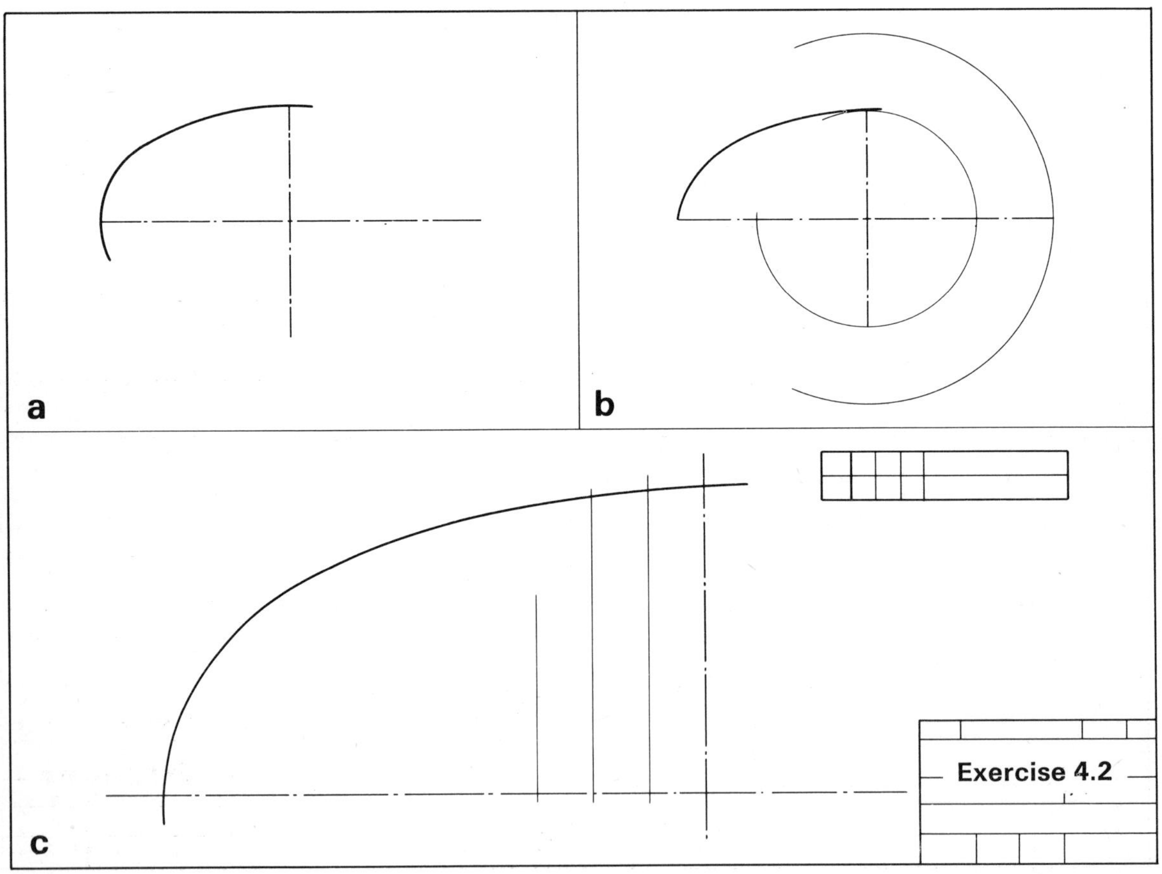

EXERCISE 4.3

Divide an A2 or a half-imperial sheet of drawing paper into four equal parts as shown, and construct the four loci from the given information.

(a) A focus is positioned 30 mm from a directrix. Draw the parabola which satisfies these conditions.

(b) A deep scratch is made on the circumference of a skate wheel of 50 mm diameter. Plot the path described by the scratch in one complete revolution of the wheel and name the curve.

(c) Draw the involute curve which is generated from a base circle of 50 mm diameter.

(d) Draw the Archimedean spiral which has a sideways displacement of 48 mm in one complete revolution. The outside end of the curve lies on the circumference of a circle having a diameter of 140 mm.

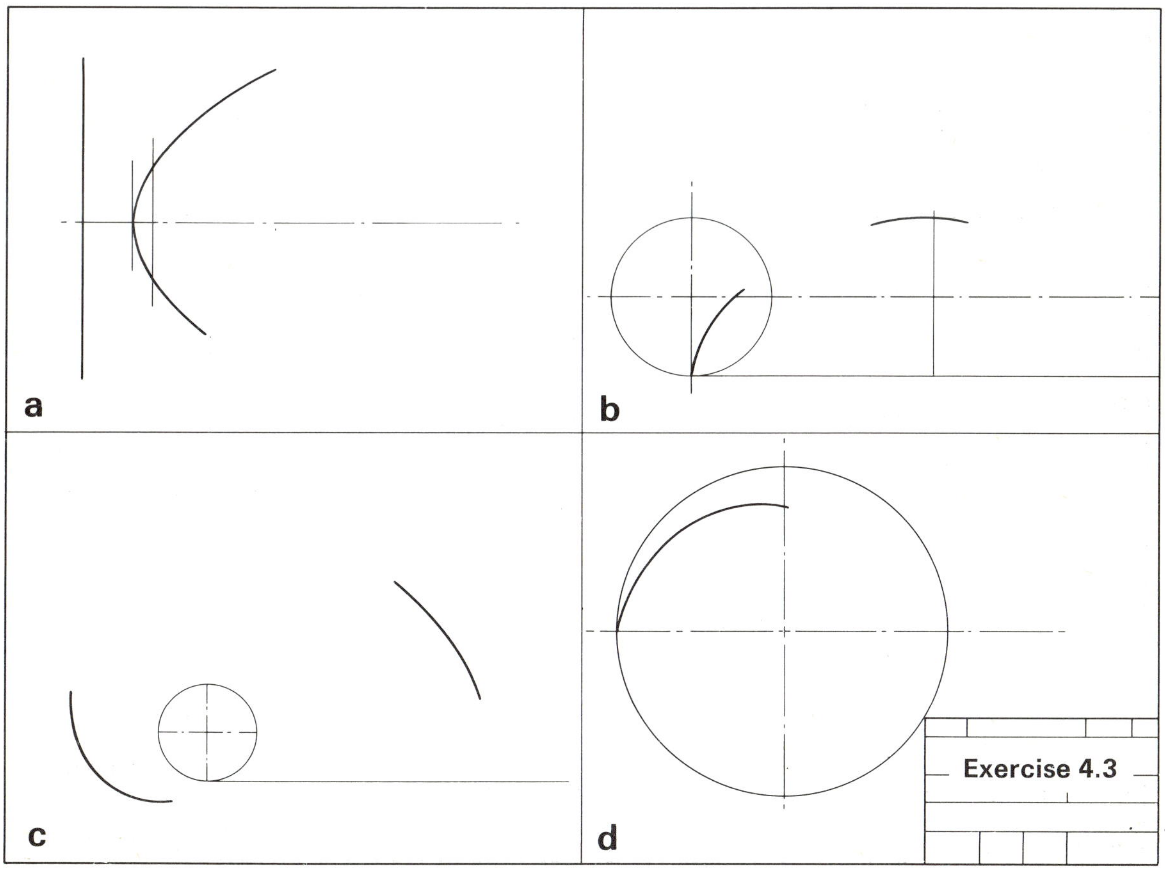

Practical tangency

The industrial draughtsman has often to produce accurate scale drawings for production by blending together circular arcs and straight lines into smooth profiles. Examples can be seen on the following pages.

Too often in practice, much time is wasted in trial and error, by stabbing around with the compass point to find the centres from which the blending curves are struck. To draw such contours accurately requires an understanding of three very simple tangency constructions which produce the *exact* centres from which the blending arcs are struck; these always occur at the *intersection of two* construction lines. The *exact blending points* of adjoining curves can also be established by joining the centres of the two curves.

The three constructions are illustrated below, and can be remembered as (R), ($R + r$) and ($R - r$) where R is the radius of the blending curve.

Problem

To draw an arc of radius R to blend with a given straight line PQ.

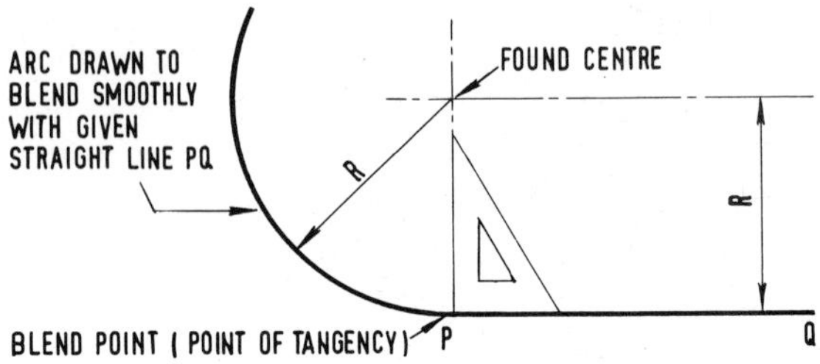

Method

1. Draw given line PQ.
2. Draw parallel line distance R from PQ.
3. Erect a perpendicular from one end of line PQ.
4. Intersection of construction lines is centre for drawing arc.

5.1 CONSTRUCTION (R)

To blend circular arcs to straight lines. Figure 5.1.

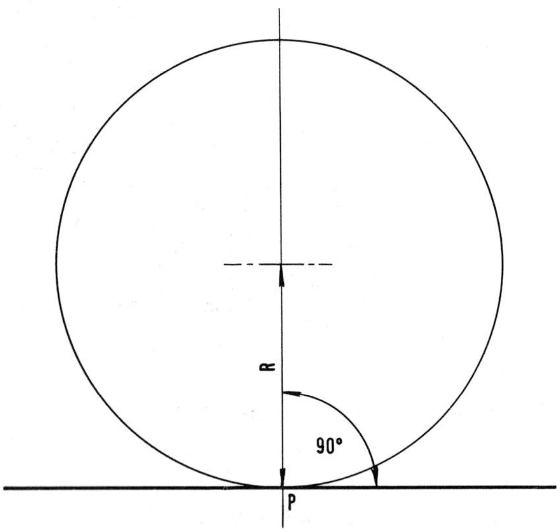

Figure 5.1

1. Straight line is tangent to circle.
2. P is the point of tangency.
3. Radius R is perpendicular to tangent.

Examples showing the method used to blend corner radii to lines meeting at an angle.

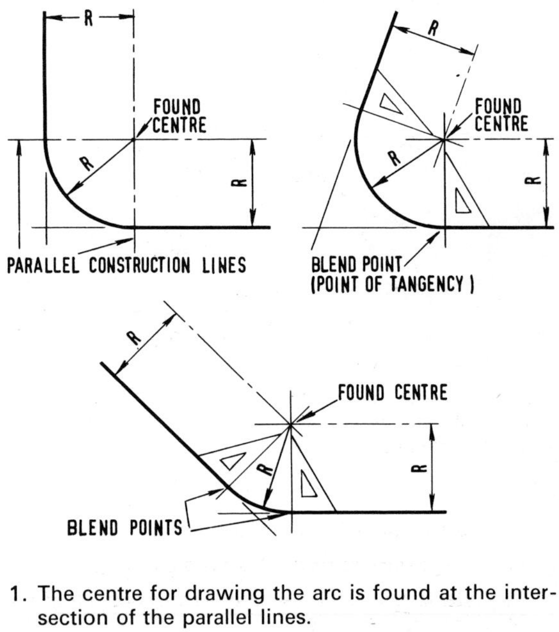

1. The centre for drawing the arc is found at the intersection of the parallel lines.
2. Perpendiculars drawn through the centre establish the blend points.

5.2 CONSTRUCTION ($R+r$)

To blend external circular arcs. Figure 5.2.

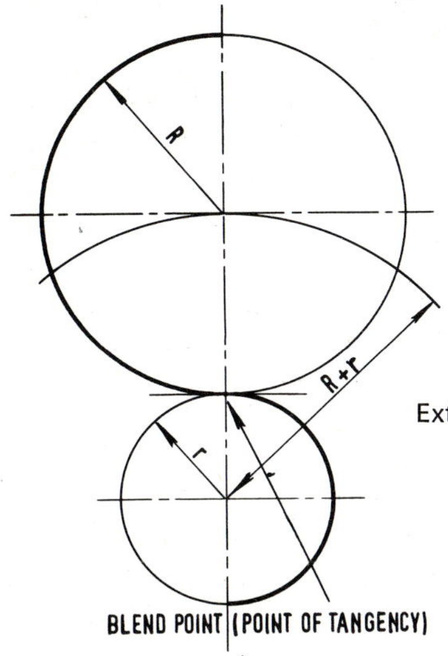

External Circles

Figure 5.2

1. The distance between the centres of the circles is the sum of the radii ($R + r$).
2. The blend point of the arcs lies on a line drawn between their centres.

Problem

To join two given circles A and B or radii r_1 and r_2 with external blending arc of radius R.

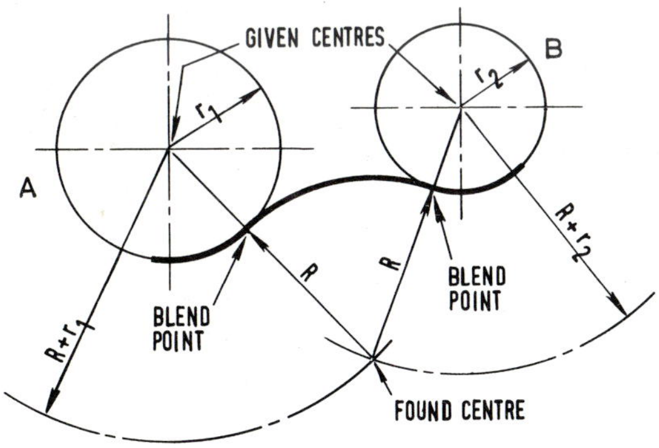

Note: To obtain the position of the found centre for drawing the blending arc of radius R, start construction from the given centres.

Method

1. From the centre of circle A draw arc $R + r_1$.
2. From the centre of circle B draw arc $R + r_2$.
3. The intersection of the arcs locates the centre of the blending arc of radius R.
4. Join the given centres to the found centre to establish the exact blend points of joining arcs.

CONSTRUCTION ($R-r$)

To blend internal circular arcs.

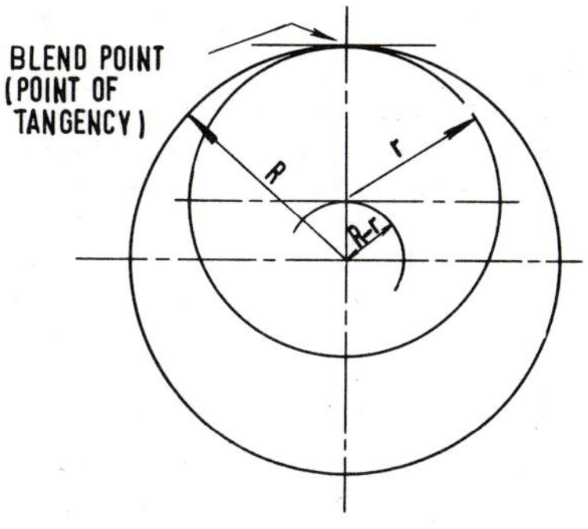

Internal Circles

Note:

1. Distance between centres of circles is the difference beween the radii (R-r)
2. The blend point lies at the intersection of the arcs, by a line drawn through the centres of the circles.

Figure 5.3

Problem

To join two given circles C and D having Radii r_1 and r_2 respectively with an internal blending arc of radius 'R'.

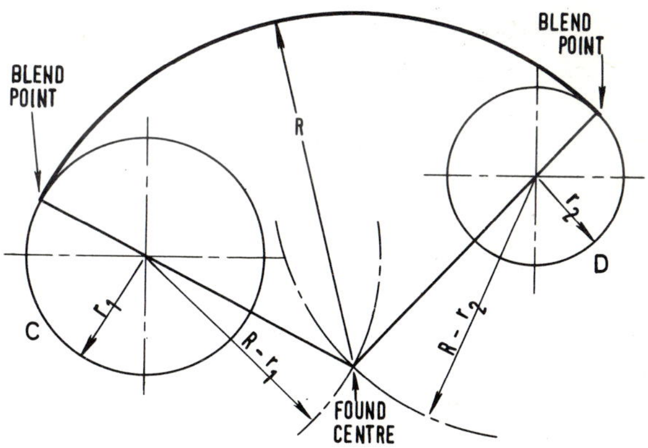

Note:

Construction starts from centres of given circles.

Method

1. From the centre of circle C draw arc R-r_1
2. From the centre of circle D draw arc R-r_2
3. Intersection of arcs locate the centre of blending arc of radius R.
4. Join the found centre to the given centres, and produce lines beyond to intersect circles C and D to establish the exact position of blend points.

Practical tangency

EXERCISES

EXERCISE 5.1. See page 57

Divide an A2 or a half-imperial sheet of drawing paper into five sections as shown, and draw the objects in alphabetical sequence.

Figures (a), (b) and (c) are simple geometric exercises in tangency; figures (d), (e) and (f) are practical examples of tangency in the design of common workshop tools.

Use the dimensions given to draw and position the objects on the drawing sheet. Show all centre lines as chain-dot lines. Leave in all construction lines used to find centres, and blend points of arcs.

Outlines to be sharp, dense, and between 0·75 mm and 1·0 mm wide. Do not add dimensions to the finished drawing.

EXERCISE 5.2. See page 58

Divide an A2 or a half-imperial sheet of drawing paper into seven parts as shown, and draw the objects in alphabetical sequence.

(a) An aluminium extrusion.
(b) Curved link.
(c) Stylised fish for pottery decoration.
(d) Cross-section through a stoneware egg cup.
(e) Cork gasket.
(f) Handle of smoothing plane.
(g) Crane hook.

Use the dimensions given to draw and position the objects on the drawing sheet. Show all centre lines as chain-dot lines. Leave in all construction lines used to find centres, and blend points of arcs.

Outlines to be sharp, dense, and between 0·75 mm and 1·0 mm. Do not add dimensions to the finished drawings.

EXERCISE 5.3. See page 59

Using the dimensions given and the correct construction for blending circular arcs, draw the profile of the helicopter. Estimate any dimensions not given.

EXERCISE 5.4. See page 60

Draw the outline of the sports car shown, using the dimensions given and the geometric construction for blending circular arcs. Estimate any dimensions not given.

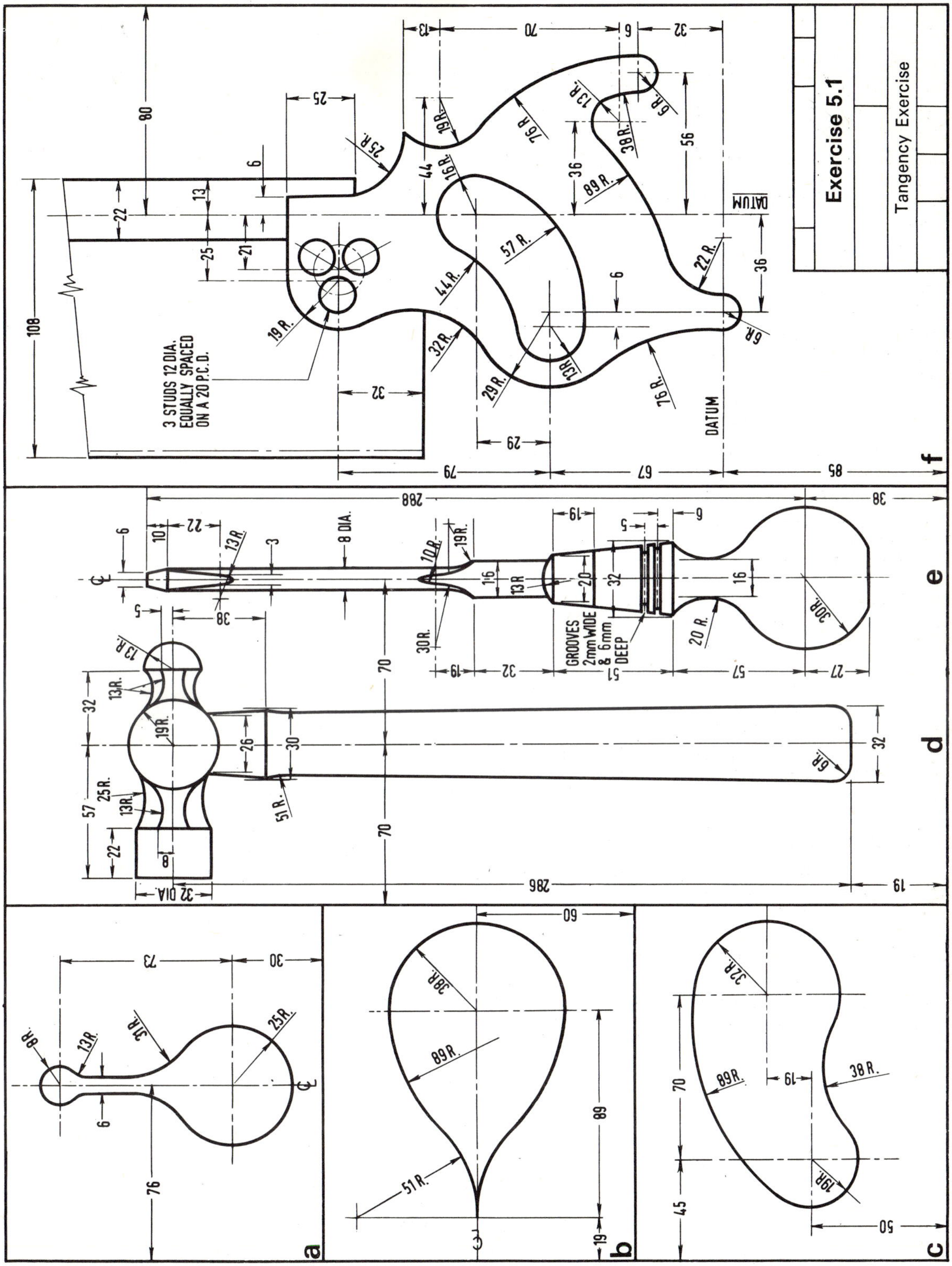

Exercise 5.1
Tangency Exercise
3 STUDS 12 DIA. EQUALLY SPACED ON A 20 P.C.D.
GROOVES 2mm WIDE & 6mm DEEP
DATUM
a
b
c
d
e
f

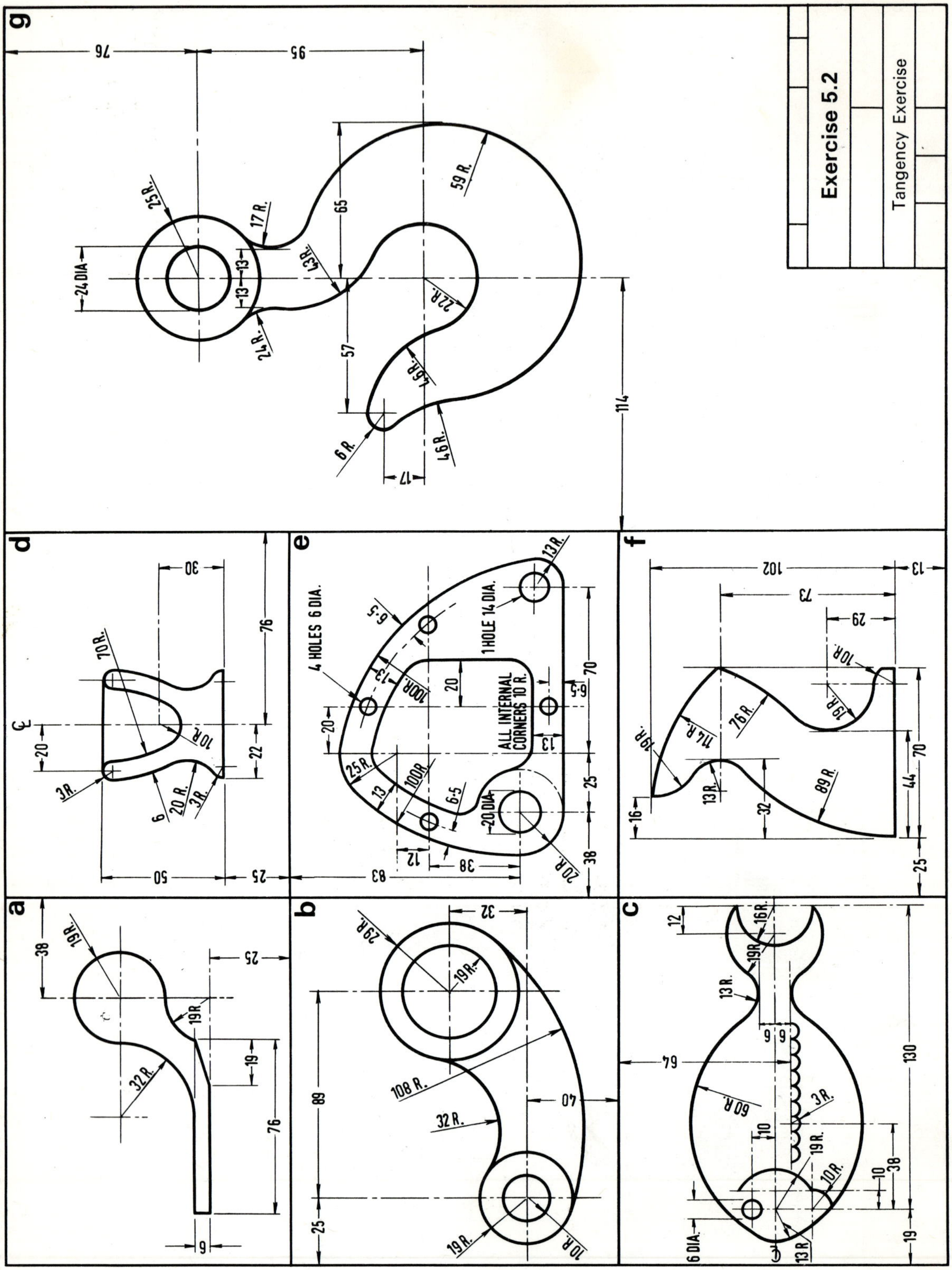
Exercise 5.2
Tangency Exercise
a
b
c
d
e
f
g
4 HOLES 6 DIA.
1 HOLE 14 DIA.
ALL INTERNAL CORNERS 10 R.

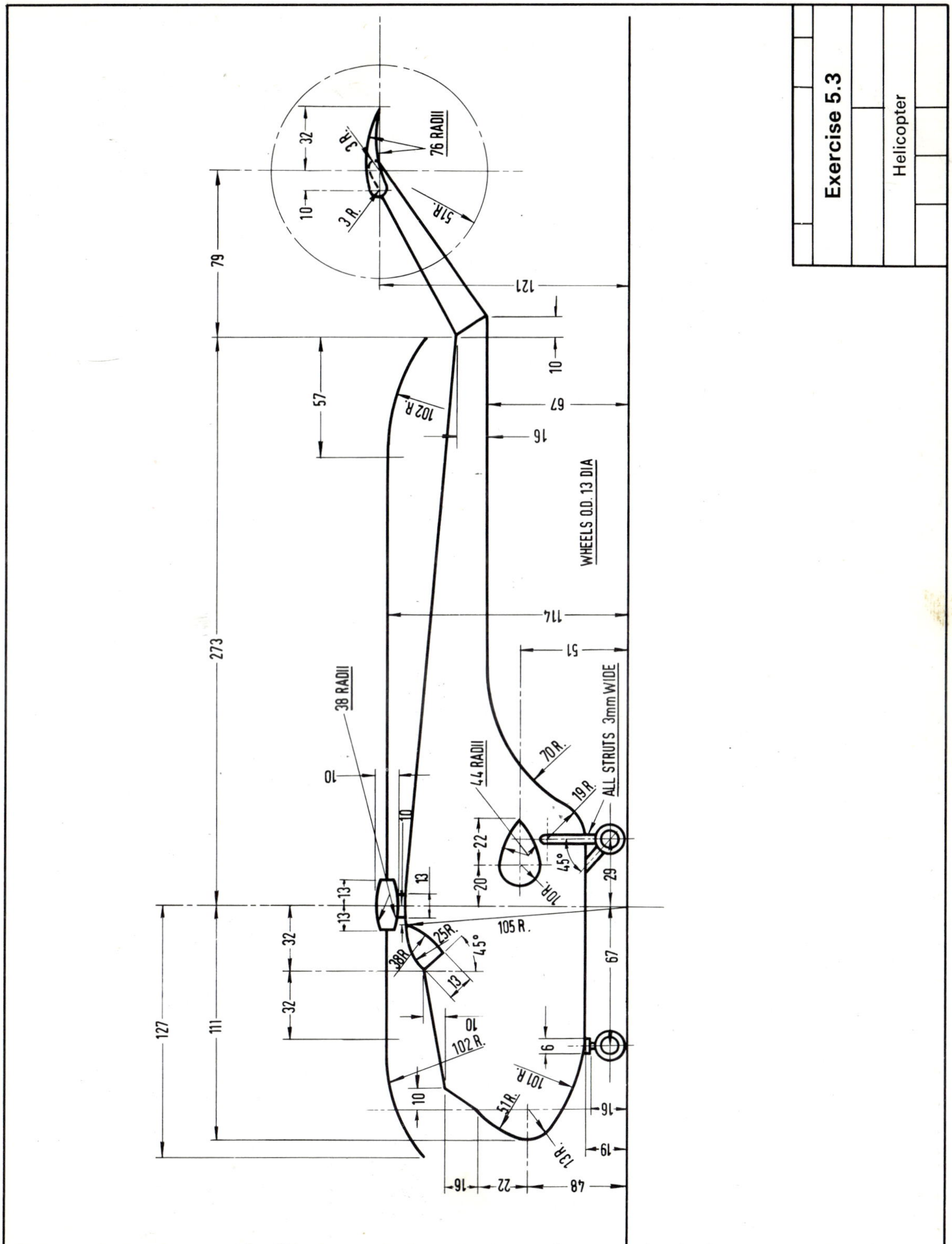
Exercise 5.3
Helicopter
76 RADII
3 R.
3 R.
51 R.
102 R.
WHEELS O.D. 13 DIA
ALL STRUTS 3mm WIDE
38 RADII
44 RADII
70 R.
19 R.
10 R.
45°
105 R.
25 R.
38 R.
102 R.
101 R.
51 R.
13 R.

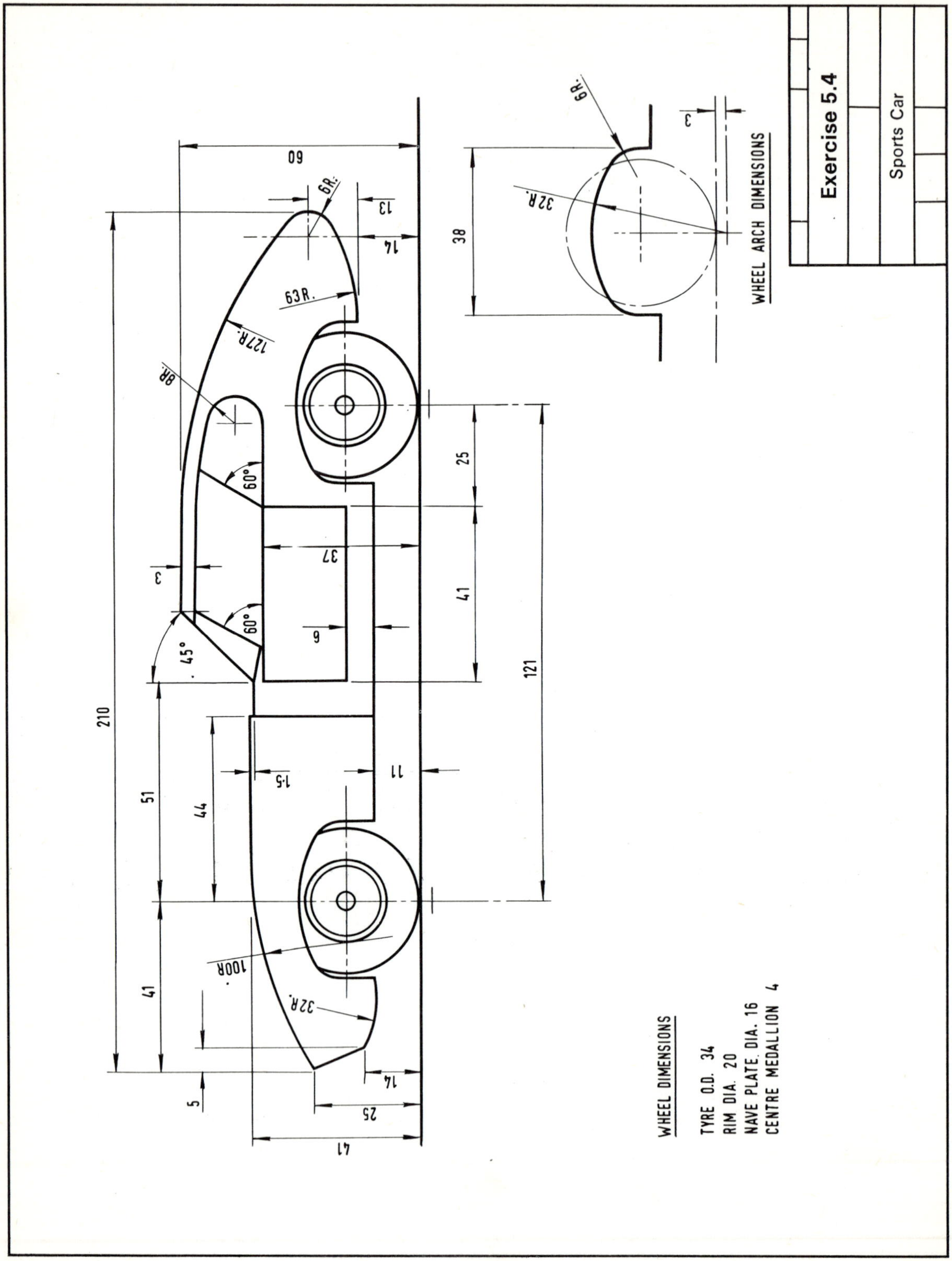
Exercise 5.4
Sports Car
WHEEL ARCH DIMENSIONS
WHEEL DIMENSIONS
TYRE O.D. 34
RIM DIA. 20
NAVE PLATE. DIA. 16
CENTRE MEDALLION 4

Orthographic projection

As explained in Chapter 2, this is a multi-view system of drawing by which a designer's precise requirements are conveyed to a manufacturing organisation. It has the important advantage of allowing any number of views of the same object, and is widely used in engineering.

In orthographic projection the shape of an object is defined by a system of separated but related views. The basic views, which are drawn true to scale, are taken at 90° to each other and when set on paper are said to be drawn in the principal planes.

It was explained in Chapter 2 that there are at present two separate methods of relating the principal views to each other, namely *first-angle projection* (used in Britain and in some European countries) and *third-angle projection* (used in Canada and the United States). Since World War II, however, third-angle projection has been increasingly used and may well become the universally accepted system.

WHAT IS MEANT BY FIRST-ANGLE AND THIRD-ANGLE PROJECTION?

The terms *first-angle* and *third-angle* are derived from the mathematician's convention of annotating the four right angles which make up the 360° of a circle, the first angle being 0° to 90° and the third angle from 180° to 270° (see Figure 6.1).

HOW IS A DRAWING MADE IN FIRST-ANGLE PROJECTION?

The object of an engineering drawing is to define precisely a solid three-dimensional object on flat paper. In the first angle (Figure 6.1), three planes of projection have been erected, and a solid object has been suspended between the planes.

The planes are actually drawing surfaces, and the views drawn on the two vertical surfaces are usually referred to as elevations, while the view on the horizontal surface is referred to as the plan.

Figure 6.2 shows the object being viewed from direction A, which is at right angles to a vertical principal plane; the view seen is drawn on the plane *beyond* the object. In this view the following information concerning the object has been given:

1. The length of the angle, *l*.
2. The height of the angle, *h*.
3. The thickness of the horizontal flange, t_1.
4. The thickness of the triangular web, t_2.
5. The position of the triangular web from the ends of the angle, *x*.

Figure 6.3 shows the object being viewed from direction B at right angles to the end vertical principal plane, and the view seen is drawn on the plane *beyond* the object. To assist in drawing this view, the height (*h*) and the

Figure 6.1

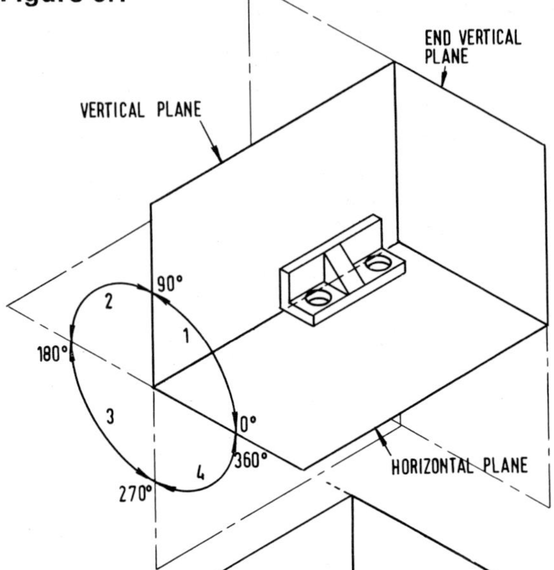

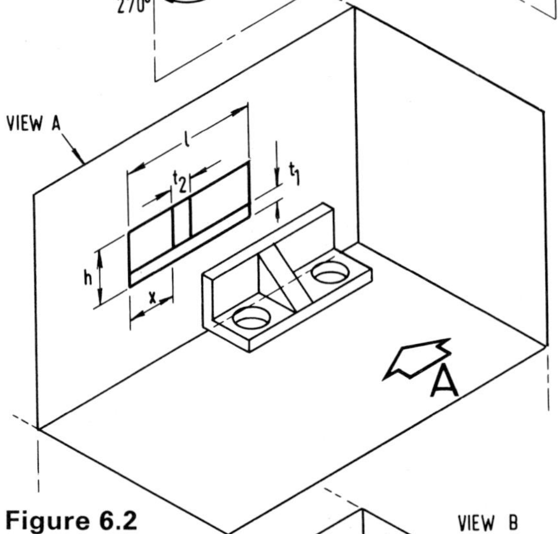

Figure 6.2

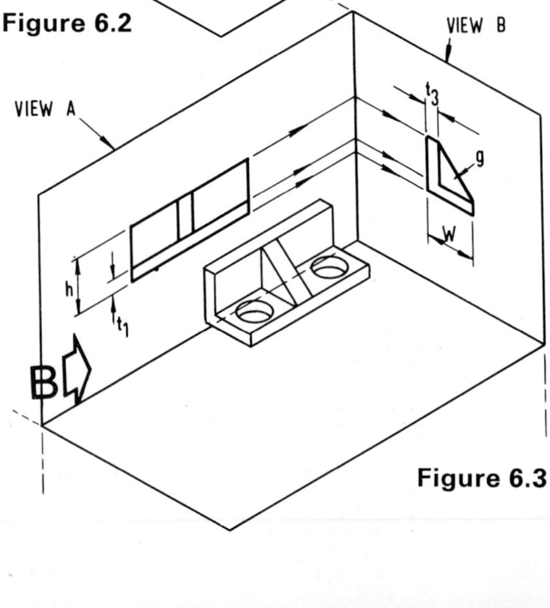

Figure 6.3

Figure 6.4

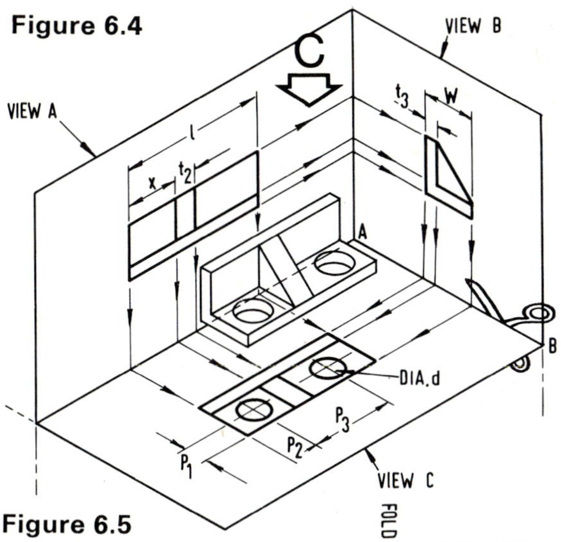

Figure 6.5

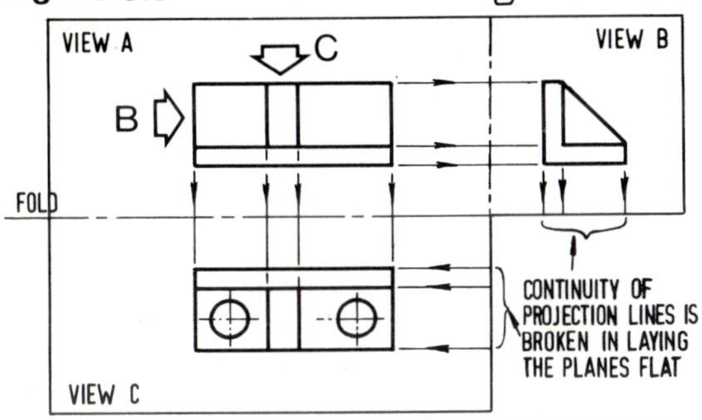

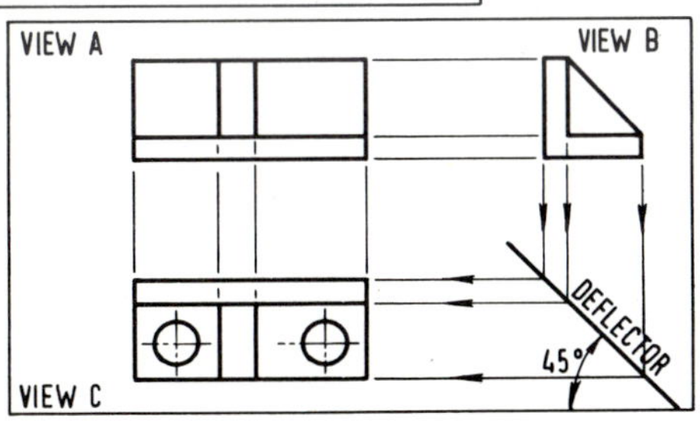

Figure 6.6

Figure 6.7

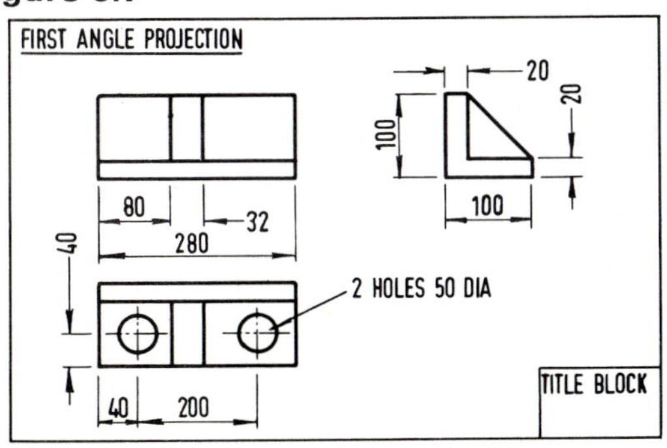

thickness of the horizontal flange (t_1) have been transferred from view A by means of horizontal lines known as projection lines. From this view further information concerning the object is given:

6. The width of the horizontal flange of the angle, *w*.
7. The triangular shape of the web, *g*.
8. The thickness of the vertical flange of the angle, t_3.

In the two views now drawn, the holes in the lower flange cannot be shown, so a third view is required.

Figure 6.4 shows the object being viewed from direction C, which is directly down, at right angles to the horizontal principal plane; the view seen is drawn on the plane *beyond* the object.

To assist in drawing this view, the length *l*, position *x*, and thickness t_2 of the triangular web have been projected down from the view A. From view B the width of the horizontal flange, *w*, and the thickness of the vertical flange, t_3, have been projected. On this view the final information can be given.

9. The size of the holes, diameter *d*.
10. The position of the holes relative to the sides of the horizontal flange, P_1, P_2 and P_3.

The three views now drawn completely define the shape, but drawing on three planes at right angles in the shape of a corner is inconvenient. In Figure 6.4 the three connected planes are cut along line AB, and when laid flat (as in Figure 6.5) result in three quarters of a flat sheet with three separated views, each one related to the other.

In laying the planes flat the *projection lines* transferring the sizes and points between the related views are seen to be horizontal or vertical parallel lines, but those between view B and plan view C have become broken, and the disjointed parts are at 90° to each other.

To enable sizes and points to be projected from view B to plan view C, a deflector line is drawn at 45° below view C bisecting the angle through which the projection lines are to be turned (see Figure 6.6). Lines are then continued vertically down from view B to the 45° *deflector* line, and from the point of contact drawn horizontally across to plan view C, thus transferring the information between views B and C.

Figure 6.7 shows the three views in *first-angle projection* as they would be drawn on a drawing sheet with dimensions added.

HOW A DRAWING IS MADE IN THIRD-ANGLE PROJECTION

The views drawn by this method differ from those in first-angle projection only in the position of the views relative to each other. Chapter 2 has a short explanation of the difference between the two systems—see page 8.

In Figure 6.8 three transparent planes of projection have been erected in the third angle, and a solid object has been suspended between the planes. The object is viewed through the transparent planes, and the view of the object seen is drawn on the transparent surface through which it is viewed. As in first-angle projection, the views on the vertical surfaces are referred to as elevations and the view on the horizontal plane is referred to as the plan.

Figure 6.9 shows the object being viewed from direction A, at right angles and through the vertical transparent principal plane. The view seen is drawn on this transparent plane. Note that the plane on which the view is drawn is *adjacent* to the object, and not *beyond* the object as in first-angle projection.

In this view the following information concerning the object has been given:

1. The length of the angle, *l*.
2. The height of the angle, *h*.
3. The thickness of the horizontal flange, t_1.
4. The thickness of the triangular web, t_2.
5. The position of the triangular web from the ends of the angle, *x*.

As this single view A does not completely define the object, other views must be drawn.

Figure 6.10 shows the object being viewed from direction B, at right angles and through the transparent end vertical plane. The view seen is drawn on the transparent end vertical plane, which is adjacent to the object.

To assist in drawing this view, the height *h* and thickness t_1 of the lower flange has been transferred from view A by means of horizontal projection lines. This view supplies further information:

6. The width of the horizontal flange, *w*.
7. The triangular shape of the web, *g*.
8. The thickness of the vertical flange of the angle, t_3.

In the two views now drawn the holes in the lower flange cannot be specified, so a third view is required.

Figure 6.11 shows the object being viewed through the principal plane from above, in the direction C; the view seen is drawn on the transparent plane.

To assist in drawing and relating the third view C to the two views already drawn, the length *l*, the position *x* and the thickness t_2 of the triangular web have been projected round from view A. From view B the width of the horizontal flange (*w*) and the thickness of the vertical flange (t_3) have also been projected round to the horizontal plane.

On this view the final information concerning the holes can be given:

Figure 6.8

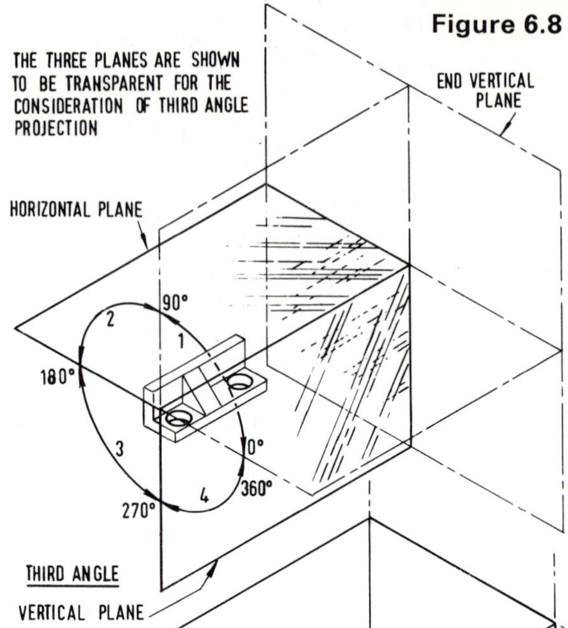

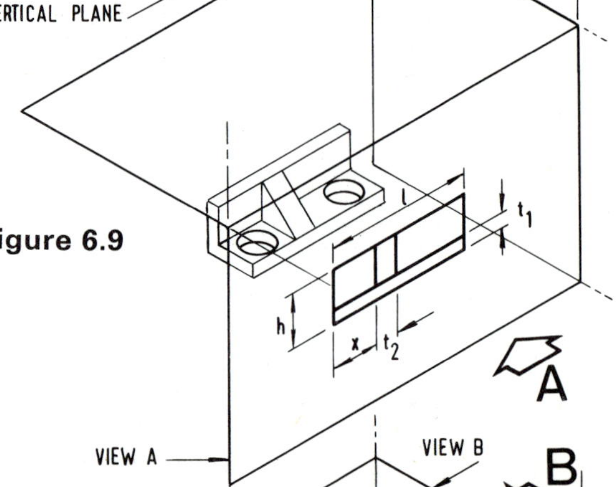

Figure 6.9

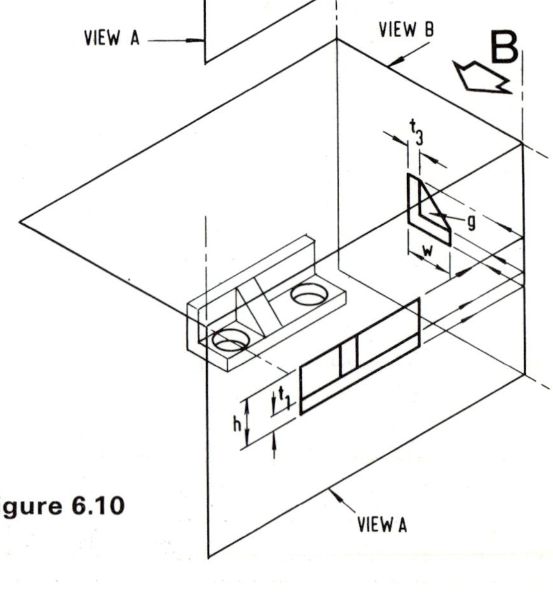

Figure 6.10

Figure 6.11

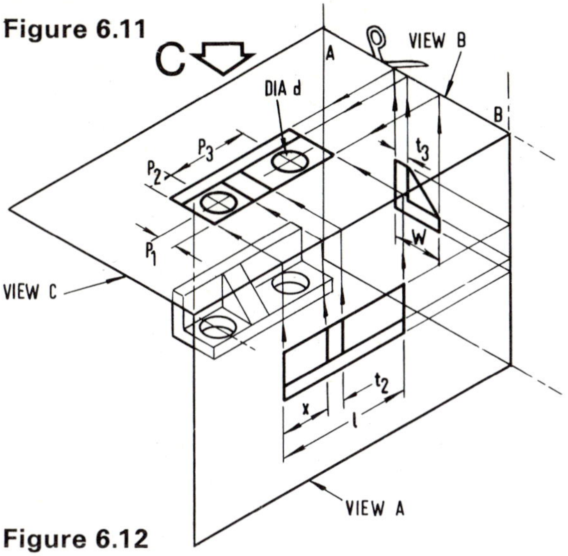

Figure 6.12

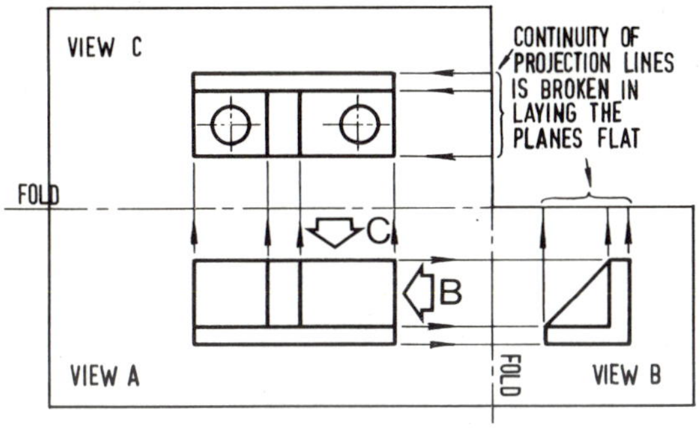

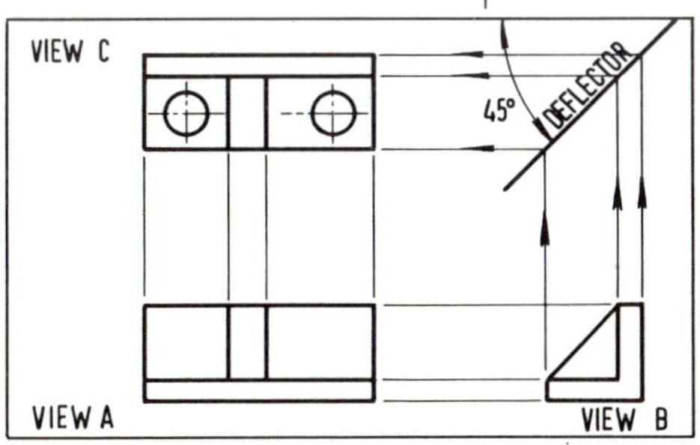

Figure 6.13

Figure 6.14

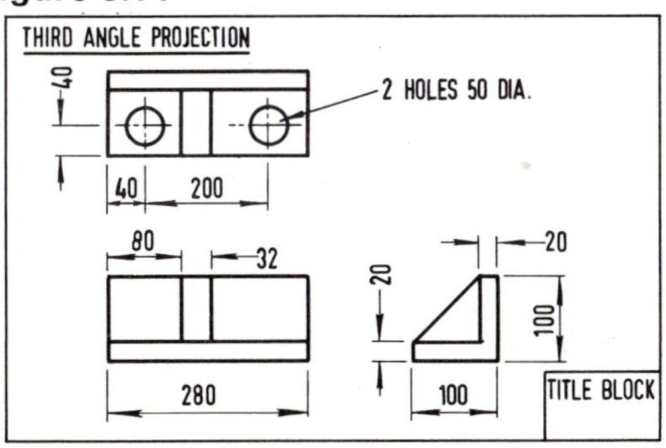

9. The size of the holes, dia. *d*.
10. The positions of the holes in the horizontal flange, P_1, P_2, P_3.

The three views now drawn completely define the shape and, by cutting along edge AB (Figure 6.11), the views are laid out flat, resulting in three quarters of a flat sheet with three separated views related to each other (Figure 6.12).

When drawing the views on a drawing sheet, the information can be transferred between view B and view C by means of a 45° *deflector* line (Figure 6.13) as in first-angle projection.

Figure 6.14 shows the three views in *third-angle projection* as they would be drawn on a drawing sheet. Note that the plan view C is now positioned above view A and that view B appears reversed. In third-angle projection this is because views are positioned *adjacent* to the surfaces being illustrated, whereas in first-angle projection the views of surfaces illustrated are positioned on the opposite side, beyond the surface being viewed.

POINTS TO NOTE. Figures 6.7 and 6.14.

1. The three views are well spaced on the drawing sheet to leave space for dimensioning between views.
2. Visible outlines of the object are thick, so that the object stands out clearly from the dimension lines.
3. Projection lines from features which are dimensioned are thin. A small gap is left between the projection line and the object.
4. Dimension lines are thin, terminating in arrow heads that are slim, solid and about 3/16″ long.
5. Dimensions are placed on the view where the information is most relevant. Dimensions are read from the bottom of the drawing or from the left. Dimensions are not placed close to visible outline but far enough away to give an uncrowded appearance. They are of good size and upright.
6. Centre lines end a small distance beyond the visible outlines of the circles, except where they are produced for dimensioning.
7. Notes are printed between guide lines.

Orthographic projection

WHY PROJECTION IS IMPORTANT IN ENGINEERING DRAWING

An engineering drawing in its proper use is a communication from a designer to manufacturing units. The object as shown on the paper does not exist when the drawing leaves the design office, so drawings cannot be made by copying the object.

The designer, using the technique known as orthographic projection, has precisely placed his ideas for a three-dimensional object on flat paper as a series of related views. In making the drawing, the designer has used two techniques, visualisation and projection between views. The basis of visualisation is thinking in three dimensions and setting down on paper what is seen in the mind's eye. Although all drawing starts by visualisation, projection must soon take over on the drawing board to produce correct orthographically projected views of the object.

Basically, the only thing that can be projected is a *point,* and to plot a point on a flat plane distances from two reference points must be known. The point is usually established by the intersection of two lines projected from existing views where the information can be stated, or by a projection line and a dimension.

Views on engineering drawings are defined by drawing the boundaries or sharp edges of an object. A straight edge will appear in all views as a straight line except when it is viewed end on; it will then appear as a dot. It will never appear as a curved line. When the two end points of a straight line are established in a given view, the line can be drawn as it appears by joining the two points.

When an edge is curved it appears in all views as a curved line except when the plane in which it lies is presented edge-on to the direction of viewing; it then appears as a straight line.

To project a curved edge to another view, the end points and several selected points along the curve are established by projection in the required view. The points are then joined by a smooth curve.

As it is impossible to visualise the shape in all the necessary views, projection is necessary to provide the information.

Figures 6.15 to 6.20, if carefully followed, should give an understanding of the techniques of projection.

Figure 6.15 shows three views of a short hollow cylinder. These views are simple enough to visualise, but projection is necessary to position one view relative to another. It is well to remember that these three views are continually arising in sheet-metal development work; for accuracy the views should be drawn in the following sequence.

Method

1. Draw the centre lines in view A. Then, taking the radius *r* measured along the centre line in the compasses, draw the plan consisting of a circle.
2. Project the centre line and the width of the circle into view B, measure off the height *h*, and complete the rectangle which represents the elevation of the cylinder.
3. View C should now be drawn by projection or by transfer of dimensions by dividers. *No further measuring by ruler* should be attempted. Project height *h* of cylinder to view C. Draw the centre line to establish the axis of the cylinder in view C, and produce it to cut the horizontal axis of view A. Through the intersection draw a 45° deflector line, and project the width of the cylinder from view A to view C. Line in the rectangle which represents the cylinder.

By this method the views are accurately related dimensionally.

Figure 6.15

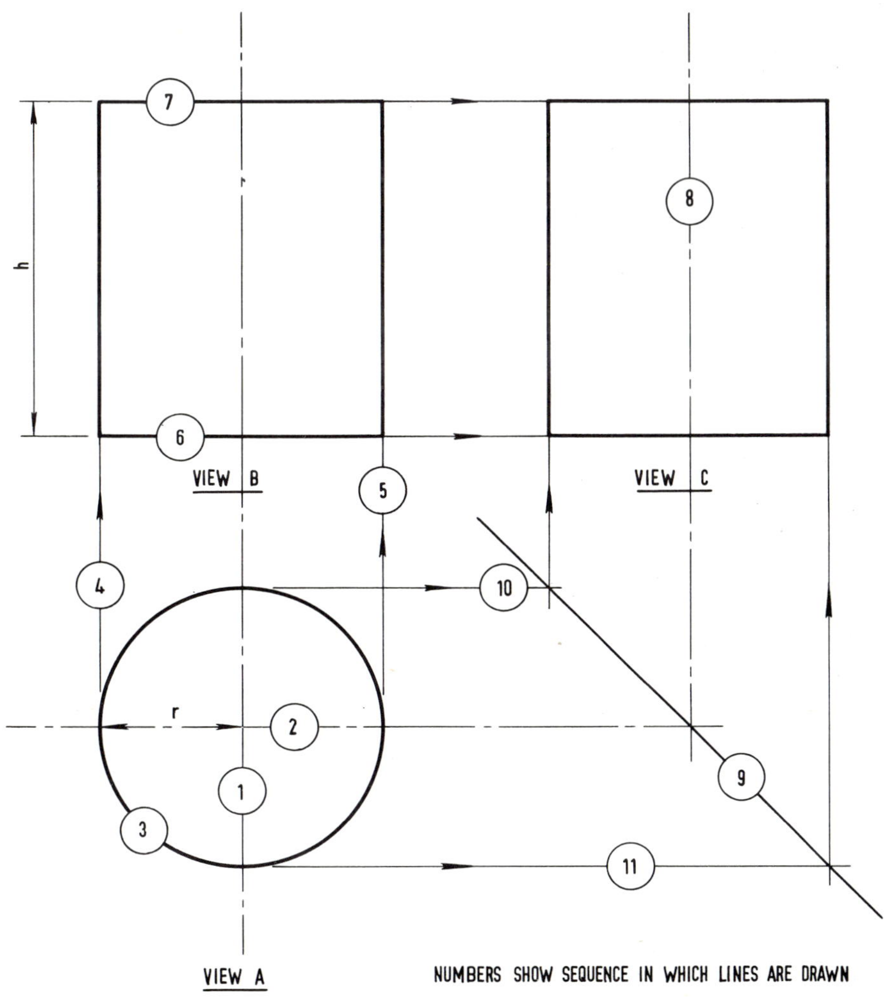

Figure 6.16 shows three views of a cylinder with three points marked as crosses on view B. It is required to complete view C showing the crosses correctly positioned. This is achieved by projection.

Method

1. Give each point in view B a number to identify it.
2. Establish the positions of the points in the plan view A, giving them the corresponding numbers for identification. Having established the points in the two views A and B, they can now be positioned by projection into view C.
3. Project the points one at a time.
4. From point 1 in view B draw a horizontal projection line to view C. This establishes the vertical height of point 1 above the base.
5. From point 1 in view A draw a projection line via the deflector line to view C. This establishes the horizontal distance of point 1 from the axis.
6. The intersection of the projection lines establishes point 1 in view C.
7. Repeat for points 2 and 3.
8. The view is that seen when viewing in the direction of arrow C. Thus points 1 and 2 will be seen, and point 3 will be shown in hidden detail because it is out of sight round the back.

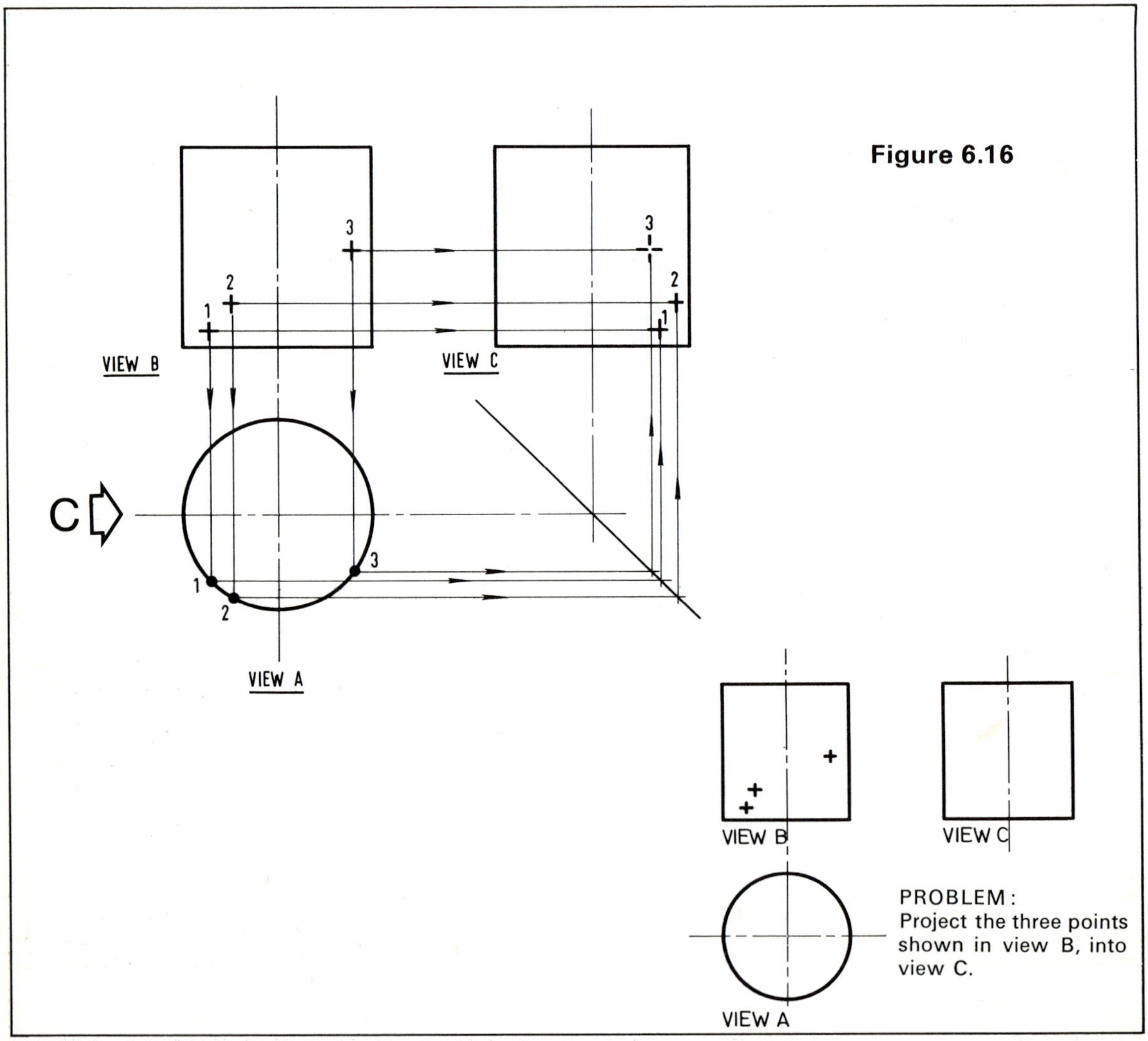

Figure 6.16

Figure 6.17 shows three views of a cylinder with a line, which appears to be straight, drawn on view C. This line is actually a curve; it runs round the cylinder. It is required to complete view B showing the line correctly positioned.

To project a curve, the end points and several points along its length must be projected; these points are then joined by a smooth curve.

Method

1. Identify by numbers the end points and several other points evenly spaced along the line in view C.
2. Establish by projection, via the deflector line, the points in the plan view A (knowing they will be positioned on the outside surface) and give them corresponding numbers for identification. Having now established the points in the two views A and C, they can be projected into view B.
3. Project one point at a time to avoid confusion.
4. From point 1 in view C draw a horizontal projection line to view B. This establishes the vertical height of point 1 above the base.
5. From point 1 in plan view A draw a vertical projection line up to view B. This establishes the horizontal distance of point 1 from the axis.
6. The intersection of the projection lines establishes point 1 in view B.
7. Repeat for the remaining points 2, 3, 4 and 5.
8. The plotted points in view B are connected with a smooth curve to produce the line as it appears in view B. Half the line, between points 1 and 3, is hidden from view, so this appears as hidden detail.

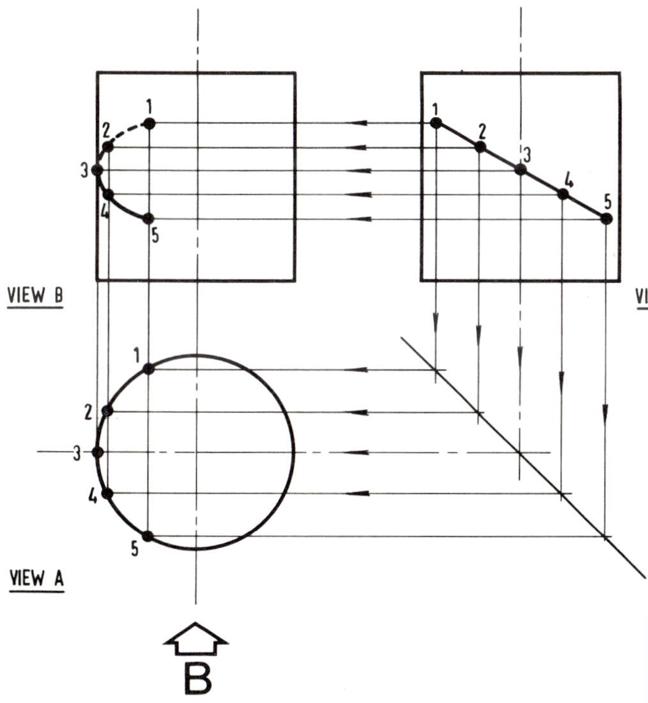

Figure 6.17

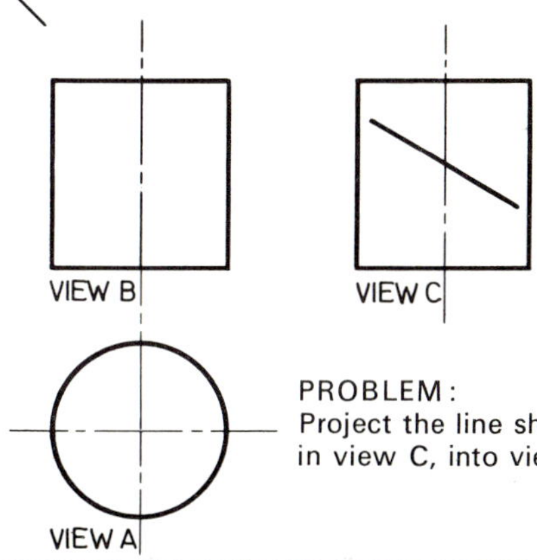

PROBLEM:
Project the line shown in view C, into view B.

Figure 6.20*a* shows views of a horizontal cylinder intersecting a right cone. It is required to complete views A and B showing the true shape of intersection. The line of intersection will be the mating shape assumed by the cylinder on attachment to the larger solid, the cone.

Figure 6.20b

This shows the series of numbered points established by the simplified technique using attached half end views.

In view B projection lines are drawn horizontally from the numbered points to the far side of the cone. These lines represent cutting lines. The sections (or shapes) shown in the sectioned pictorial view are drawn one on top of the other in view A. This is known as a 'nest of sections'.

Sections are drawn in view A by connecting projection lines from the numbered points on the cylinder with the relevant circular section, which is drawn using the radius *r* projected down from view B.

Figure 6.20c

This shows the completed nest of sections in view A, and the resulting series of points which lie on the intersection line. These points are projected into view B to plot the line of intersection in this view.

Figure 6.20d

This shows the two views completed.

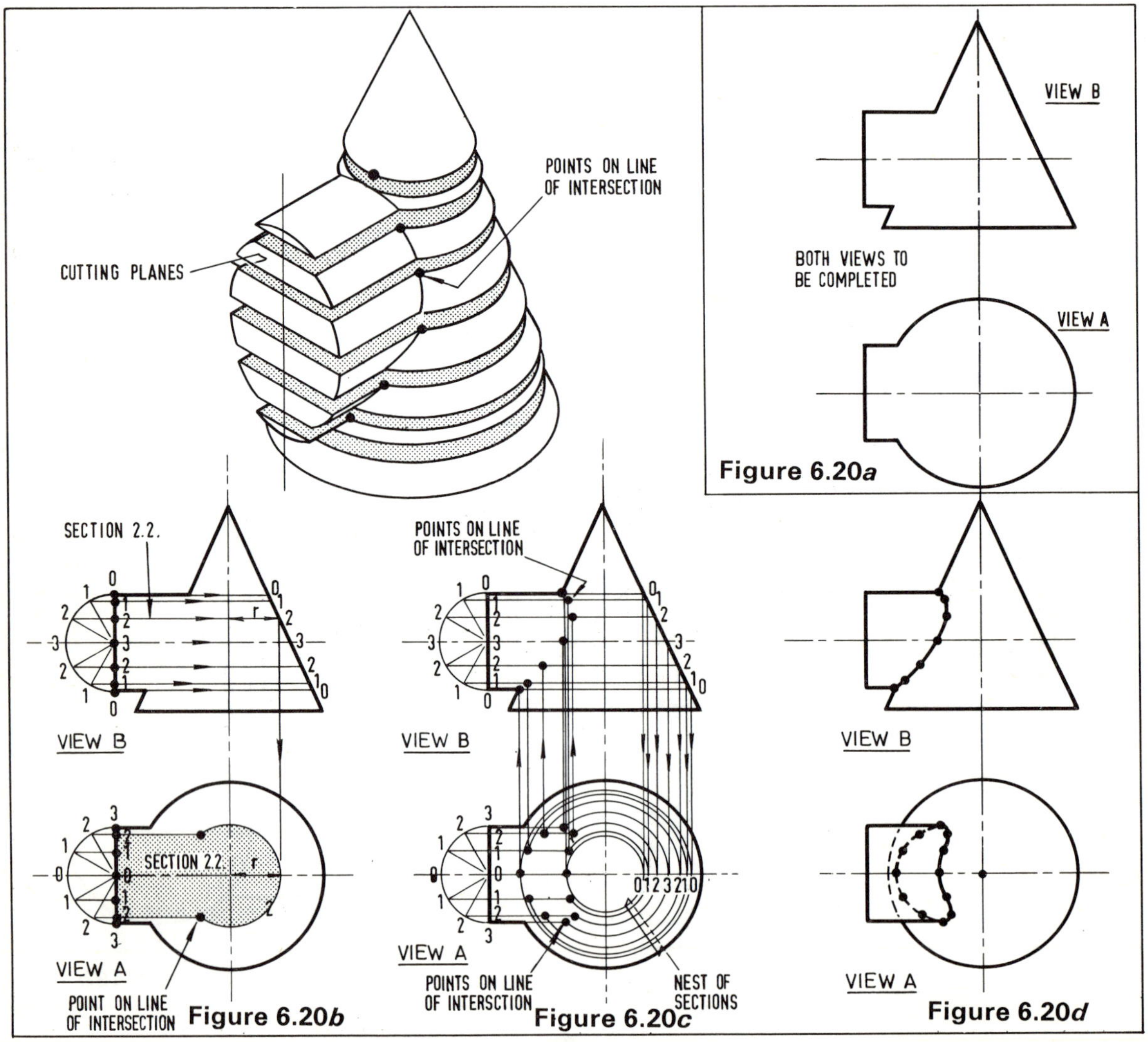

Figure 6.20*a*

Figure 6.20*b*

Figure 6.20*c*

Figure 6.20*d*

Figure 6.17 shows three views of a cylinder with a line, which appears to be straight, drawn on view C. This line is actually a curve; it runs round the cylinder. It is required to complete view B showing the line correctly positioned.

To project a curve, the end points and several points along its length must be projected; these points are then joined by a smooth curve.

Method

1. Identify by numbers the end points and several other points evenly spaced along the line in view C.
2. Establish by projection, via the deflector line, the points in the plan view A (knowing they will be positioned on the outside surface) and give them corresponding numbers for identification. Having now established the points in the two views A and C, they can be projected into view B.
3. Project one point at a time to avoid confusion.
4. From point 1 in view C draw a horizontal projection line to view B. This establishes the vertical height of point 1 above the base.
5. From point 1 in plan view A draw a vertical projection line up to view B. This establishes the horizontal distance of point 1 from the axis.
6. The intersection of the projection lines establishes point 1 in view B.
7. Repeat for the remaining points 2, 3, 4 and 5.
8. The plotted points in view B are connected with a smooth curve to produce the line as it appears in view B. Half the line, between points 1 and 3, is hidden from view, so this appears as hidden detail.

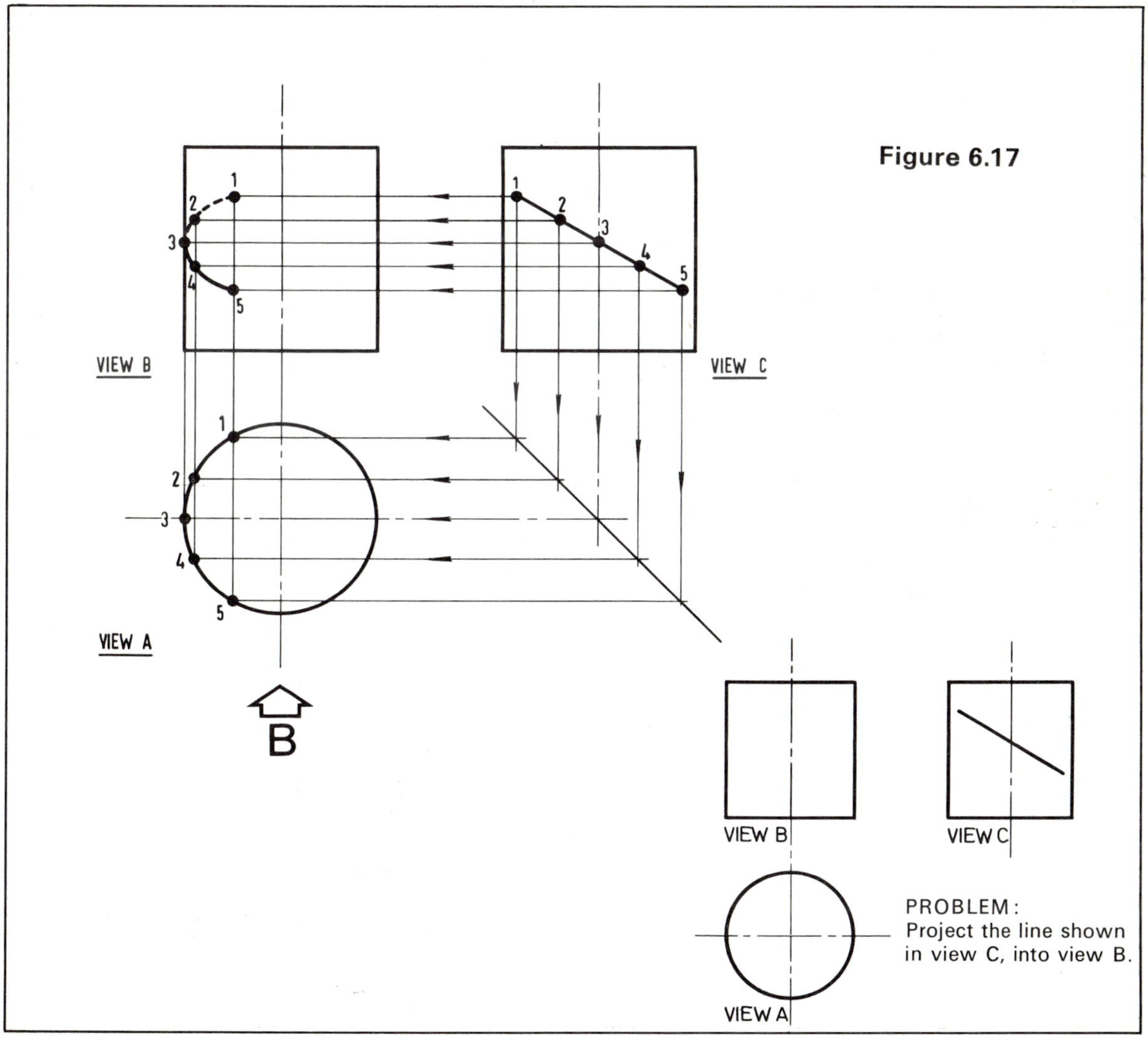

Figure 6.17

Figure 6.18 shows two views of a square bar with two V-shaped grooves cut at right angles in the top surface. It is required to project a plan view A from the given views.
The object is made up of straight edges which will appear in the views as straight lines.

Method

1. Systematically identify the end points of all the straight edges with a letter or number. Each point as it appears in both views must carry the same number. In this example there are 26 points to identify as follows:

   16 points on the top surface
   4 points at the bottom of the shallow V
   2 points at the bottom of the deep V
   4 points at the corners of the base.

2. Individually, or in groups, systematically plot all points by projection on the plan view A as follows:
   (a) Draw a common projection line down from points A, E, I, M, U, W and Y in view C to the deflector, and then horizontally across to view A.
   (b) Draw vertical lines down from points A, E, I, M, U, W and Y in view B to intersect the projection line drawn in (a) in order to plot points A, E, I, M, U, W and Y on view A.
3. Continue in the same manner by plotting groups B, F, J, N; Q, R, S, T; C, G, K, O; D, H, L, P, V, X, Z.
4. To complete plan view A, join the plotted points in the same sequence as they appear in the other two views, e.g. W to Y, Y to Z, Z to X, X to W. This will give the shape of the base.
   Join A to B (already drawn), B to F, F to E, E to A (already drawn). Join D to C, C to G, G to H, H to D. Join F to R, R to G. Join Q to R and U to V. The remaining points give a mirror image of the shape obtained.

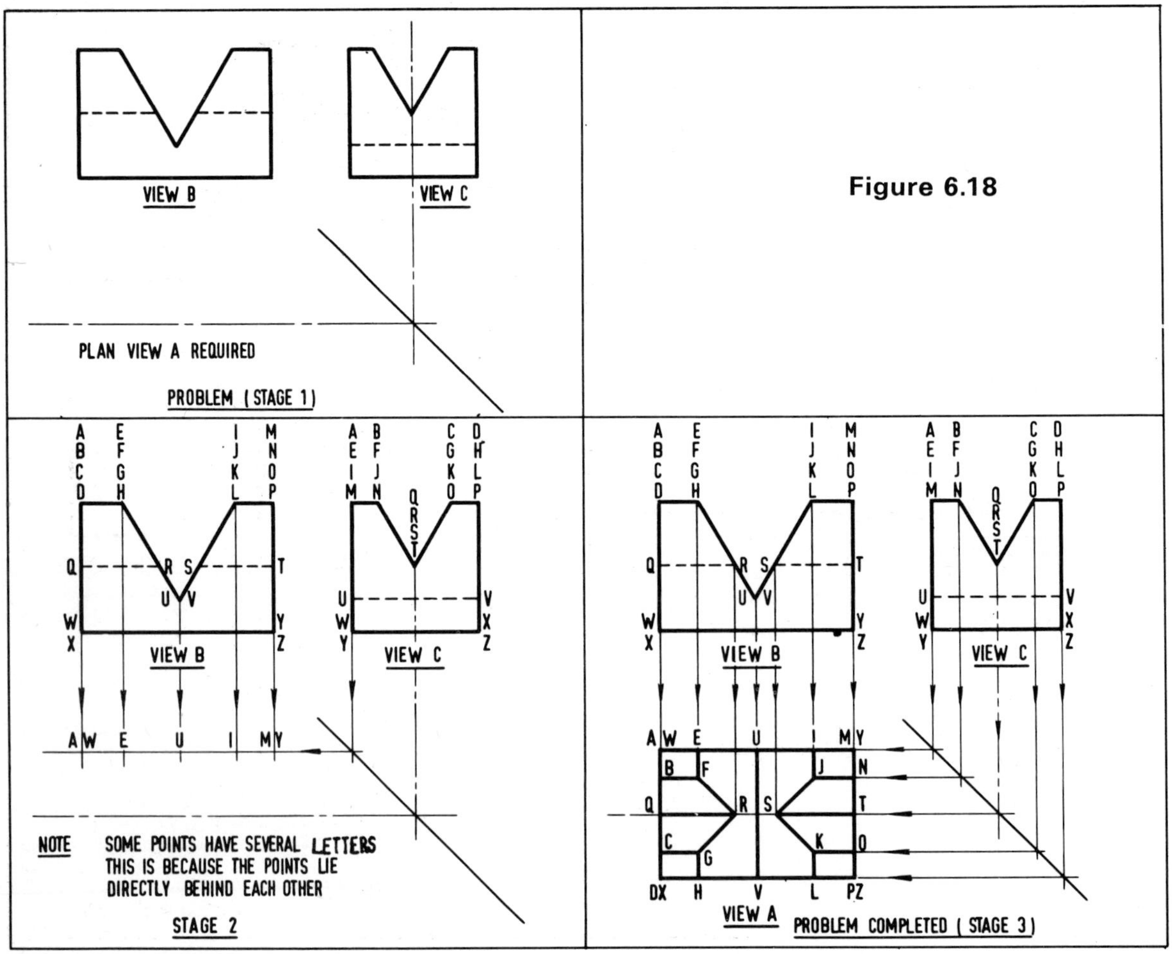

Figure 6.18

Figure 6.19 shows views of two cylinders intersecting at right angles. It is required to complete view B showing the true shape of the line of intersection. Where solid shapes intersect, the smaller one should be regarded as being attached to the larger, and the line of intersection is the mating shape assumed by the smaller on attachment.

Figure 6.19a

In view C establish a series of equally spaced points around the circumference of the small cylinder by dividing the end view of the cylinder into 12 equal parts, using a 60°–30° set square. Project the points into views A and B. Note the system of numbering the points. Corresponding points on both sides of the cylinder carry the same number to prevent excess numbering and confusion during projection.

Figure 6.19b

1. The figure shows a simplified technique for establishing points. View C is not required and is replaced by half views of the circumference drawn attached to the smaller cylinder in views A and B. The half views are divided into 6 equal parts using a 60°–30° set square.
2. Projection lines drawn horizontally from the numbered points in view A can be regarded as lines scribed on the cylinder surface. It will be seen that the length of these lines finishes on contact with the circle representing the surface of the larger cylinder. Number the ends of the lines in contact with the circle with the same numbers as at their other end.
3. The same series of projection or scribed lines is drawn horizontally in view B and the lengths are established by projection from view A.
4. The points plotted in view B are points on the intersection line. A smooth curve joining the points in view B will produce the line of intersection.

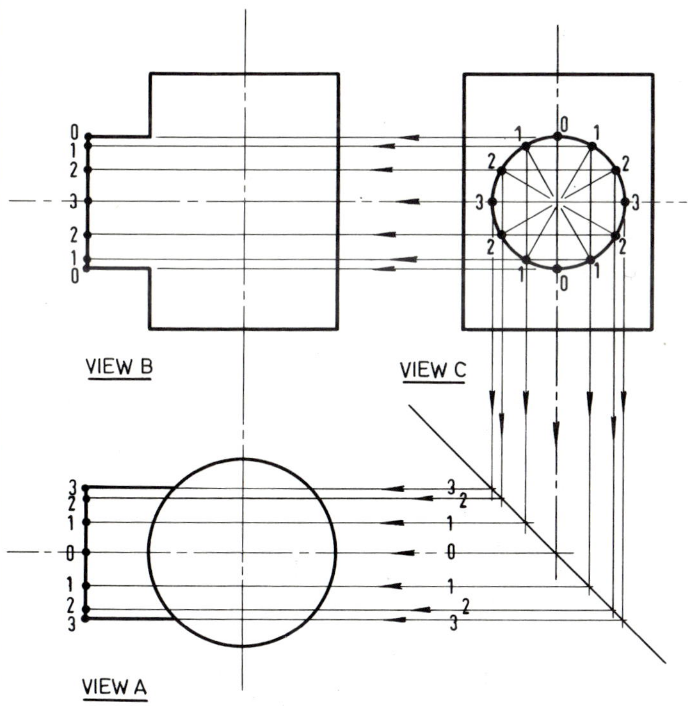

Figure 6.19*a*

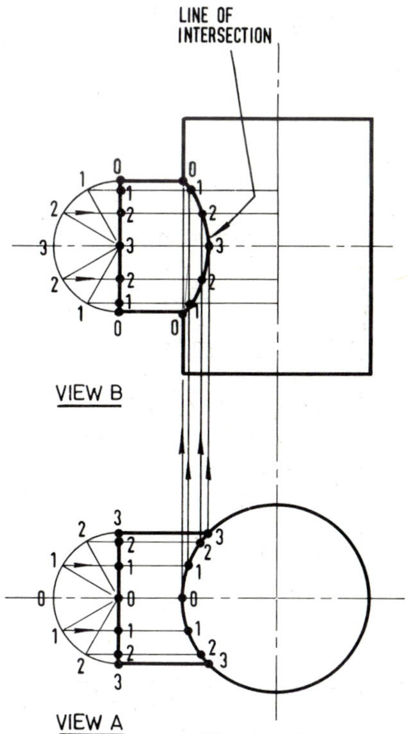

Figure 6.19*b*

Figure 6.20*a* shows views of a horizontal cylinder intersecting a right cone. It is required to complete views A and B showing the true shape of intersection. The line of intersection will be the mating shape assumed by the cylinder on attachment to the larger solid, the cone.

Figure 6.20b

This shows the series of numbered points established by the simplified technique using attached half end views.

In view B projection lines are drawn horizontally from the numbered points to the far side of the cone. These lines represent cutting lines. The sections (or shapes) shown in the sectioned pictorial view are drawn one on top of the other in view A. This is known as a 'nest of sections'.

Sections are drawn in view A by connecting projection lines from the numbered points on the cylinder with the relevant circular section, which is drawn using the radius *r* projected down from view B.

Figure 6.20c

This shows the completed nest of sections in view A, and the resulting series of points which lie on the intersection line. These points are projected into view B to plot the line of intersection in this view.

Figure 6.20d

This shows the two views completed.

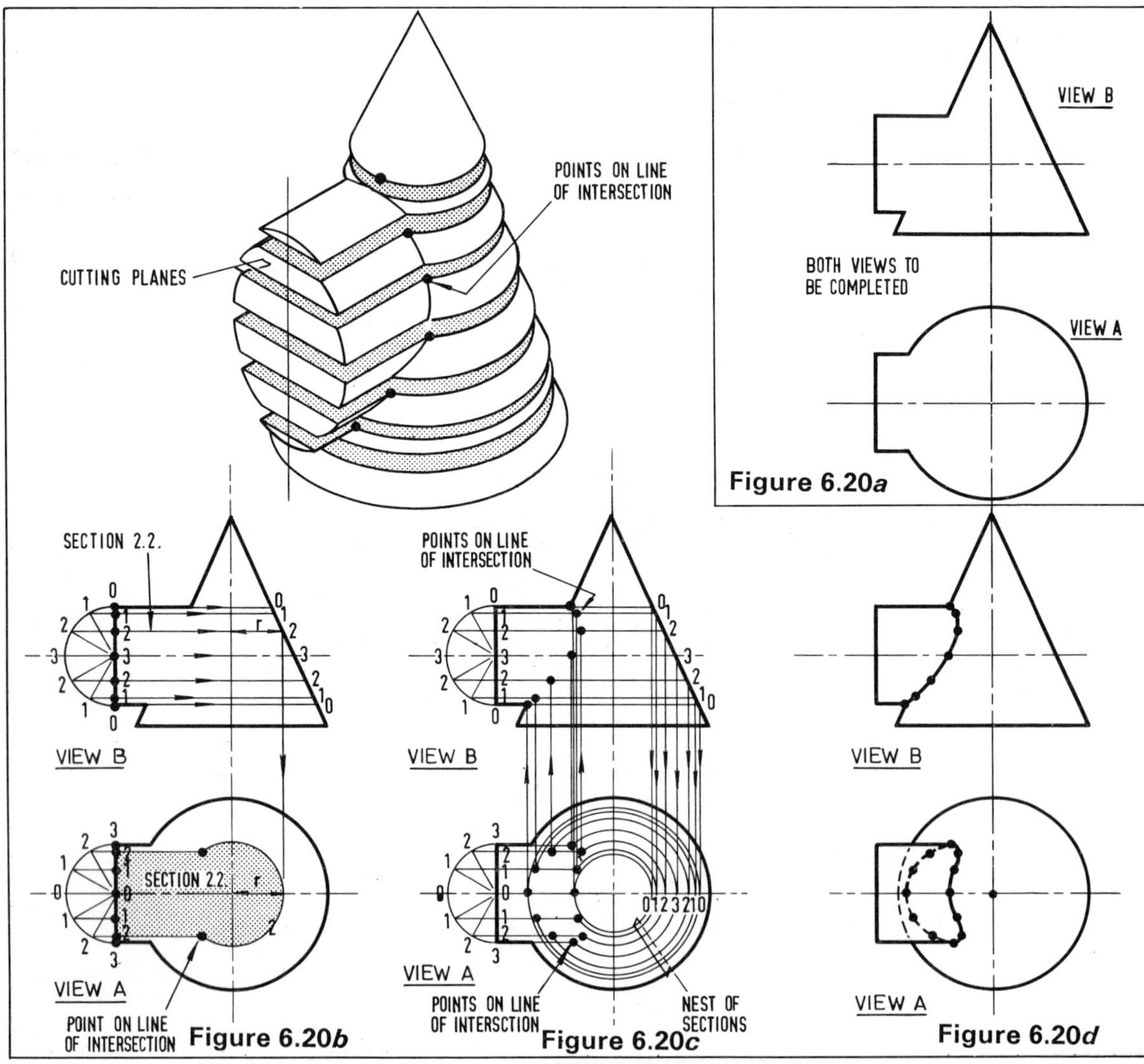

Figure 6.20*a*

Figure 6.20*b*

Figure 6.20*c*

Figure 6.20*d*

AUXILIARY PROJECTION

When making an engineering drawing, instances often arise where the object has surfaces which cannot be clearly defined or dimensioned in the principal planes of drawing. An example is where objects have inclined surfaces with complex visible outlines and machine details. To present the true shape and the precise manufacturing information, additional views are required showing the surface as it appears to an observer when looking directly at that surface. These views are called *auxiliary views.*

Auxiliary views may be projected from elevations or plans, and are produced either by complete projection from existing views, or by part projection plus transfer of dimensions by dividers. In practice, it is usual to draw only the inclined surface and not the entire object. The use of the partial view decreases the risk of confusion when reading the drawing. The remainder of the object in the view is usually of pictorial interest only, being non-informative and often an unnecessary, time-wasting, complex exercise.

However, when learning to draw it is advisable to draw several complete objects in auxiliary projection to increase one's ability in drawing technique.

Figure 6.21 shows an auxiliary plane erected in the first angle to receive the true view of the inclined surface containing the blind hole. The true width *w* of the inclined face and other points are projected at right angles across the horizontal plane, then vertically up the auxiliary plane. The true height is projected around from the elevation. The hole is then positioned on the auxiliary view; where necessary dimensioning would appear on a dimensioned drawing.

Figure 6.21

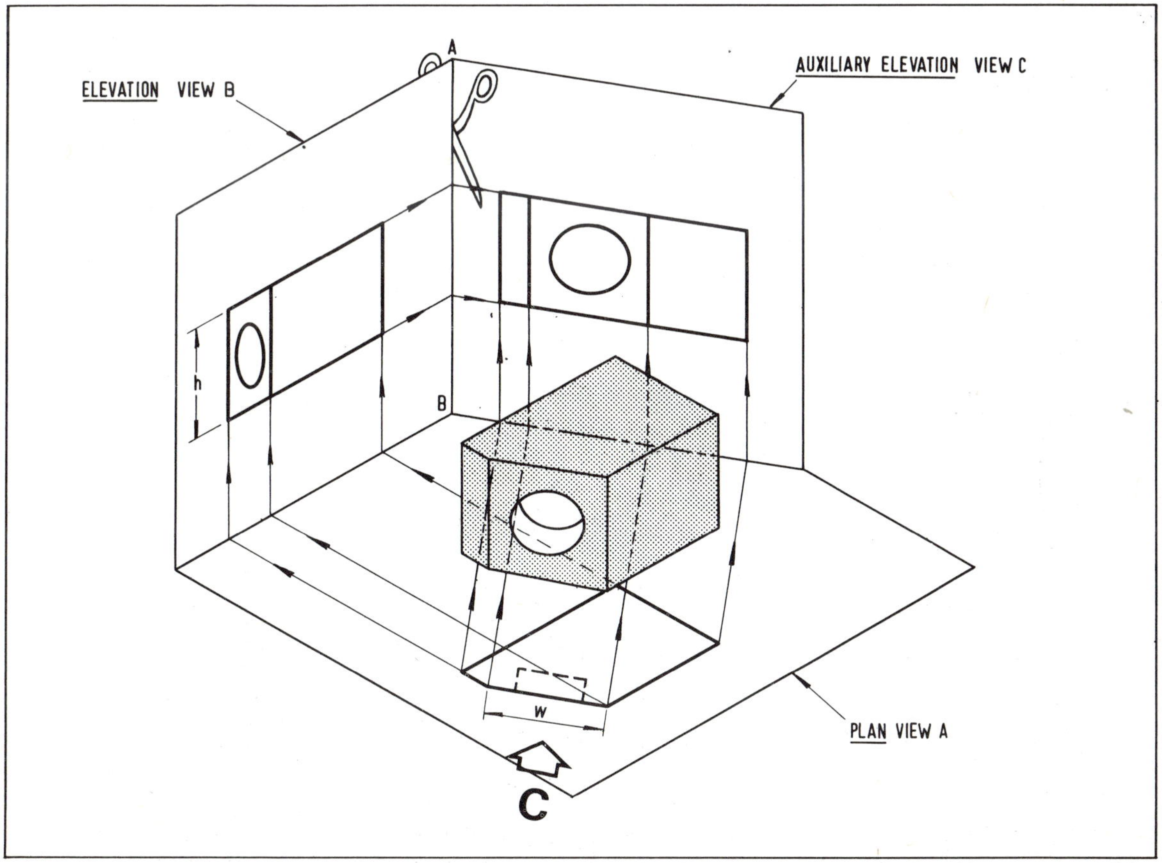

Figure 6.22 shows the three planes laid out flat after cutting along line AB. It will be seen that the projection lines from the elevation to the auxiliary view have become disjointed. When projecting on a drawing sheet, the height *h* is transferred via a deflector line which is set up by bisecting the angle through which the dimensions have to be transferred.

The true width *w* of the inclined face is transferred to the auxiliary views by projection lines at right lines to the inclined face.

Figure 6.22

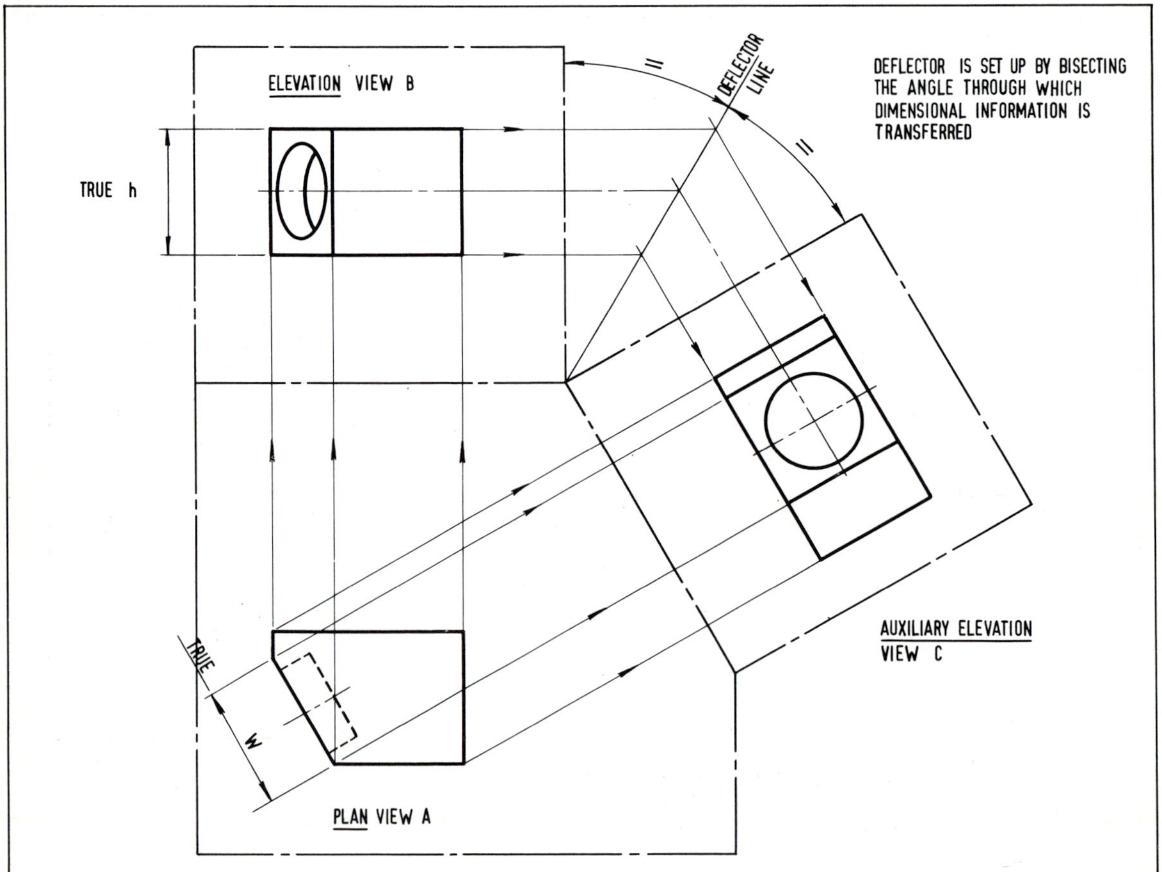

In Figure 6.23 the three views are shown on a drawing sheet where lack of space prevents full projection of the auxiliary view. In such circumstances the following method of projection, plus transfer of dimensions by dividers, is used.

Method

1. Complete the outline of elevation and plan.
2. Parallel to the inclined face set up a datum line from which dimensions can be located. The datum in this instance represents the base of the elevation.
3. In the plan view, project width *w* and other points at right angles to the inclined surface on to the auxiliary view.
4. Using dividers, transfer the true height *h* from the elevation to the auxiliary view. Then line in the view as necessary.
5. On completion of the visible outline, position the hole, using exact dimensions.
6. Transfer dimensional information concerning the hole back to the other views.

Figure 6.23

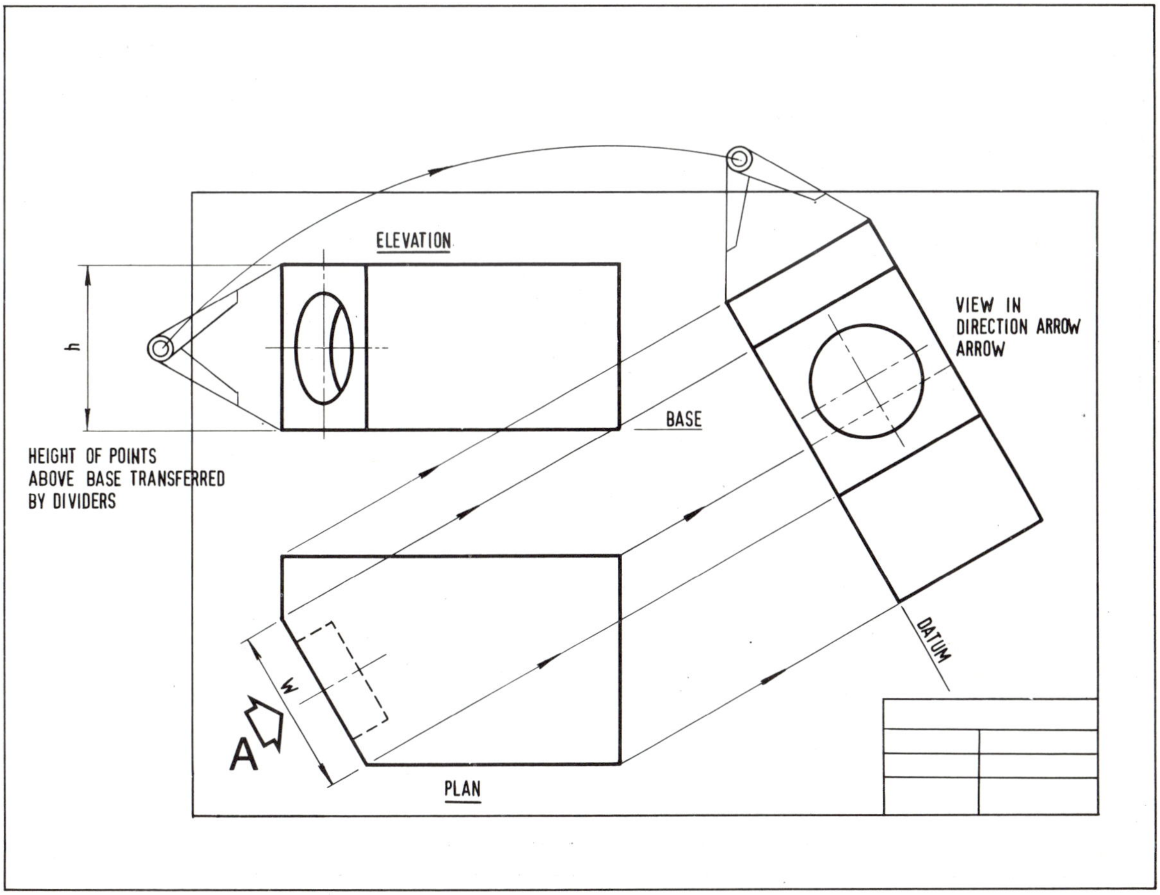

PROBLEM Figure 6.24

A square pyramid with a portion machined away is shown in plan and elevation. Project an auxiliary elevation in the direction of arrow A.

Method

1. In the plan and elevation identify the end points of all the straight lines with a letter, corresponding points to carry the same letter in both views.
2. Set up a datum or ground line at right angles to the direction of viewing.
3. Draw projection lines from all points in the *plan view* on to the position of the auxiliary view, parallel to the direction of viewing.
4. Transfer the vertical height dimensions of the points above the ground line in the *elevation* to the relevant projection line in the *auxiliary view* as shown.
5. Join points in the *auxiliary view* in the same sequence as they appear in the *elevation* and *plan* to produce the required view, i.e. J to K, K to M, M to L, L to J; then J to G, G to H, H to K, H to F, E to G etc.

Figure 6.24

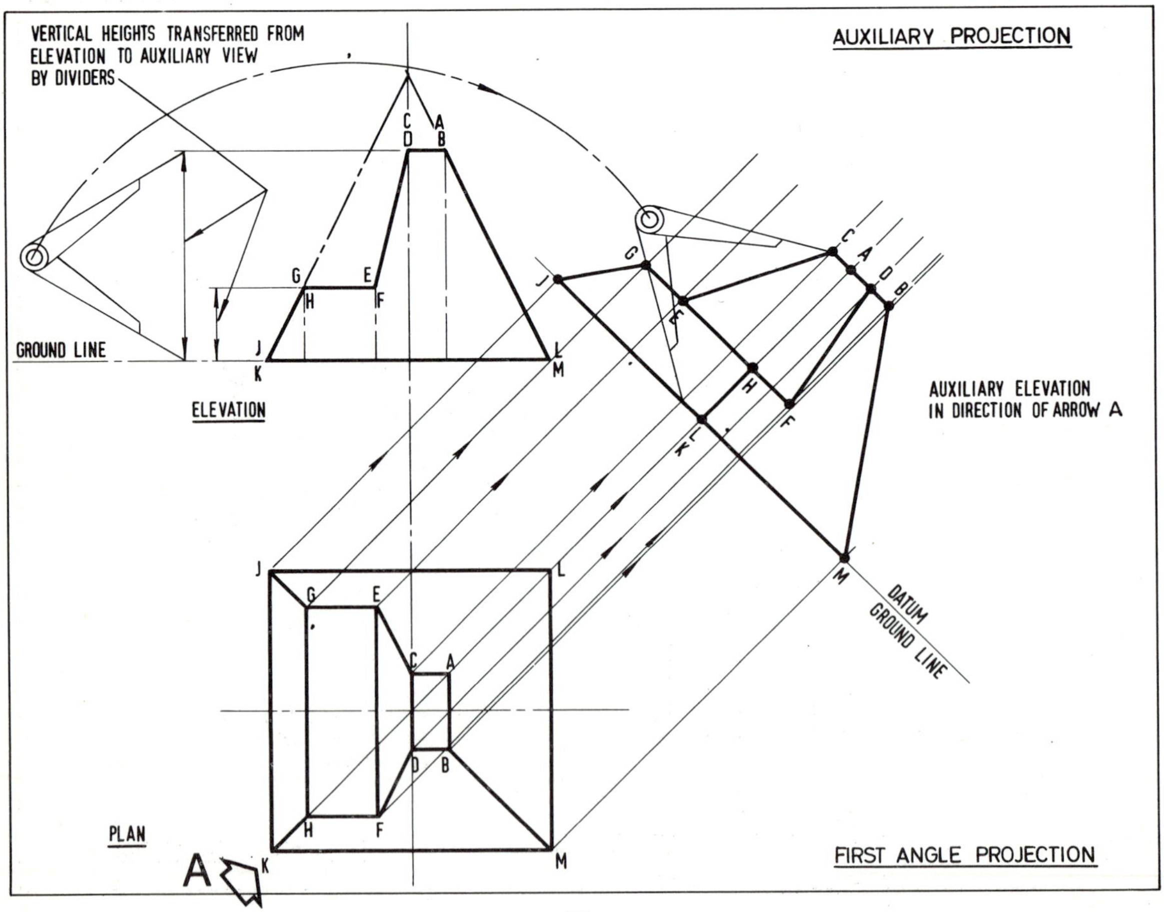

PROBLEM Figure 6.25

A square pyramid with a portion machined away is shown in plan and elevation. Project an auxiliary plan in the direction of arrow A.

Method

1. In the plan and elevation identify the end points of all the straight lines with a letter, corresponding points to carry the same letter in both views.
2. Set up a datum at right angles to the direction of viewing, and establish the position of the datum in the plan view. The datum in this case is the vertical plane containing the edge JL.
3. Draw projection lines from all points in the *elevation* on to the position of the *auxiliary view,* parallel to the direction of viewing.
4. Transfer the perpendicular distances of the points from the datum in the *plan view* to the relevant projection lines in the *auxiliary view* as shown.
5. Join the points in the *auxiliary view* in the same sequence as they appear in the *elevation* and *plan* to produce the required view, i.e. J to K, G to H, J to G, H to K, G to E, H to F, E to F etc.

Note that edge LM is out of sight from the direction of viewing, and is shown by hidden detail.

Figure 6.25

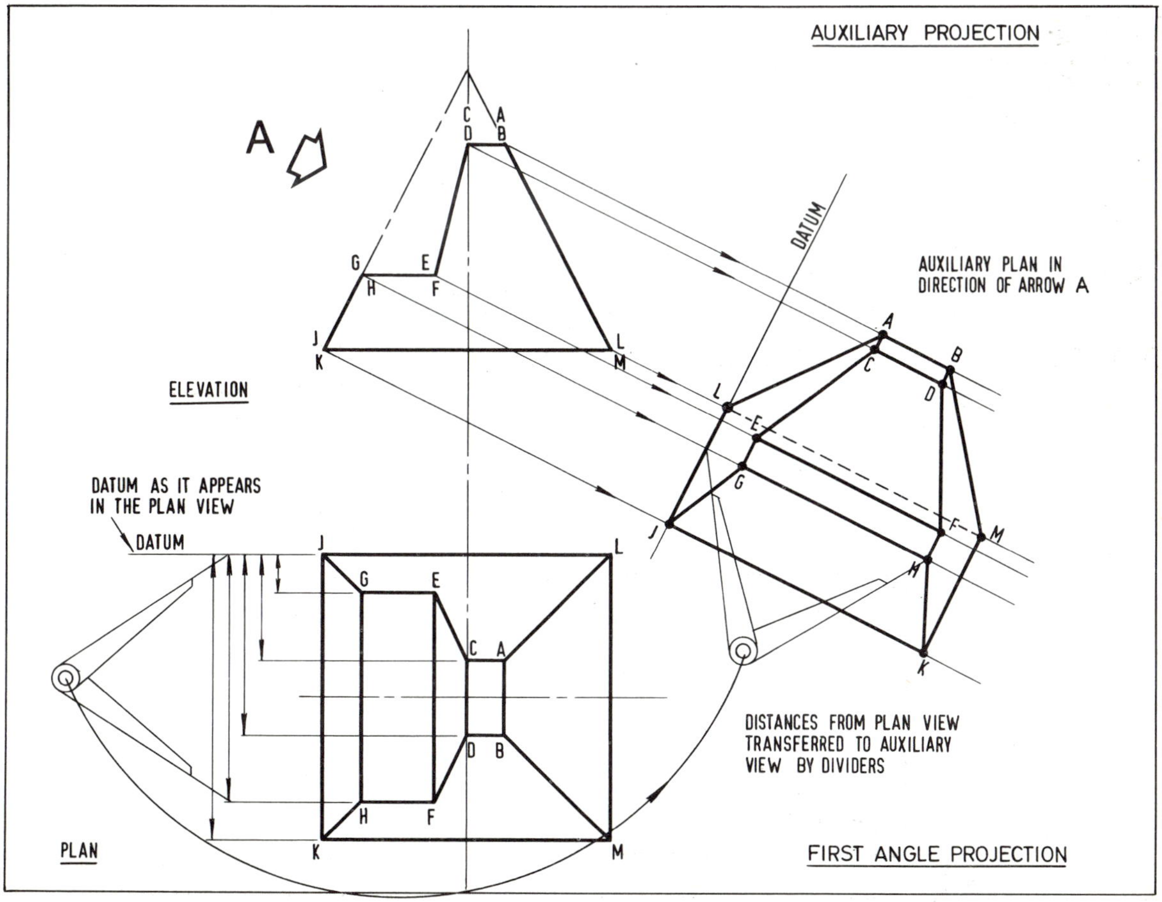

PROBLEM Figure 6.26

A square prism with a sloping top surface, and a hole drilled through it diagonally across the corners, is shown in plan and elevation drawn in third-angle projection. Project an auxiliary elevation in the direction of arrow A. This problem is divided into two parts:

Part 1: Project the prism only.

Part 2: Project the shape of the hole on to the prism.

Method: Part 1

1. In the *plan* and *elevation* identify all the end points of the straight lines with a letter, corresponding points to carry the same letter in both views.
2. Set up a datum or ground line at right angles to the direction of viewing and establish the position of the datum in the *elevation.*
3. Draw projection lines from all points in the *plan* on to the position of the *auxiliary view,* parallel to the direction of viewing.
4. Transfer the perpendicular distances of the points in the *elevation* to the relevant projection lines in the *auxiliary view* as shown.
5. Join the points in the auxiliary view in the same sequence as they appear in the *elevation* and *plan* to produce the required view of the prism.

Method: Part 2

1. Establish 12 equally spaced points around the hole in the *elevation.*
2. Project these points on to both ends of the hole in the *plan view.*
3. Draw projection lines from these points in the *plan* on to the *auxiliary view.*
4. Transfer the perpendicular distances of the points in the *elevation* to the relevant projection lines in the auxiliary view with dividers as shown.
5. Join the points in numerical sequence.

Figure 6.26

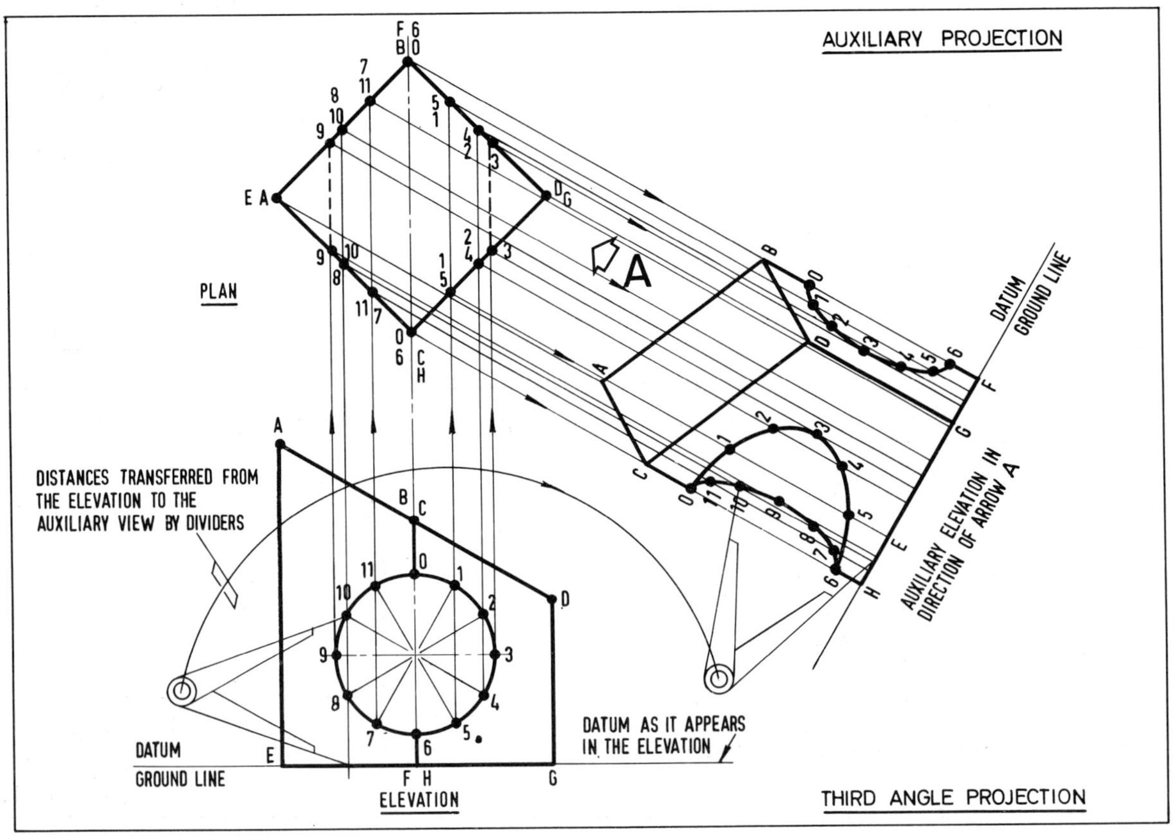

EXERCISES

EXERCISE 6.1

Divide an A2 or half imperial sheet into four parts as shown. Using the dimensions given copy the views shown in the four exercises and complete the unfinished views by projection.

Number all points for projection and show all projection and construction lines. Visible outlines should be thick and bold. Projection lines to be thin. Do not dimension the exercise.

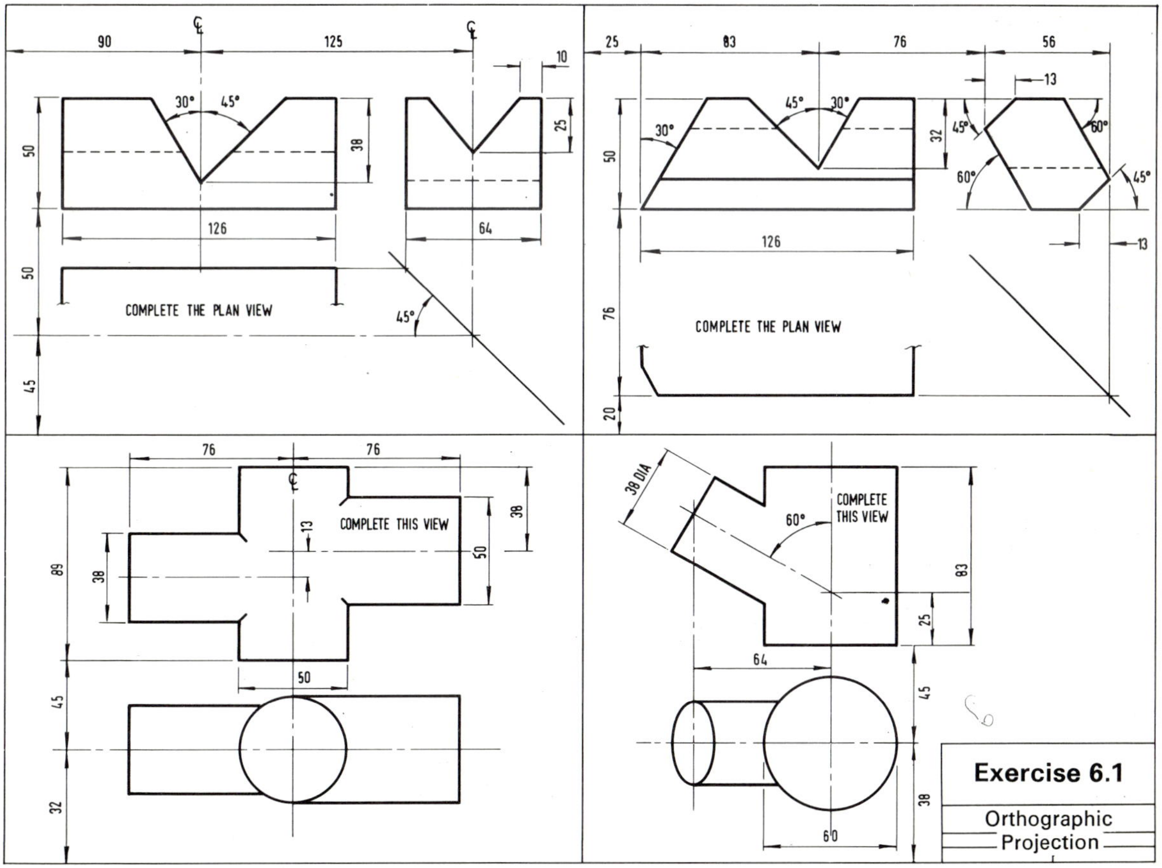

Fits and clearances

WHY DO WE HAVE TO BOTHER ABOUT FITS AND CLEARANCES?

Up to the beginning of the twentieth century mechanical engineers had no universal standard to which they could refer when producing mating components. What size hole was required to give a satisfactory fit, and a satisfactory working life, was the individual knowledge of skilled craftsmen who gained this knowledge by long practice.

If a shaft had to fit a hole, the shaft size was often determined by rule of thumb, by previous experience, or by what material was available. Having decided on the shaft size, the problem was to fit the shaft into the hole. The function of the shaft was known. It could fit so tightly that on driving it in it did not come apart in use. Alternatively, the shaft could be a nice tight fit, but could be taken apart without difficulty, or it could rotate freely.

HOW WAS THIS CARRIED OUT?

The hole was drilled and the shaft machined down very carefully until the desired fit was achieved by making several trial assemblies. Alternatively, finished machined shafts were fitted to bearing journals made of bronze or white metal, which were cast as near to size as possible, and then painstakingly scraped until at least an 80% contact between shaft and bearing was produced. This was the heyday of the skilled fitter, whose long and practical experience enabled him to use hand tools to produce mating parts of the required fit.

However, fitting was a time-consuming and tedious practice which nowadays, with high wages and overheads, would be uneconomic. Also, because all parts were individually matched by hand fitting, there was no interchangeability whereby parts could be purchased or ordered from stock and be fitted without further work being done on them.

The need to produce standard parts to achieve interchangeability was recognised by Sir Joseph Whitworth (the man who gave us the first standardised thread form). Between 1840 and 1860 he introduced in his works a system of gauging to which parts were manufactured, thus producing a measure of standardisation. The parts were still made by trial and error, however, and although Whitworth's work gave a great impetus to accurate mechanical measurement no reference standard of limits and fits was produced until the beginning of the nineteenth century.

It is interesting to note here that the outside micrometer caliper as used by engineers was first manufactured for industry by Brown and Sharpe in 1867, although some earlier models had been produced. The improved model of 1885 produced by the same company differs very little in external appearance from the modern micrometer. However, this accurate measuring device took a number of years before it became universally available and in use in industry. The method of measuring by inside and outside calipers to produce mating parts was more the order of the day.

MASS PRODUCTION

With the increase in population, and society growing more affluent, there was an increasing demand for manufactured goods at prices which ordinary people could afford. Also, there were the requirements of the armed forces for large supplies of standardised equipment with interchangeability. To produce complex engineering goods cheaply, a large labour force had to be employed, with labour broken down into individual tasks. Each person in this large force was engaged on a simple task in the overall production, which was spread over a large area, often in many factories. Today production is so specialised that it is not only spread over many factories but over many independent companies, often in different countries.

At the final assembly stage, thousands of individual parts and small assemblies have to be fitted together to produce the complete machine or structure. Also an excess has to be retained in stock for replacements made necessary by wear or breakage. To enable this to be done, the individual parts have to be made to precise sizes within the limits of a fixed manufacturing tolerance, and then inspected with measuring instruments and gauges at each stage of production to ensure that only parts of correct size are passed on to the assembly stages.

HOW THIS WAS ACHIEVED

The problem was first tackled on a rational basis in 1902 by J. W. Newall, who collected and collated a large amount of material from many engineering sources and laid down a tolerance system on the 'hole' basis, by which all holes were produced as far as commercially possible to exact size. The system was a bilateral one in which the limits for the manufacture of holes were fixed above and below the nominal hole size. It provided for two grades of hole designated Class A and Class B. Class A holes were for fine tool work and had a smaller tolerance than class B holes. The slightly increased tolerance in Class B made this commercially more economical for general use.

For shafts, tables for the production of various fits were compiled. Tables F, D and P gave tolerances for the manufacture of force, driving and push fits; Tables X, Y and Z gave tolerances for three grades of running fits.

Since this first Standard of limits and fits, which is still in use but fast becoming obsolete, the British Standards Institution, with recommendations from the Institution of Production Engineers, have issued two standards on limits and fits. The first (BS 164) was issued in 1906 and revised in 1924 and 1941. A new Standard (BS 1916) was produced in 1953 and is in current use.

Fits and clearances

Chapter 7

BS 1916: 1953 Limits and fits for engineering

It is advisable for the student technician who is to be working in design or production to familiarise himself with the tolerancing of mating parts, and he should either buy copies of Parts 1 and 2 of BS 1916* or at least read through the Standard, which should be found in the library of all design and production units.

Part 1 covers 21 fits between holes and mating shafts in both inch and metric measures. For the *holes* the varying degrees of size above or below nominal size are designated by a capital letter; for the *shafts* they are designated by a small letter, e.g. Hole A, Shaft a.

All letters of the alphabet are used, with the exception of I, L, O, Q and W. The 21 fits are graded, and range from an extreme clearance fit at the beginning of the alphabet to extreme interference at the end of the alphabet. Figure 7.1 shows this diagrammatically.

The Standard also gives 16 grades of accuracy for each of the 21 grades of fit to which holes and shafts can be manufactured; these are designated by number 1 to 16. To give a complete description of a hole or shaft, the fit and grade of manufacturing accuracy is stated, as, for example, Hole H 7, Shaft e 8.

A particular fit is stated by the description of the hole followed by the description of the shaft, for example H 7–e 8. This particular combination would give an easy-running fit.

The Standard contains a selection of suitable combinations of the 21 fits and 16 grades of manufacturing tolerance. Figure 7.1 shows this selection diagrammatically, and also a smaller selection of 3 holes and 17 shafts from which most of the required fits in mechanical engineering can be satisfied.

* Part 1: *Limits and Tolerances;* Part 2: *Guide to the Selection of Fits in BS 1916; Part 1.*

HOLES

Hole H. From Figure 7.1 it will be seen that the hole H has for its low limit (i.e. smallest diameter) the nominal size of the hole, and the various grades of manufacturing tolerance give the high limit (i.e. diameters which are larger than nominal). This H hole is the one recommended to be used in practice. It gives a tolerance which is *unilateral*, from nominal size to slightly larger. This allows for simple manufacture of holes by using standard-size drills and reamers.

Holes A–G have a fixed low limit which is larger than nominal, with manufacturing tolerances always increasing the hole diameter still further.

Hole J has tolerance zones which lie on either side of the nominal size, producing holes which are slightly smaller or slightly larger than nominal

Holes K–Z are toleranced to produce holes which are always smaller than nominal size.

Shafts are toleranced in the reverse manner.

Shafts a–g have a fixed high limit which is smaller than the nominal diameter, with the various grades of manufacturing tolerance decreasing the size of the shaft still further.

Shaft h has a high limit (largest diameter) which is the nominal size of the shaft, with the various grades of tolerance decreasing the shaft diameter for the low limit.

Shaft j has tolerance zones which lie either side of the nominal size, allowing for shafts slightly larger or slightly smaller than nominal diameter.

Shafts k–z are all toleranced to produce shafts which are larger than nominal diameter.

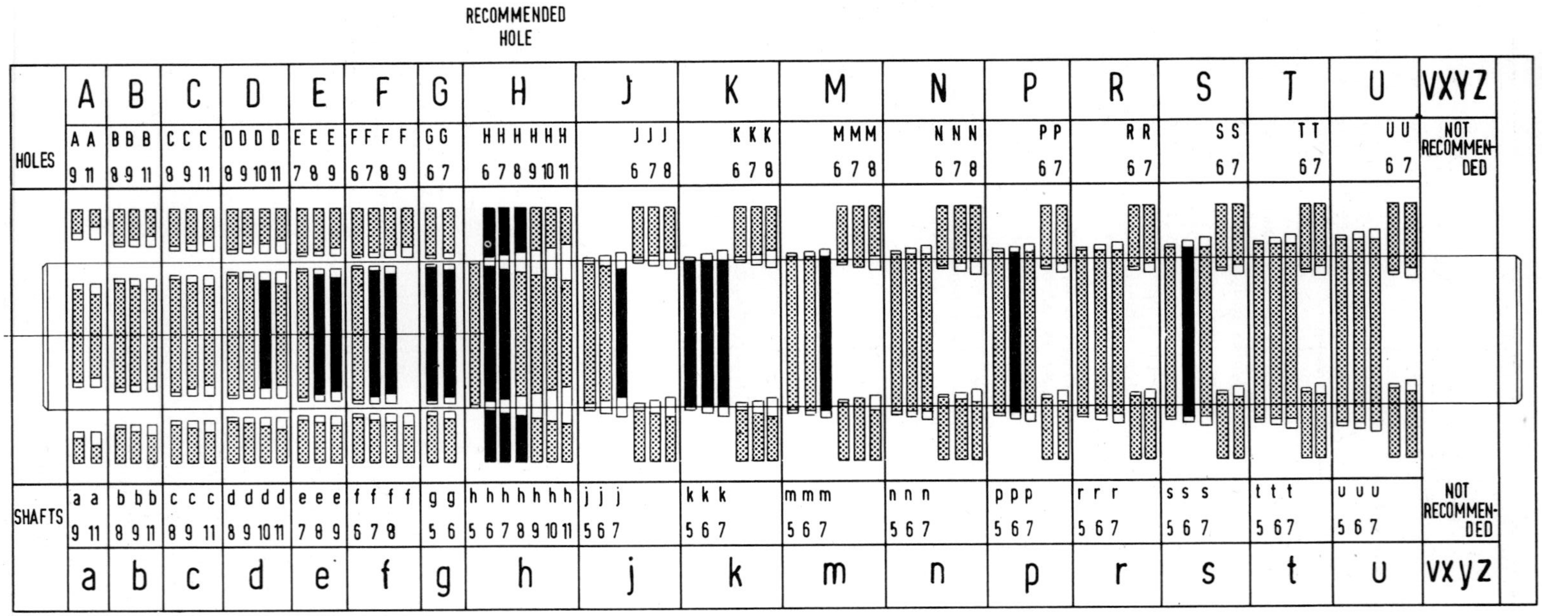

Figure 7.1

A recommended selection from BS 1916 of the twenty-one fits between shaft and hole, and a selection of the sixteen grades of tolerance diagrammatically arranged according to symbol about a nominal-sized shaft. Note that for engineering purposes only a small number of possible fits are required. Fits produced between three types of hole and sixteen types of shaft cover most engineering requirements.

		HOLES			TYPE OF FIT
		H. 6	H. 7	H. 8	
CLEARANCE	SHAFTS			d. 10	LOOSE RUNNING
			e. 8	e. 9	EASY RUNNING
			f. 7	f. 8	RUNNING
		g. 5	g. 6		CLOSE RUNNING
			h. 6	h. 7	SLIDE
				j. 7	PUSH FIT
		k. 5	k. 6	k. 7	EASY KEYING
				m. 7	TIGHT KEYING
			p. 6		PRESS FIT (FERROUS PART)
INTERFERENCE			s. 6		PRESS FIT (NON-FERROUS PARTS)

Figure 7.2 Table showing combination of holes with shafts to produce types of fit

SELECTION FOR FITS

The practice usually recommended is to select an H hole with a manufacturing tolerance to suit economic production capacity, and then match it to a shaft from the whole a–z range to produce the required fit. For most fits between mating parts, a surprisingly small selection of holes and shafts satisfies most practical requirements. This selection is shown in Figure 7.2 giving the type of fit for various combinations of hole and shaft. Reference to Figure 7.1 shows where this selection has been extracted from the full recommendation.

WHO DECIDES THE TYPE OF FIT REQUIRED

When a new project is begun, the designer draws layouts and schemes of the mechanisms and structures required, to illustrate his thoughts and ideas. The object is to prove a design on paper, and layout and schemes are drawn, altered, scrapped and redrawn until the new project becomes a workable proposition. During these layout stages, each individual part is considered for strength, material and functional requirements; this includes the fit of mating components, and a good designer will note, either in words or symbols, the type of fit he requires.

A section of a designer's layout is shown in Figure 7.3, which gives an overall picture of the mechanism, with the desired fits shown as hole and shaft symbols. It is essentially a communication between designer and detailing draughtsman.

It is the responsibility of the detailing draughtsman to produce from the designer's layout detail drawings of each part, fully dimensioned for manufacture. Figure 7.4 shows one method by which the draughtsman would interpret the designer's instructions with regard to the toleranced dimensions on the individual detail drawings. The machining tolerance is indicated by stating the acceptable limits for manufacture as the maximum and minimum diameters for both holes and shafts. This is a precise method of instruction which is easily understood by production and inspection staff.

BS 308 (see page 39) shows other acceptable methods of dimensioning which can be used on drawings for tolerancing mating components.

Figure 7.3 Designer's layout showing nominal diameter and types of fit required

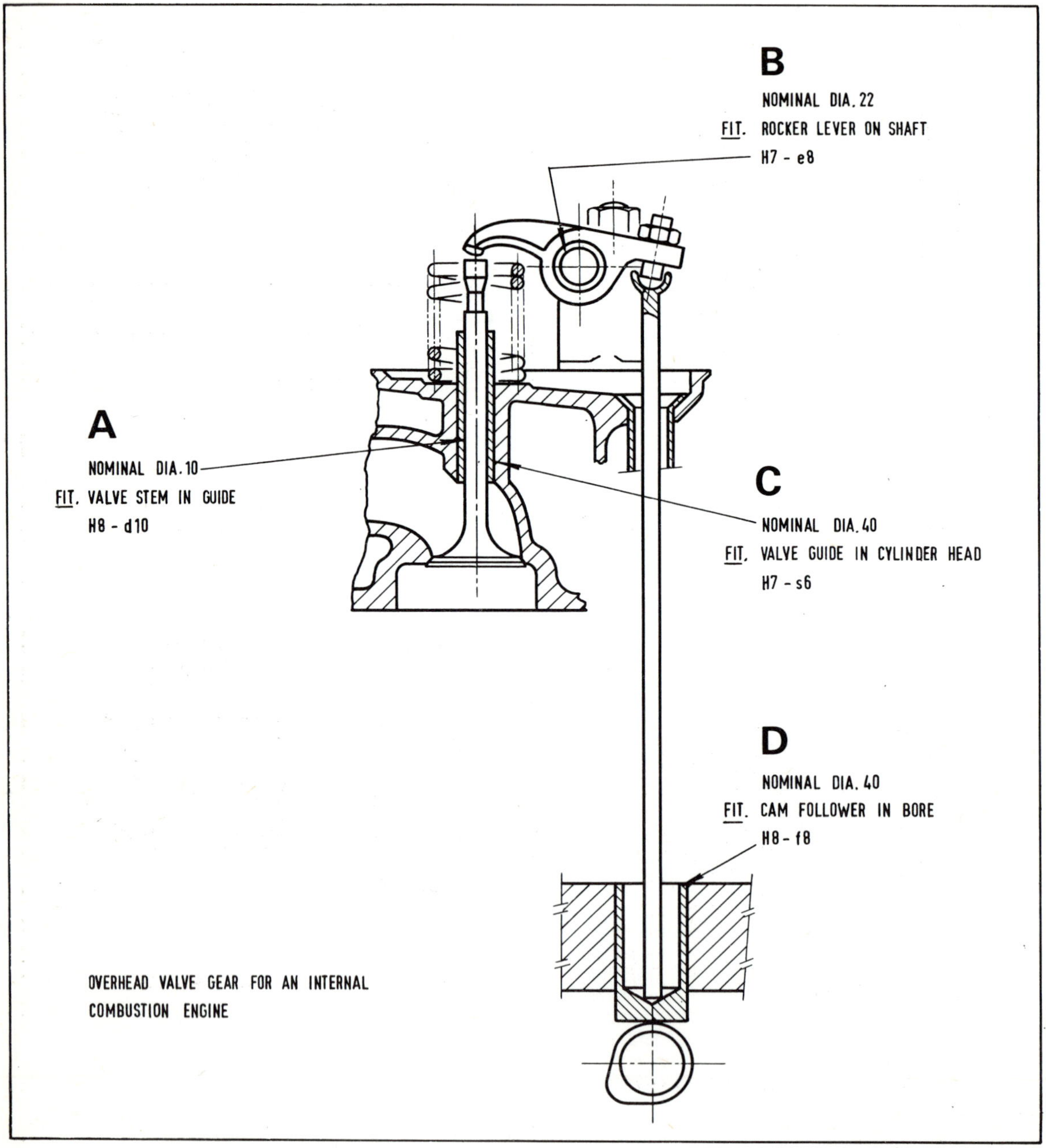

Figure 7.4 Draughtsman's interpretations of designer's instructions when making detail drawings for production

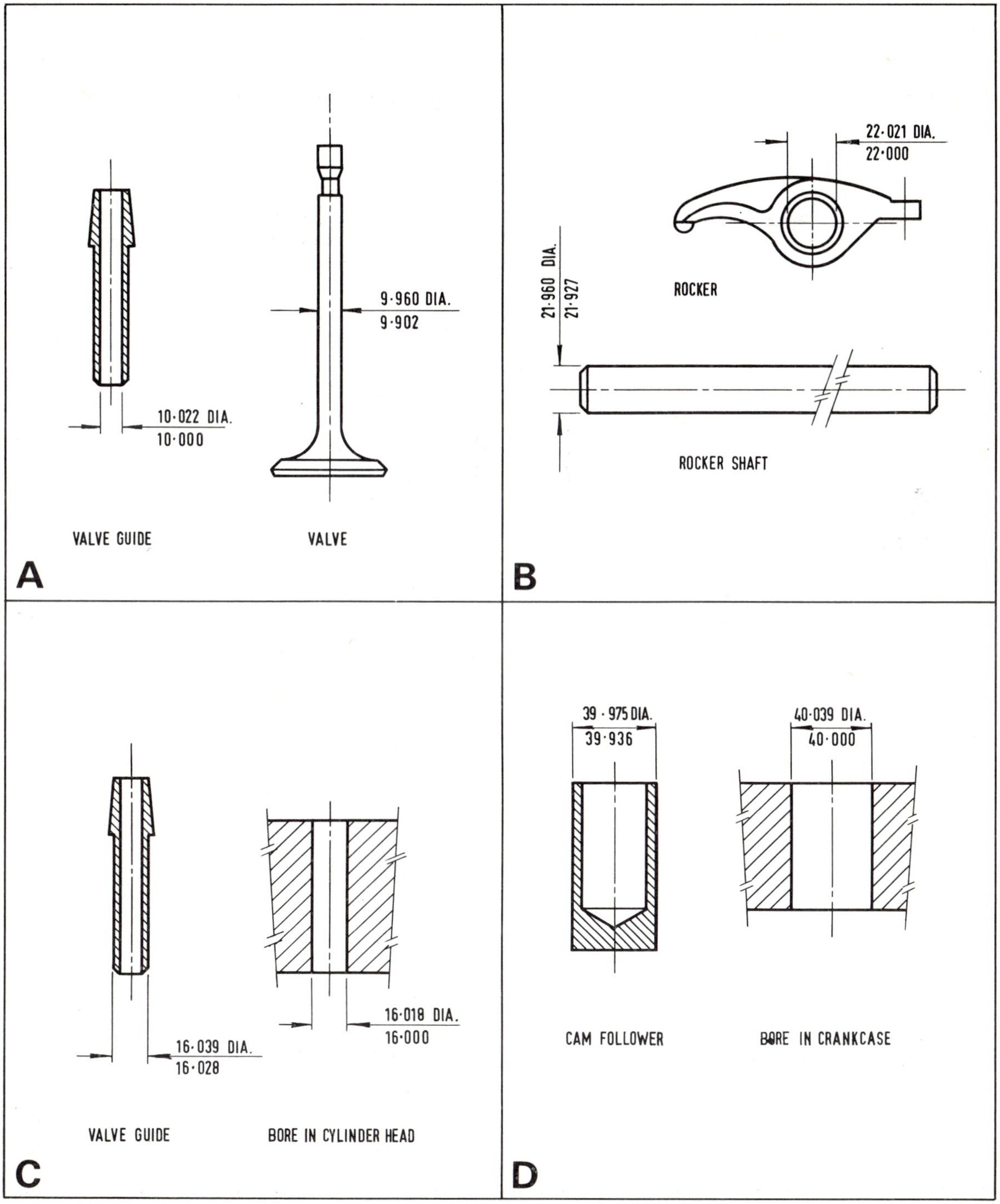

Limits for 3 selected holes and 16 shafts as shown in Figure 2. (B.S. 1916 limits and fits)

Tolerances in units of 0·001 inch

Nominal diameter in inches	HOLES						SHAFTS															
	H 6		H 7		H 8		d 10		e 8		e 9		f 7		f 8		g 5		g 6		h 6	
	High limit	*Low limit*	*High limit*	*Low limit*	*High limit*	*Low limit*	*High limit*	*Low limit*	*High limit*	*Low limit*	*High limit*	*Low limit*	*High limit*	*Low limit*	*High limit*	*Low limit*	*High limit*	*Low limit*	*High limit*	*Low limit*	*High limit*	*Low limit*
0·04–0·12	+0·25	+0	+0·4	+0	+0·6	+0	−1·0	−2·6	−0·6	−1·2	−0·6	−1·6	−0·3	−0·7	−0·3	−0·9	−0·1	−0·3	−0·1	−0·35	−0	−0·25
0·12–0·24	+0·3	+0	+0·5	+0	+0·7	+0	−1·2	−3·0	−0·8	−1·5	−0·8	−2·0	−0·4	−0·9	−0·4	−1·1	−0·15	−0·35	−0·15	−0·45	−0	−0·3
0·24–0·40	+0·4	+0	+0·6	+0	+0·9	+0	−1·6	−3·8	−1·0	−1·9	−1·0	−2·4	−0·5	−1·1	−0·5	−1·4	−0·2	−0·45	−0·2	−0·6	−0	−0·4
0·40–0·56	+0·4	+0	+0·7	+0	+1·0	+0	−2·0	−4·8	−1·2	−2·2	−1·2	−2·8	−0·6	−1·3	−0·6	−1·6	−0·25	−0·55	−0·25	−0·65	−0	−0·4
0·56–0·71	+0·4	+0	+0·7	+0	+1·0	+0	−2·0	−4·8	−1·2	−2·2	−1·2	−2·8	−0·6	−1·3	−0·6	−1·6	−0·25	−0·55	−0·25	−0·65	−0	−0·4
0·71–0·95	+0·5	+0	+0·8	+0	+1·2	+0	−2·5	−6·0	−1·6	−2·8	−1·6	−3·6	−0·8	−1·6	−0·8	−2·0	−0·3	−0·7	−0·3	−0·8	−0	−0·5
0·95–1·19	+0·5	+0	+0·8	+0	+1·2	+0	−2·5	−6·0	−1·6	−2·8	−1·6	−3·6	−0·8	−1·6	−0·8	−2·0	−0·3	−0·7	−0·3	−0·8	−0	−0·5
1·19–1·58	+0·6	+0	+1·0	+0	+1·6	+0	−3·0	−7·0	−2·0	−3·6	−2·0	−4·5	−1·0	−2·0	−1·0	−2·6	−0·4	−0·8	−0·4	−1·0	−0	−0·6
1·58–1·97	+0·6	+0	+1·0	+0	+1·6	+0	−3·0	−7·0	−2·0	−3·6	−2·0	−4·5	−1·0	−2·0	−1·0	−2·6	−0·4	−0·8	−0·4	−1·0	−0	−0·6
1·97–2·56	+0·7	+0	+1·2	+0	+1·8	+0	−4·0	−8·5	−2·5	−4·3	−2·5	−5·5	−1·2	−2·4	−1·2	−3·0	−0·4	−0·9	−0·4	−1·1	−0	−0·7
2·56–3·15	+0·7	+0	+1·2	+0	+1·8	+0	−4·0	−8·5	−2·5	−4·3	−2·5	−5·5	−1·2	−2·4	−1·2	−3·0	−0·4	−0·9	−0·4	−1·1	−0	−0·7
3·15–3·94	+0·9	+0	+1·4	+0	+2·2	+0	−5·0	−10·0	−3·0	−4·2	−3·0	−6·5	−1·4	−2·8	−1·4	−3·6	−0·5	−1·1	−0·5	−1·4	−0	−0·9
3·94–4·73	+0·9	+0	+1·4	+0	+2·2	+0	−5·0	−10·0	−3·0	−5·2	−3·0	−6·5	−1·4	−2·8	−1·4	−3·6	−0·5	−1·1	−0·5	−1·4	−0	−0·9
4·73–5·52	+1·0	+0	+1·6	+0	+2·5	+0	−6·0	−12·0	−3·5	−6·0	−3·5	−7·5	−1·6	−3·2	−1·6	−4·1	−0·6	−1·3	−0·6	−1·6	−0	−1·0
5·52–6·30	+1·0	+0	+1·6	+0	+2·5	+0	−6·0	−12·0	−3·5	−6·0	−3·5	−7·5	−1·6	−3·2	−1·6	−4·1	−0·6	−1·3	−0·6	−1·6	−0	−1·0
6·30–7·09	+1·0	+0	+1·6	+0	+2·5	+0	−6·0	−12·0	−3·5	−6·0	−3·5	−7·5	−1·6	−3·2	−1·6	−4·1	−0·6	−1·3	−0·6	−1·6	−0	−1·0

Nominal diameter in inches	HOLES						SHAFTS															
	H 6		H 7		H 8		h 7		j 7		k 5		k 6		k 7		m 7		p 6		s 6	
	High limit	*Low limit*	*High limit*	*Low limit*	*High limit*	*Low limit*	*High limit*	*Low limit*	*High limit*	*Low limit*	*High limit*	*Low limit*	*High limit*	*Low limit*	*High limit*	*Low limit*	*High limit*	*Low limit*	*High limit*	*Low limit*	*High limit*	*Low limit*
0·04–0·12	+0·25	+0	+0·4	+0	+0·6	+0	−0	−0·4	+0·3	−0·1	—	—	—	—	—	—	+0·55	+0·15	+0·65	+0·4	+0·85	+0 6
0·12–0·24	+0·3	+0	+0·5	+0	+0·7	+0	−0	−0·5	+0·4	−0·1	—	—	—	—	—	—	+0·7	+0·2	+0·8	+0·5	+1·0	+0·7
0·24–0·40	+0·4	+0	+0·6	+0	+0·9	+0	−0	−0·6	+0·4	−0·2	+0·35	+0·1	+0·5	+0·1	+0·7	+0·1	+0·8	+0·2	+1·0	+0·6	+1·4	+1·0
0·40–0·56	+0·4	+0	+0·7	+0	+1·0	+0	−0	−0·7	+0·5	−0·2	+0·4	+0·1	+0·5	+0·1	+0·8	+0·1	+1·0	+0·3	+1·1	+0·7	+1·6	+1·2
0·56–0·71	+0·4	+0	+0·7	+0	+1·0	+0	−0	−0·7	+0·5	−0·2	+0·4	+0·1	+0·5	+0·1	+0·8	+0·1	+1·0	+0·3	+1·1	+0·7	+1·6	+1·2
0·71–0·95	+0·5	+0	+0·8	+0	+1·2	+0	−0	−0·8	+0·5	−0·3	+0·5	+0·1	+0·6	+0·1	+0·9	+0·1	+1·1	+0·3	+1·3	+0·8	+1·9	+1·4
0·95–1·19	+0·5	+0	+0·8	+0	+1·2	+0	−0	−0·8	+0·5	−0·3	+0·5	+0·1	+0·6	+0·1	+0·9	+0·1	+1·1	+0·3	+1·3	+0·8	+1·9	+1·4
1·19–1·58	+0·6	+0	+1·0	+0	+1·6	+0	−0	−1·0	+0·6	−0·4	+0·5	+0·1	+0·7	+0·1	+1·1	+0·1	+1·4	+0·4	+1·6	+1·0	+2·4	+1·8
1·58–1·97	+0·6	+0	+1·0	+0	+1·6	+0	−0	−1·0	+0·6	−0·4	+0·5	+0·1	+0·7	+0·1	+1·1	+0·1	+1·4	+0·4	+1·6	+1·0	+2·4	+1·8
1·97–2·56	+0·7	+0	+1·2	+0	+1·8	+0	−0	−1·2	+0·7	0·5	+0·6	+0·1	+0·8	+0·1	+1·3	+0·1	+1·7	+0·5	+2·1	+1·4	+2·7	+2·0
2·56–3·15	+0·7	+0	+1·2	+0	+1·8	+0	−0	−1·2	+0·7	−0·5	+0·6	+0·1	+0·8	+0·1	+1·3	+0·1	+1·7	+0·5	+2·1	+1·4	+2·9	+2·2
3·15–3·94	+0·9	+0	+1·4	+0	+2·2	+0	−0	−1·4	+0·8	−0·6	+0·7	+0·1	+1·0	+0·1	+1·5	+0·1	+1·9	+0·5	+2·5	+1·6	+3·7	+2·8
3·94–4·73	+0·9	+0	+1·4	+0	+2·2	+0	−0	−1·4	+0·8	−0·6	+0·7	+0·1	+1·0	+0·1	+1·5	+0·1	+1·9	+0·5	+2·5	+1·6	+3·9	+3·0
4·73–5·52	+1·0	+0	+1·6	+0	+2·5	+0	−0	−1·6	+0·9	−0·7	+0·8	+0·1	+1·1	+0·1	+1·7	+0·1	+2·2	+0·6	+2·8	+1·8	+4·5	+3·5
5·52–6·30	+1·0	+0	+1·6	+0	+2·5	+0	−0	−1·6	+0·9	−0·7	+0·8	+0·1	+1·1	+0·1	+1·7	+0·1	+2·2	+0·6	+2·8	+1·8	+5·0	+4·0
6·30–7·09	+1·0	+0	+1·6	+0	+2·5	+0	−0	−1·6	+0·9	−0·7	+0·8	+0·1	+1·1	+0·1	+1·7	+0·1	+2·2	+0·6	+2·8	+1·8	+5·5	+4·5

TABLE 7.1 (inches)

Limits for 3 selected holes and 16 shafts as shown in Figure 2. (B.S. 1916 limits and fits)

Tolerances in units of 0·001 millimetres

Nominal diameter in millimetres	HOLES						SHAFTS															
	H 6		H 7		H 8		d 10		e 8		e 9		f 7		f 8		g 5		g 6		h 6	
	High limit	Low limit	High limit	Low limit	High limit	Low limit	High limit	Low limit	High limit	Low limit	High limit	Low limit	High limit	Low limit	High limit	Low limit	High limit	Low limit	High limit	Low limit	High limit	Low limit
1– 3	+7	+0	+9	+0	+14	+0	−20	−60	−14	−28	−14	−39	−7	−16	−7	−21	−3	−8	−3	−10	−0	−7
3– 6	+8	+0	+12	+0	+18	+0	−30	−78	−20	−38	−20	−50	−10	−22	−10	−28	−4	−9	−4	−12	−0	−8
6– 10	+9	+0	+15	+0	+22	+0	−40	−98	−25	−47	−25	−61	−13	−28	−13	−35	−5	−11	−5	−14	−0	−9
10– 14	+11	+0	+18	+0	+27	+0	−50	−120	−32	−59	−32	−75	−16	−34	−16	−43	−6	−14	−6	−17	−0	−11
14– 18	+11	+0	+18	+0	+27	+0	−50	−120	−32	−59	−32	−75	−16	−34	−16	−43	−6	−14	−6	−17	−0	−11
18– 24	+13	+0	+21	+0	+33	+0	−65	−149	−40	−73	−40	−92	−20	−41	−20	−53	−7	−16	−7	−20	−0	−13
24– 30	+13	+0	+21	+0	+33	+0	−65	−149	−40	−73	−40	−92	−20	−41	−20	−53	−7	−16	−7	−20	−0	−13
30– 40	+16	+0	+25	+0	+39	+0	−80	−180	−50	−89	−50	−112	−25	−50	−25	−64	−9	−20	−9	−25	−0	−16
40– 50	+16	+0	+25	+0	+39	+0	−80	−180	−50	−89	−50	−112	−25	−50	−25	−64	−9	−20	−9	−25	−0	−16
50– 65	+19	+0	+30	+0	+46	+0	−100	−220	−60	−106	−60	−134	−30	−60	−30	−76	−10	−23	−10	−29	−0	−19
65– 80	+19	+0	+30	+0	+46	+0	−100	−220	−60	−106	−60	−134	−30	−60	−30	−76	−10	−23	−10	−29	−0	−19
80–100	+22	+0	+35	+0	+54	+0	−120	−260	−72	−126	−72	−159	−36	−71	−36	−90	−12	−27	−12	−34	−0	−22
100–120	+22	+0	+35	+0	+54	+0	−120	−260	−72	−126	−72	−159	−36	−71	−36	−90	−12	−27	−12	−34	−0	−22
120–140	+25	+0	+40	+0	+63	+0	−145	−305	−85	−148	−85	−185	−43	−83	−43	−106	−14	−32	−14	−39	−0	−25
140–160	+25	+0	+40	+0	+63	+0	−145	−305	−85	−148	−85	−185	−43	−83	−43	−106	−14	−32	−14	−39	−0	−25
160–180	+25	+0	+40	+0	+63	+0	−145	−305	−85	−148	−85	−185	−43	−83	−43	−106	−14	−32	−14	−39	−0	−25

Nominal diameter in millimetres	HOLES						SHAFTS															
	H 6		H 7		H 8		h 7		j 7		k 5		k 6		k 7		m 7		p 6		s 6	
	High limit	Low limit	High limit	Low limit	High limit	Low limit	High limit	Low limit	High limit	Low limit	High limit	Low limit	High limit	Low limit	High limit	Low limit	High limit	Low limit	High limit	Low limit	High limit	Low limit
1– 3	+7	+0	+9	+0	+14	+0	−0	−9	+7	−2	—	—	—	—	—	—	—	—	+16	+9	+22	+15
3– 6	+8	+0	+12	+0	+18	+0	−0	−12	+9	−3	—	—	—	—	—	—	—	—	+20	+12	+27	+19
6– 10	+9	+0	+15	+0	+22	+0	−0	−15	+10	−5	+7	+1	+10	+1	+16	+1	+21	+6	+24	+15	+32	+23
10– 14	+11	+0	+18	+0	+27	+0	−0	−18	+12	−6	+9	+1	+12	+1	+19	+1	+25	+7	+29	+18	+39	+28
14– 18	+11	+0	+18	+0	+27	+0	−0	−18	+12	−6	+9	+1	+12	+1	+19	+1	+25	+7	+29	+18	+39	+28
18– 24	+13	+0	+21	+0	+33	+0	−0	−21	+13	−8	+11	+2	+15	+2	+23	+2	+29	+8	+35	+22	+48	+35
24– 30	+13	+0	+21	+0	+33	+0	−0	−21	+13	−8	+11	+2	+15	+2	+23	+2	+29	+8	+35	+22	+48	+35
30– 40	+16	+0	+25	+0	+39	+0	−0	−25	+15	−10	+13	+2	+18	+2	+27	+2	+34	+9	+42	+26	+59	+43
40– 50	+16	+0	+25	+0	+39	+0	−0	−25	+15	−10	+13	+2	+18	+2	+27	+2	+34	+9	+42	+26	+59	+43
50– 65	+19	+0	+30	+0	+46	+0	−0	−30	+18	−12	+15	+2	+21	+2	+32	+2	+41	+11	+51	+32	+72	+53
65– 80	+19	+0	+30	+0	+46	+0	−0	−30	+18	−12	+15	+2	+21	+2	+32	+2	+41	+11	+51	+32	+78	+59
80–100	+22	+0	+35	+0	+54	+0	−0	−35	+20	−15	+18	+3	+25	+3	+38	+3	+48	+13	+59	+37	+93	+71
100–120	+22	+0	+35	+0	+54	+0	−0	−35	+20	−15	+18	+3	+25	+3	+38	+3	+48	+13	+59	+37	+101	+79
120–140	+25	+0	+40	+0	+63	+0	−0	−40	+22	−18	+21	+3	+28	+3	+43	+3	+55	+15	+68	+43	+117	+92
140–160	+25	+0	+40	+0	+63	+0	−0	−40	+22	−18	+21	+3	+28	+3	+43	+3	+55	+15	+68	+43	+125	+100
160–180	+25	+0	+40	+0	+63	+0	−0	−40	+22	−18	+21	+3	+28	+3	+43	+3	+55	+15	+68	+43	+133	+108

TABLE 7.2 (millimetres)

EXERCISES

Exercise 7.1

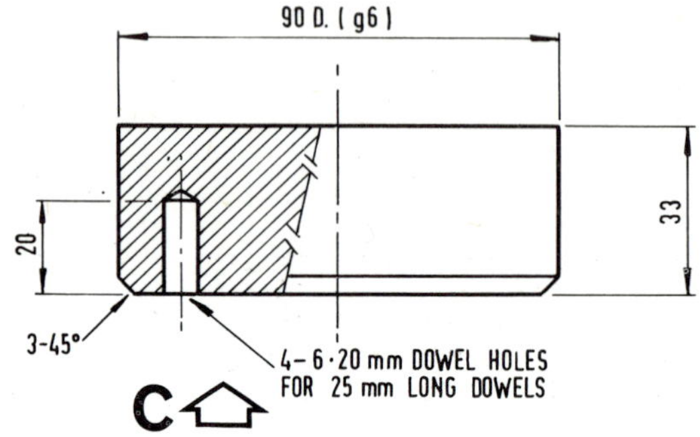

MATRIX MATERIAL. 60% Pb. 40% Sn.ALLOY FINISH. NATURAL, TOP SURFACE POLISHED

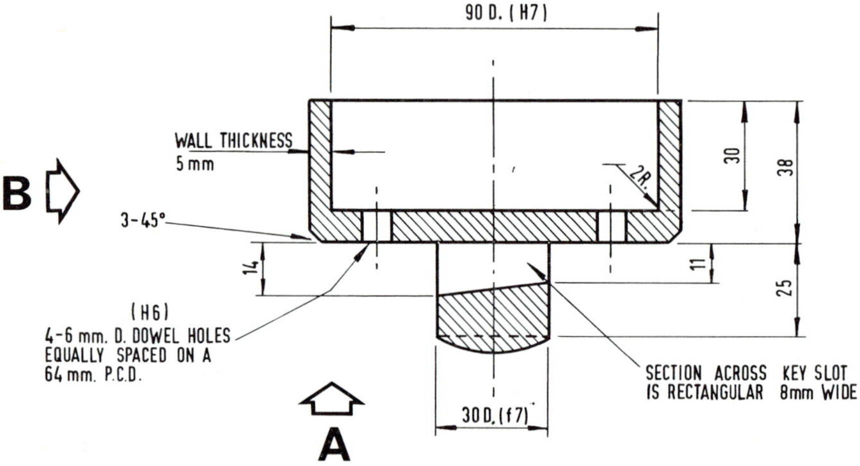

MATRIX HOLDER. MATERIAL. STEEL En. 32 FINISH. CASE HARDEN. CAD. PL.& PASSIVATE.

PROBLEM: THIRD-ANGLE

From the given information produce manufacturing drawings fully dimensioned and showing manufacturing limits, where indicated, in accordance with recommendations given in BS 308 and BS 1916.

Matrix holder. Draw as follows:

(a) Sectioned view as shown.

(b) Plan in direction of arrow A.

(c) Elevation in direction of arrow B.

Matrix. Draw as follows:

(a) Half-sectioned view.

(b) Plan in direction of arrow C.

Add title and details of material and finish.

Finally, draw a half-sectional assembly drawing of the two parts dowelled together, and add a plan view. Balloon reference parts, and add a parts list and a title block showing title, date, scale, name, projection and drawing number.

Exercise 7.2

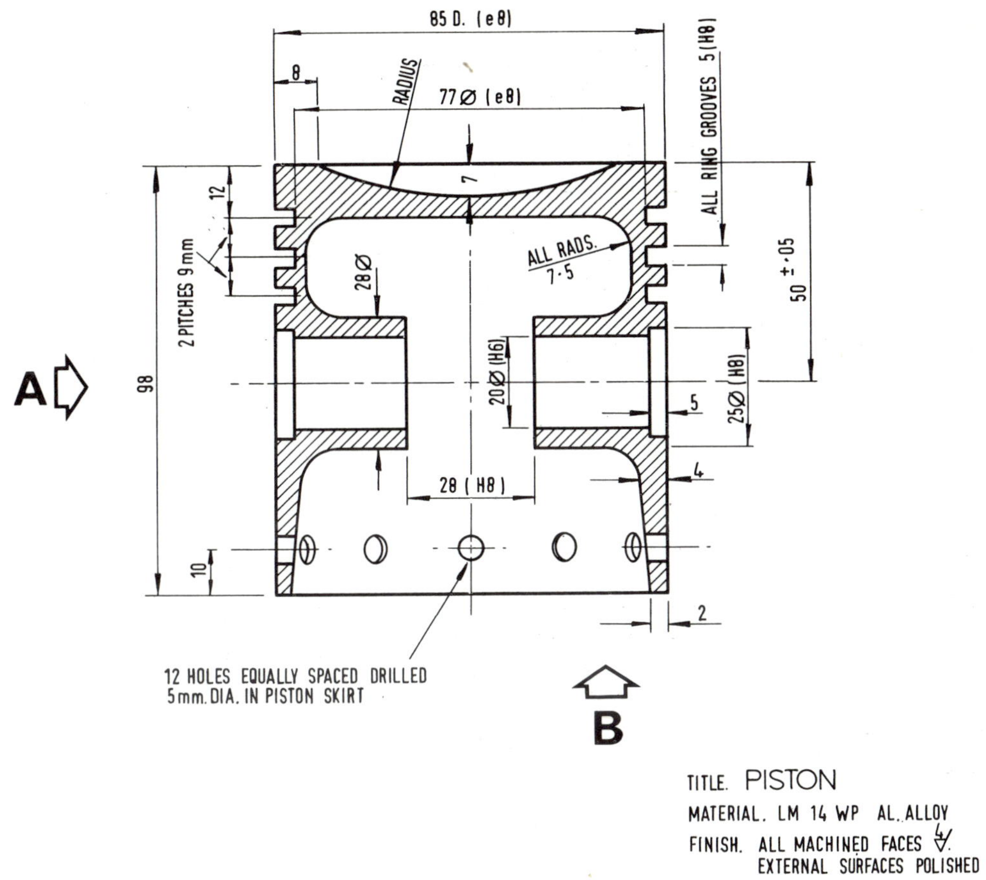

PROBLEM: THIRD-ANGLE

From the sketch of a piston make a manufacturing drawing as follows:

(a) Copy the sectioned view.

(b) Add an elevation in direction of arrow A.

(c) Add a plan view in direction of arrow B.

Do not copy the dimensions as shown but fully dimension for manufacture in accordance with the recommendations shown in BS 308. On dimensions as indicated, add upper and lower limits for manufacture.

Show machining symbols on all machined faces.

Add a title block showing title, material, finish, date, scale, projection, name and drawing number.

Exercise 7.3

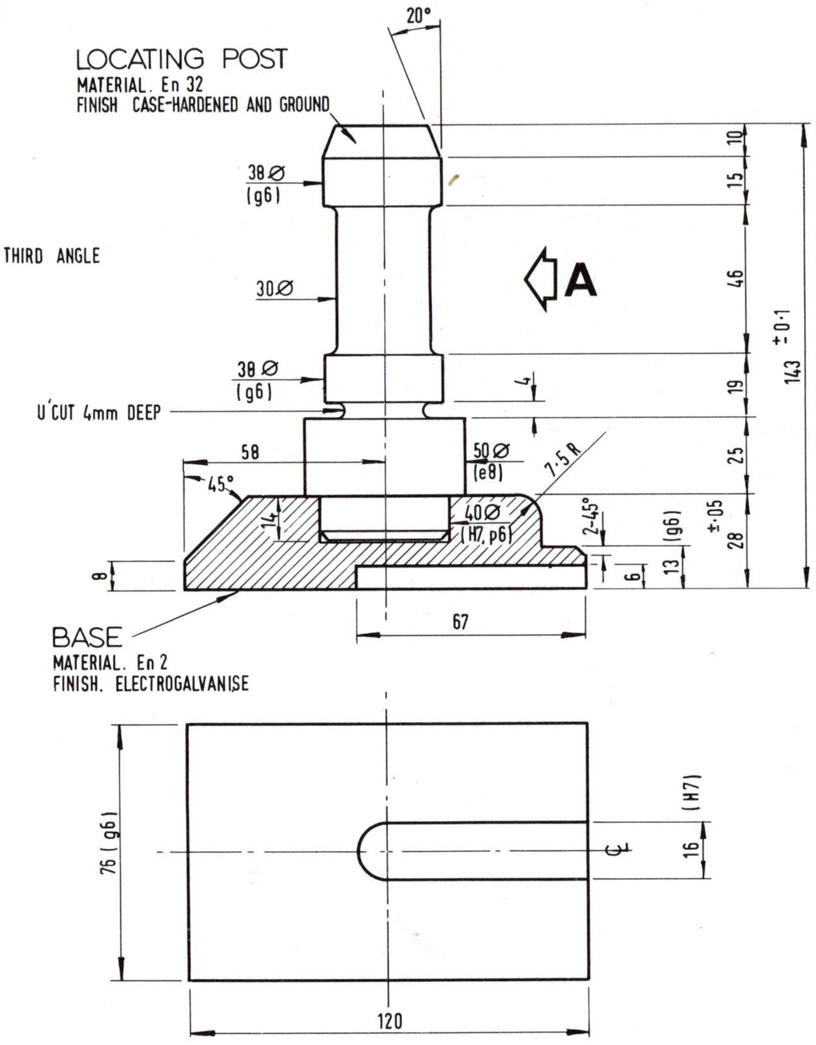

PROBLEM

The designer's layout shows a dimensioned assembly of a locating post for a checking fixture. From the information given make the following drawings:

(a) A fully dimensioned detail drawing of the locating post.

(b) A fully dimensioned detail drawing of the base.

Drawings are to be dimensioned in accordance with the recommendations given in BS 308. Dimensions requiring machining tolerance should state upper and lower dimensions. Limits to be taken from Tables shown in BS 1916.

Add title, details of material and finish, and scale.

(c) A three-view assembly of the two parts, consisting of:

1. Sectional view as shown.
2. Elevation in the direction of arrow A.
3. Plan.

Exercise 7.4

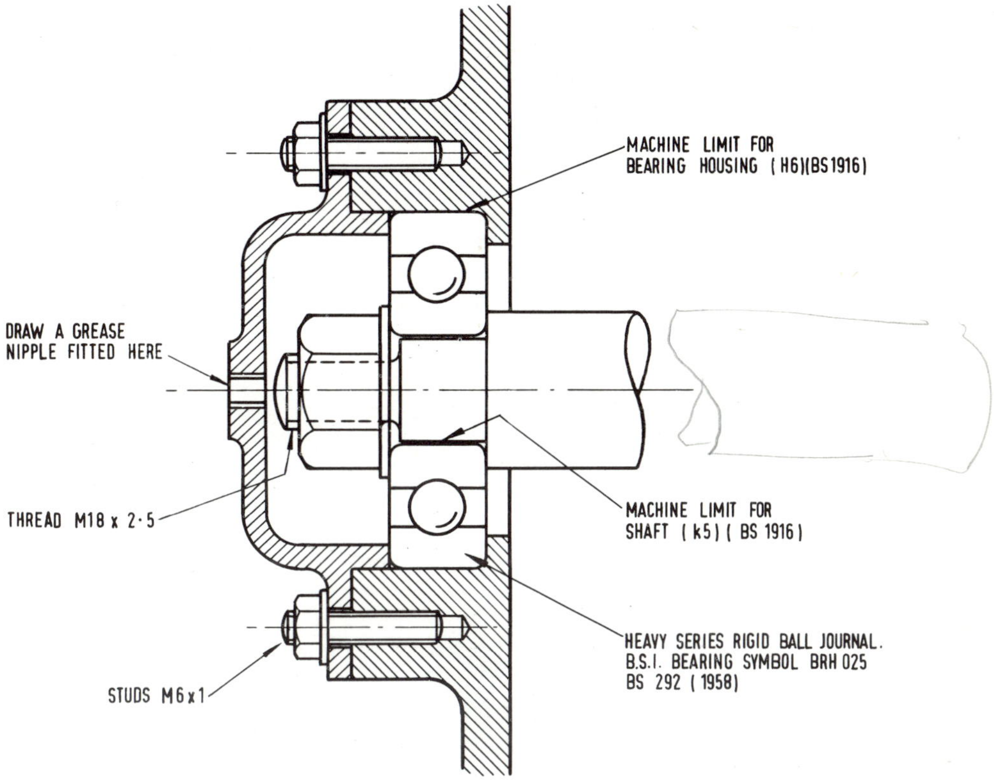

PROBLEM

The drawing shows a bearing arrangement for a ball journal BRH 025. Proceed as follows:

(a) From BS 292 or from a bearing catalogue, find the dimensions for bearing BRH 025, and draw the bearing.

(b) Draw the bearing arrangement around the bearing, estimating sizes and adding a grease nipple.

(c) Draw a scrap view of the bearing housing, and add machining limits for bore.

(d) Draw a scrap view of the end of the shaft and fully dimension with machining limits to fit bearing.

All dimensions to be in millimetres.

Drawings are to be dimensioned in accordance with the recommendations given in BS 308. Dimensions requiring machining tolerance should state upper and lower dimensions. Limits to be taken from Tables shown in BS 1916.

Add title, details of material and finish, and scale.

FASTENERS

The term *fasteners* is used to describe the multitude of nuts, bolts, rivets, pins, keys, clips and proprietary devices for joining engineering materials together.
The most common devices are screws, nuts and bolts. Millions of these are used daily, and it is advisable for the student to commit to memory a simple method of drawing these common fasteners, so as to give a clear representation of the device without waste of valuable time.

On a drawing, the representation of a fastener does not give dimensional information but shows production and assembly staff the type of fastener required and the method of fitting. The fasteners are fully described in the parts list, which gives type, size, thread, length, material and finish, or by a code reference which covers the same information.

Sometimes, due to lack of space or as a result of modification, a fastener is fitted close to other parts, and in such cases it should be drawn accurately (even several times full size) to dimensions taken from manufacturers' data sheets in order to ascertain whether fouling will occur between fasteners and components on assembly. However, where fasteners are fitted so close that clearance has to be proved by accurate drawing, the design should be reappraised; manufacturing time will be lost when such fasteners have to be fitted, and special tools may have to be used.

Figure 8.1 shows a draughtsman's simplified dimensions for drawing a hexagon nut and bolt, and common screwheads and rivets involving in general two basic dimensions:

(a) the diameter D of the shank of the bolt, screw or rivet, this dimension also being used for head thickness and most radii.
(b) the width 2D of all heads of bolts, screws, nuts and rivets.

Note also that it is practice to draw nuts and bolts across the corners in all views. Drawing hexagon heads across flats can give a false impression of clearance.

Figure 8.1

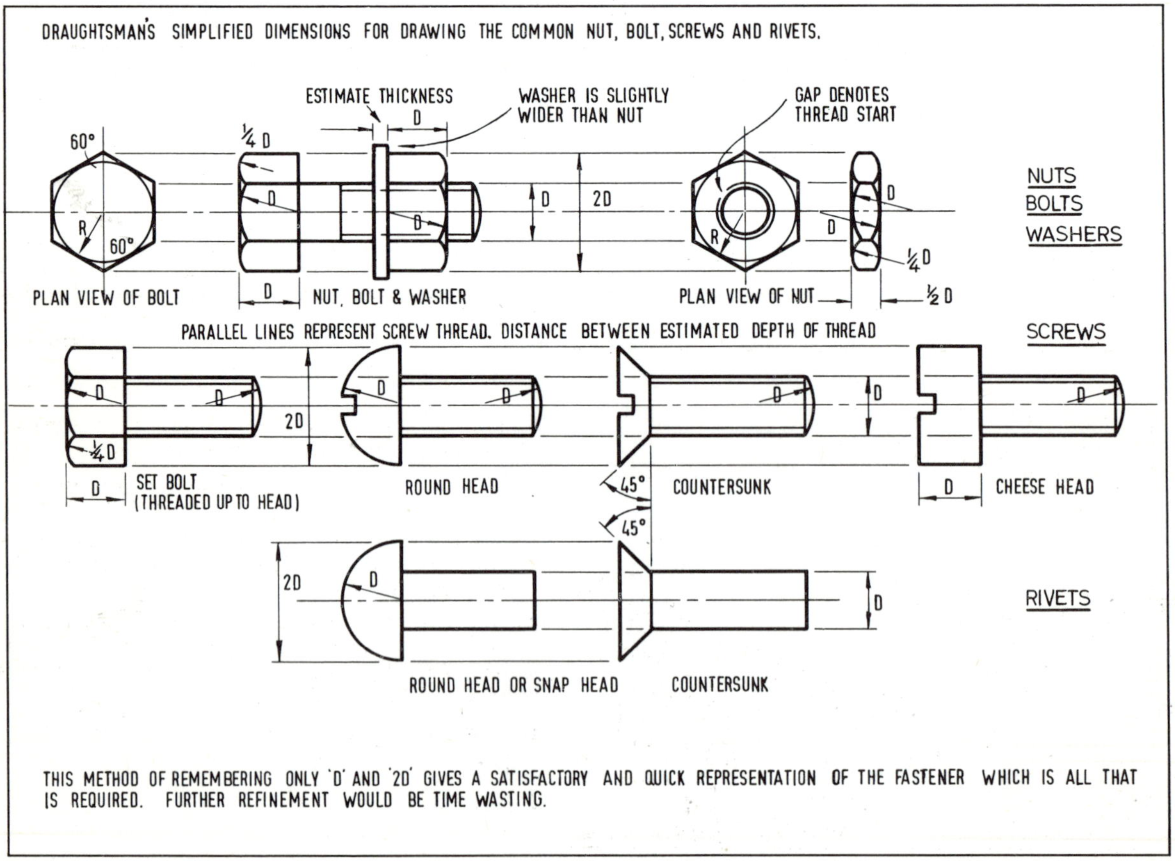

SCREW THREADS: DEFINITIONS

A *screw thread* is a *ridge of uniform section* formed on the outside or inside of a cylinder. When outside, it is the shape on a bolt; when inside, it is the shape on a nut.

A ridge of uniform section similarly formed as a conical spiral on a cone produces a tapered thread.

The helix

Figure 8.2 shows an inclined plane or slope. Engineers have long made use of slopes or ramps to raise heavy loads from one level to a higher level; this method was used, for example, in building the pyramids and huge temples of ancient times that still stand today.

On either side of the slope in Figure 8.2 a cylinder is shown with the slope wrapped around it in the form of a *helix*. Note that in equal angular distance the line rises equal linear distance.

The screw thread is no more than the inclined plane in a more compact form which allows engineers to apply a force through a distance to do useful work. A thread, although basically an inclined plane wrapped around a cylinder, has far more versatility in this form. Fitted to a nut and turning device, it can apply compressive and tensile forces for lifting, wedging, pushing, stretching etc., and can be used for applying power to move and control many forms of mechanical equipment and machine tools.

Right-hand thread. A thread is right-handed if, when viewed axially, it moves in a receding direction when turned *clockwise*.

Left-hand thread. A thread is left-handed if, when viewed axially, it moves in a receding direction when turned *anticlockwise*.

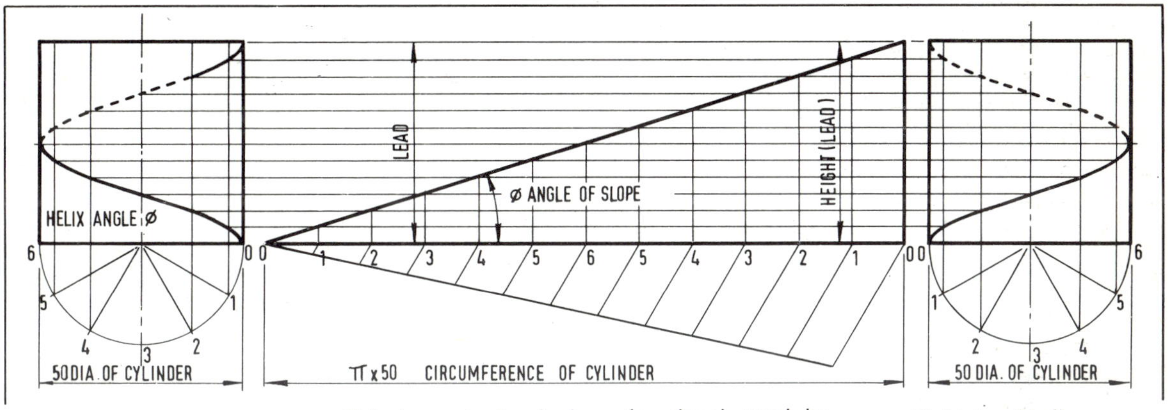

Left-hand helix
This is an inclined plane wrapped round a cylinder.

This is an inclined plane (or slope) used by engineers in early times to raise loads from one level to a higher level.
The helix is the basic shape of all screw threads.

Right-hand helix
This is an inclined plane wrapped round a cylinder in the opposite direction.

Figure 8.2

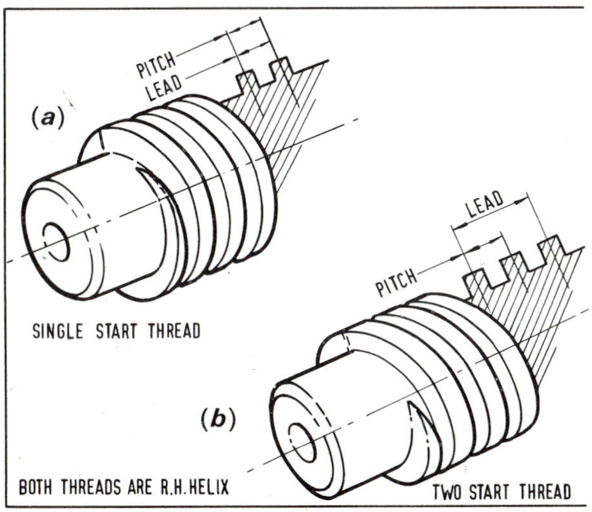

Pitch is the distance between the centres of adjacent screw thread-forms parallel to the axis.

Lead is the distance a threaded part moves axially with respect to the fixed mating part in one complete rotation.

Single thread. A single-start thread is one continuous thread in the form of a helix, having the lead equal to the pitch (Figure 8.3a).

Multiple thread. A multiple (multi-start) thread is one where a number or separate threads are formed simultaneously side by side; it has a lead which is a multiple of the pitch (Figure 8.3b).

LEAD = PITCH × NUMBER OF THREADS (OR STARTS)

Figure 8.3

Major diameter. In bolts the diameter over the crests of the threads; in nuts the diameter between the roots of the threads.

Minor diameter. In bolts the diameter between the roots of the threads; in nuts the diameter between the crests of the threads.

Pitch diameter (*effective diameter*) is the diameter between lines drawn through the thread profile on both sides of the screw where the groove width equals half-pitch (Figure 8.4).

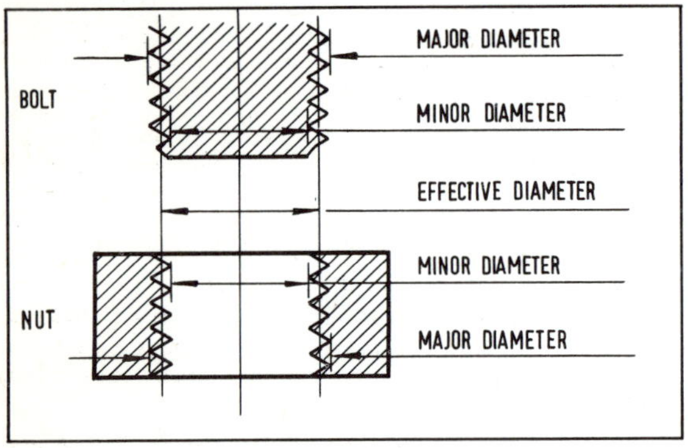

Figure 8.4

THREAD FORMS

In the mid-nineteenth century the position regarding screw threads was chaotic. Every manufacturer made screw threads to suit his own requirements, and any interchangeability that existed was by chance and not design. It was about this period that Sir Joseph Whitworth realised that standardisation of screw-thread fasteners for the rapidly expanding engineering industry in Great Britain was essential for ease and economy of manufacture, and he therefore introduced the 55° Whitworth thread form which is still in use today.

At about the same time, William Sellers introduced in America the Sellers 60° thread from which developed the American National Thread System which forms the basis of the modern Unified Thread System.

Before World War II it had been recognised that a universal common standard for screw threads was required. The War itself accentuated the problem; vast quantities of engineering products, of which Britain and America were the main suppliers, had to be handled in many countries.

In 1945 a conference on the subject was held in Ottawa between representatives of governments and industry from the United States, Britain and Canada, and in 1949 the British Standards Institution published the first specification for Unified Screw Threads (BS 1580). In 1953 this was revised in the light of experience gained by industry in applying the Standard. Later the International Organisation for Standardisation (ISO) decided to recommend the Unified Thread System in inch units as an international standard, in parallel with a similar standard in metric units, and the system became recognised internationally.

In 1962, the British Standards Institution issued a revised edition of BS 1580 to correspond with the revised American, Canadian and ISO standard. This is referred to as the ISO Inch Screw Thread System.

In 1963 the British Standards Institution issued a specification for ISO Metric Screw Threads (BS 3643). There are therefore now two systems of screw thread which are internationally recognised; both have the same basic thread form, but the manufactured threaded components conforming to one system are not interchangeable with those of the other system.

The difference between imperial units used by America, Canada and Britain and metric units used by the rest of the world, as well as the considerable expenditure already incurred by some of the major supply industries in adopting the Unified Thread System, has so far prevented the complete changeover to a universal thread system.

However, the large proportion of British exports being sold to countries using the metric system of measurement, together with Britain's closer association with the Common Market countries, has made it clear that the time has come to discontinue the irrational system of imperial units and adopt the metric system of measurement. At a fully representative Government-sponsored conference in November 1965 it was decided that British industry should be strongly recommended to adopt the internationally-agreed ISO metric screw threads (BS 3643) as first choice on all designs, and that ISO inch screw threads (BS 1580, *Unified Screw Threads*) should be regarded as second choice. The implementation of this recommendation means that, after the necessary run-down of existing machinery and equipment, BA, BSF and BSW screw threads will become obsolete and the British Standards for these will be withdrawn.

The Government, as the largest single purchaser of engineering equipment, has recommended that all current and future designs should be produced to the metric system of measurement, and that ISO metric screw threads should be the first choice for thread forms except in those industries, such as aerospace and motor-vehicle production, which have already adopted the Unified Thread System as their standard.

It is hoped that the changeover from imperial units to metric units will be complete by 1975.

VEE-FORM THREADS (Figures 8.5a to 8.5e)

There are five thread forms in common use in engineering:

1. British Standard Whitworth (BSW)	BS 84—obsolete
2. British Standard Fine (BSF)	
3. British Association (BA)	BS 93—obsolete
4. Unified Screw Thread (UNF and UNC)	BS 1580
5. ISO Metric Screw Thread	BS 3643

BSW (Figure 8.5a) and BSF (Figure 5b) threads

The basic design form of these two threads is a symmetrical V-thread in which the angle between flanks is 55°. One-sixth of the sharp V is truncated at root and crest, the thread being rounded by circular arcs blending tangentially with the flanks.

The difference between BSW and BSF is not in form, but in the number of threads per inch. Size for size, the BSF thread has the greater number of threads per inch and appears finer. It can withstand the loosening effect of machinery vibration much better than the BSW thread.

BA thread (Figure 8.5c)

The basic thread form is a symmetrical V thread in which the angle between flanks is $47\frac{1}{2}$°. The crests and flanks are rounded with equal radii, so that the basic depth of thread is reduced to 0·6 of the pitch.

BA threads are a numbered range of parallel screw threads covering 26 graded sizes numbered 0 to 25, 0 BA being the largest and 25 BA the smallest. They are used chiefly in the instrument, electrical and radio industries.

Like the BSW and BSF threads, the BA thread is being replaced by similar sizes in ISO metric and ISO inch screw threads.

Unified screw thread, BS 1580: ISO inch thread system (Figure 8.5d)

The basic thread form is a symmetrical V thread in which the angle between flanks is 60°. The crests and roots are basically flat; however, in practice the crests and roots carry small radii.

The Unified Thread System is a constant-pitch series, i.e. has a fixed number of threads to the inch—4, 6, 8, 12, 16, 20, 28, 32. For nuts and bolts there is a graded series of threads per inch giving a selection of Coarse (UNC), Fine (UNF) and Extra Fine (UNEF) threads.

A threaded fastener is designated in the Unified System by diameter/threads per inch/grade, e.g. $\frac{1}{2}$″/20/UNF; $\frac{5}{8}$″/11/UNC.

The sizes below $\frac{1}{4}$-inch diameter are a numbered series; the replacement equivalent for the 2BA size is 10/32/UNF.

ISO metric screw thread, BS 3643 (Figure 8.5e)

The basic thread form is identical with the Unified Screw Thread, being of symmetrical V form with an angle of 60° between flanks. The crests and roots are basically flat, but in practice the crests and roots carry small radii.

The ISO Metric Screw Thread, like the Unified Screw Thread, is a constant-pitch thread series. For nuts, bolts and threaded fasteners there are two series of diameters with graded pitches, one with coarse and the other with fine threads.

A threaded fastener is designated in the ISO Metric Screw Thread system by the letter M followed by values of the nominal diameter and pitch, both in millimetres, e.g.

Diameter (mm)		*Pitch of thread* (mm)
M2	×	0·4
M6	×	1·0

METRICATION

At a fully representative conference held on 23rd November, 1965 consideration was given to the action to be taken in relation to the move to metric as far as British Standards for screw threads were concerned and it was decided that:
British industry should be strongly recommended to adopt the internationally agreed ISO metric threads * or ISO inch threads † but that the ISO inch threads be regarded as second choice. The implementation of this recommendation means that BA, BSF and BSW threads should become obsolescent and should not be used in new designs.
Accordingly it has been agreed that BS: 93 be rendered obsolescent: it will be made obsolete in due course.

B.S.W. (B.S. 84)

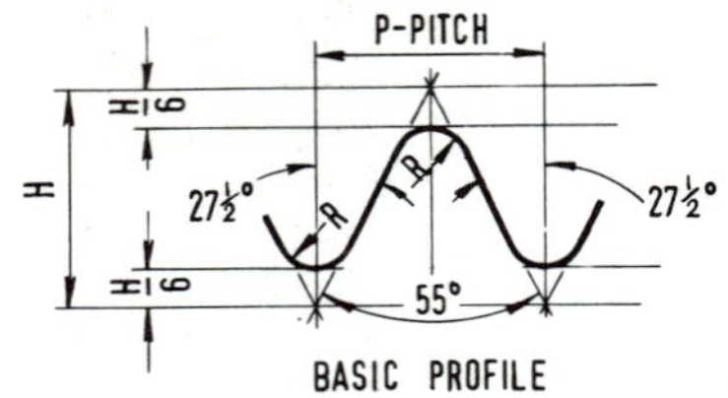

Nominal size	Threads per inch	Pitch (in.)	Depth of thread (in.)	Full diameter (in.)
$\frac{1}{4}$	20	0·050 00	0·032 0	0·250 0
$\frac{5}{16}$	18	0·055 56	0·035 6	0·312 5
$\frac{3}{8}$	16	0·062 50	0·040 0	0·375 0
$\frac{7}{16}$	14	0·071 43	0·045 7	0·437 5
$\frac{1}{2}$	12	0·083 33	0·053 4	0·500 0
$\frac{9}{16}$	12	0·083 33	0·053 4	0·562 5
$\frac{5}{8}$	11	0·090 91	0·058 2	0·625 0
$\frac{3}{4}$	10	0·100 00	0·064 0	0·750 0
$\frac{7}{8}$	9	0·111 11	0·071 1	0·875 0
1	8	0·125 00	0·080 0	1·000 0
$1\frac{1}{8}$	7	0·142 86	0·091 5	1·125 0
$1\frac{1}{4}$	7	0·142 86	0·091 5	1·250 0
$1\frac{1}{2}$	6	0·166 67	0·106 7	1·500 0
$1\frac{3}{4}$	5	0·200 00	0·128 1	1·750 0
2	$4\frac{1}{2}$	0·222 22	0·142 3	2·000 0
$2\frac{1}{4}$	4	0·250 00	0·160 1	2·250 0
$2\frac{1}{2}$	4	0·250 00	0·160 1	2·500 0
$2\frac{3}{4}$	$3\frac{1}{2}$	0·285 71	0·183 0	2·750 0
3	$3\frac{1}{2}$	0·285 71	0·183 0	3·000 0
$3\frac{1}{4}$	$3\frac{1}{4}$	0·307 69	0·197 0	3·250 0
$3\frac{1}{2}$	$3\frac{1}{4}$	0·307 69	0·197 0	3·500 0
$3\frac{3}{4}$	3	0·333 33	0·213 4	3·750 0
4	3	0·333 33	0·213 4	4·000 0

Figure 8.5*a*

B.S.F. (B.S. 84)

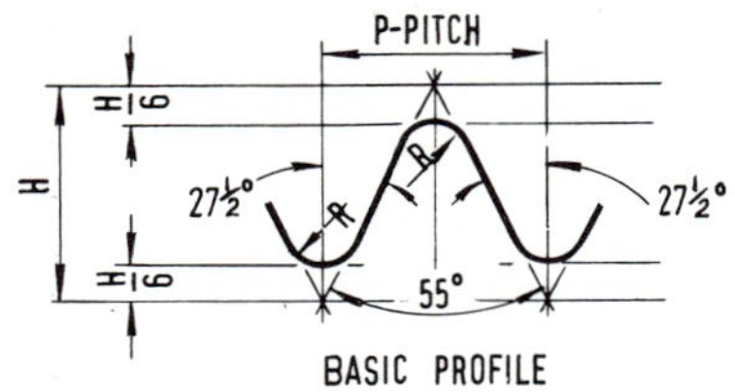

Nominal size	Threads per inch	Pitch (in.)	Depth of thread (in.)	Full diameter (in.)
$\frac{1}{4}$	26	0·038 46	0·024 6	0·250 0
$\frac{5}{16}$	22	0·045 45	0·029 1	0·312 5
$\frac{3}{8}$	20	0·050 00	0·032 0	0·375 0
$\frac{7}{16}$	18	0·055 56	0·035 6	0·437 5
$\frac{1}{2}$	16	0·062 50	0·040 0	0·500 0
$\frac{9}{16}$	16	0·062 50	0·040 0	0·562 5
$\frac{5}{8}$	14	0·071 43	0·045 7	0·625 0
$\frac{3}{4}$	12	0·083 33	0·053 4	0·750 0
$\frac{7}{8}$	11	0·090 91	0·058 2	0·875 0
1	10	0·100 00	0·064 0	1·000 0
$1\frac{1}{8}$	9	0·111 11	0·071 1	1·125 0
$1\frac{1}{4}$	9	0·111 11	0·071 1	1·250 0
$1\frac{3}{8}$	8	0·125 00	0·080 0	1·375 0
$1\frac{1}{2}$	8	0·125 00	0·080 0	1·500 0
$1\frac{3}{4}$	7	0·142 86	0·091 5	1·750 0
2	7	0·142 86	0·091 5	2·000 0
$2\frac{1}{4}$	6	0·166 67	0·106 7	2·250 0
$2\frac{1}{2}$	6	0·166 67	0·106 7	2·500 0
$2\frac{3}{4}$	6	0·166 67	0·106 7	2·750 0
3	5	0·200 00	0·128 1	3·000 0
$3\frac{1}{4}$	5	0·200 00	0·128 1	3·250 0
$3\frac{1}{2}$	$4\frac{1}{2}$	0·222 22	0·142 3	3·500 0
$3\frac{3}{4}$	$4\frac{1}{2}$	0·222 22	0·142 3	3·750 0
4	$4\frac{1}{2}$	0·222 22	0·142 3	4·000 0

Figure 8.5*b*

B.A. (B.S. 93)

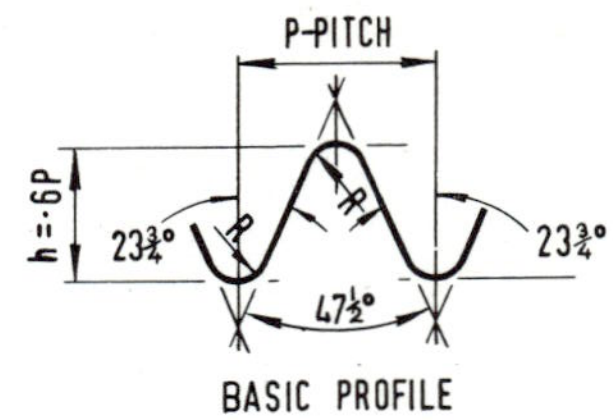

BA No.	Approximate					
	Threads per inch	Pitch (in.)	Full diameter (in.)	Pitch (mm)	Depth of thread (mm)	Full diameter (mm)
0	25·4	0·039 37	0·236 2	1·000	0·600	6·00
1	28·2	0·035 43	0·208 7	0·900	0·540	5·30
2	31·4	0·031 89	0·185 0	0·810	0·485	4·70
3	34·8	0·028 74	0·161 4	0·730	0·440	4·10
4	38·5	0·025 98	0·141 7	0·660	0·395	3·60
5	43·1	0·023 23	0·126 0	0·590	0·355	3·20
6	47·9	0·020 87	0·110 2	0·530	0·320	2·80
7	52·9	0·018 90	0·098 4	0·480	0·290	2·50
8	59·1	0·016 93	0·086 6	0·430	0·260	2·20
9	65·1	0·015 35	0·074 8	0·390	0·235	1·90
10	72·6	0·013 78	0·066 9	0·350	0·210	1·70
11	81·9					
12	90·7					
13	102·0					
14	110·0					
15	121·0					
16	134·0					
17	149·0					
18	169·0					
19	181·0					
20	212·0					
21	231·0					
22	254·0					
23	282·0					
24	318·0					
25	363·0					

BA threads smaller than 10 BA are very seldom used in general engineering.

Figure 8.5*c*

ISO-UNIFIED (B.S. 1580)

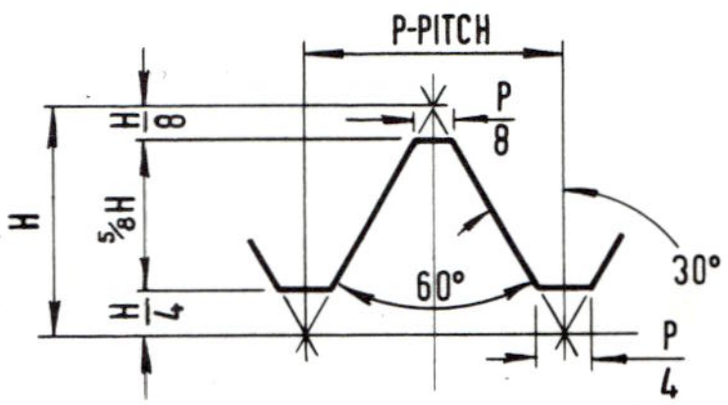

BASIC FORM ISO/ UNIFIED THREAD

First choice sizes	Full diameter (in.)	Threads per inch UNC coarse	UNF fine	UNEF extra fine
0	0·060 0		80	
1				
2	0·086 0	56	64	
3				
4	0·112 0	40	48	
5	0·125 0	40	44	
6	0·138 0	32	40	
8	0·164 0	32	36	
10	0·190 0	24	32	
1/4	0·250 0	20	28	32
5/16	0·312 5	18	24	32
3/8	0·375 0	16	24	32
7/16	0·437 5	14	20	28
1/2	0·500 0	13	20	28
9/16	0·562 5	12	18	24
5/8	0·625 0	11	18	24
3/4	0·750 0	10	16	20
7/8	0·875 0	9	14	20
1	1·000 0	8	12	20
1 1/8	1·125 0	7	12	18
1 1/4	1·250 0	7	12	18
1 3/8	1·375 0	6	12	18
1 1/2	1·500 0	6	12	18
1 5/8	1·625 0			18
1 3/4	1·750 0	5		
2	2·000 0	4 1/2		
2 1/4	2·250 0	4 1/2		
2 1/2	2·500 0	4		
2 3/4	2·750 0	4		
3	3·000 0	4		

Figure 8.5*d*

ISO-METRIC (B.S. 3643)

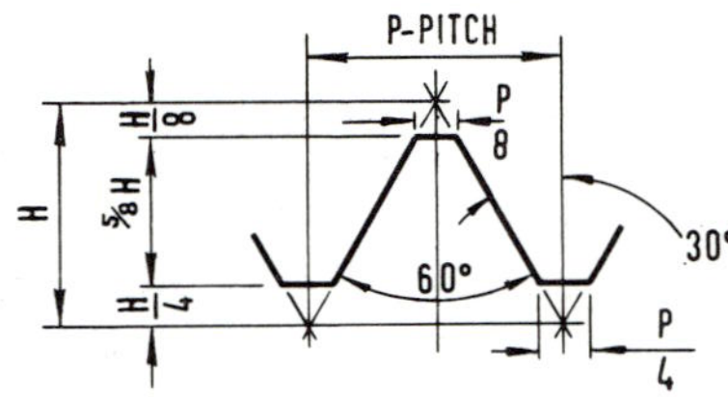

BASIC FORM ISO-METRIC THREAD

Basic major diameter (mm) First choice	Second choice	Pitch (mm) Coarse series	Fine series
1·6		0·35	
	1·8	0·35	
2·0		0·40	
	2·2	0·45	
2·5		0·45	
3·0		0·50	
	3·5	0·60	
4·0		0·70	
	4·5	0·75	
5·0		0·80	
6·0		1·00	
	7·0	1·00	
8·0		1·25	1·00
10·0		1·50	1·25
12·0		1·75	1·25
	14·0	2·00	1·50
16·0		2·00	1·50
	18·0	2·50	1·50
20·0		2·50	1·50
	22·0	2·50	1·50
24·0		3·00	2·00
	27·0	3·00	2·00
30·0		3·50	2·00
	33·0	3·50	2·00
36·0		4·00	3·00
	39·0	4·00	3·00

Note. In practice crests and roots of threads are rounded. This is to prevent fatigue cracks starting at sharp root corners, and also results from the methods used in machining.

Figure 8.5*e*

SCREWED FASTENERS

Nuts, bolts and washers

The common forms of screwed fasteners in engineering are hexagon-headed nuts and bolts, which draw the machine parts tightly together with the aid of spanners applied to the hexagons. For BSW, BSF and BA nuts and bolts, spanners are available in sizes designated by the thread size.

Unified nuts and bolts have hexagons of different size. Spanners are designated by the across-flats size of the hexagon and are known as A/F spanners. Figure 8.6 gives a table of sizes for hexagon nuts, bolts and washers.

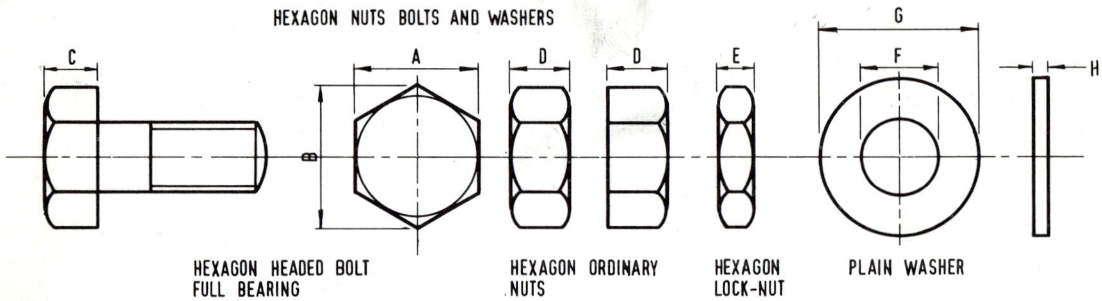

UNIFIED (*inches*)

Size (No. or in)	A	B	C	D	E	F	G	H
4	0·1875	0·216	0·060	0·098	—	0·123	0·233	0·024
6	0·2500	0·289	0·080	0·114	—	0·157	0·301	0·028
8	0·2500	0·289	0·110	0·130	—	0·177	0·341	0·028
10	0·3125	0·361	0·120	0·130	—	0·202	0·391	0·032
$\frac{1}{4}$	0·4375	0·505	0·163	0·224	0·161	0·270	0·562	0·056
$\frac{5}{16}$	0·5000	0·577	0·211	0·271	0·192	0·333	0·625	0·072
$\frac{3}{8}$	0·5625	0·650	0·243	0·333	0·244	0·395	0·750	0·072
$\frac{7}{16}$	0·6250	0·722	0·291	0·380	0·255	0·458	0·875	0·092
$\frac{1}{2}$	0·7500	0·866	0·323	0·442	0·317	0·520	1·000	0·092
$\frac{9}{16}$	0·8125	0·938	0·371	0·489	0·349	0·593	1·125	0·104
$\frac{5}{8}$	0·9375	1·083	0·403	0·552	0·380	0·656	1·250	0·116
$\frac{3}{4}$	1·1250	1·300	0·483	0·651	0·432	0·781	1·500	0·144
$\frac{7}{8}$	1·3125	1·515	0·563	0·760	0·494	0·906	1·625	0·144
1	1·5000	1·732	0·627	0·874	0·562	1·031	1·875	0·160

ISO METRIC (*millimetres*)

Size (mm)	A	B	C	D	E	F	G	H
M1·6	3·2	3·7	1·225	1·30	—	1·7	4·0	0·3
M2	4·0	4·6	1·525	1·60	—	2·2	5·0	0·3
M2·5	5·0	5·8	1·825	2·00	—	2·7	6·5	0·5
M3	5·5	6·4	2·125	2·40	—	3·2	7·0	0·5
M4	7·0	8·1	2·925	3·20	—	4·3	9·0	0·8
M5	8·0	9·2	3·650	4·00	—	5·3	10·0	1·0
M6	10·0	11·5	4·15	5·00	—	6·4	12·5	1·6
M8	13·0	15·0	5·65	6 50	5·0	8·4	17·0	1·6
M10	17·0	19·6	7·18	8·00	6·0	10·5	21·0	2·0
M12	19·0	21·9	8·18	10·00	7·0	13·0	24·0	2·5
M16	24·0	27·7	10·18	13·00	8·0	17·0	30·0	3·0
M20	30·0	34·6	13·215	16·00	9·0	21·0	37·0	3·0
M24	36·0	41·6	15·215	19·00	10·0	25·0	44·0	4·0
M30	46·0	53·1	19·26	24·00	12·0	31·0	56·0	4·0
M36	55·0	63·5	23·26	29·00	14·0	37·0	66·0	5·0

BSW and BSF (*inches*)

Size (in)	A	B	C	D	E	F	G	H
$\frac{1}{4}$	0·445	0·51	0·19	0·200	0·133	0·270	0·562	0·056
$\frac{5}{16}$	0·525	0·61	0·22	0·250	0·166	0·333	0·625	0·072
$\frac{3}{8}$	0·600	0·69	0·27	0·312	0·208	0·395	0·750	0·072
$\frac{7}{16}$	0·710	0·82	0·33	0·375	0·250	0·458	0·875	0·092
$\frac{1}{2}$	0·820	0·95	0·38	0·437	0·291	0·520	1·000	0·092
$\frac{9}{16}$	0·920	1·06	0·44	0·500	0·333	0·593	1·125	0·104
$\frac{5}{8}$	1·010	1·17	0·49	0·562	0·375	0·656	1·250	0·116
$\frac{3}{4}$	1·200	1·39	0·60	0·687	0·458	0·781	1·500	0·144
$\frac{7}{8}$	1·300	1·50	0·66	0·750	0·500	0·906	1·625	0·144
1	1·480	1·71	0·77	0·875	0·583	1·031	1·875	0·160

BA (*inches*)

No.	A	B	C	D	E	F	G	H
0	0·413	0·48	0·177	0·213	0·157	0·256	0·500	0·040
2	0·324	0·37	0·139	0·167	0·123	0·202	0·391	0·032
4	0·248	0·29	0·106	0·135	0·094	0·157	0·301	0·028
6	0·193	0·22	0·083	0·105	0·073	0·123	0·233	0 024
8	0·152	0·18	0·065	0·082	0·058	0·099	0·185	0·020
10	0·117	0·14	0·050	0·064	—	—	—	—

Figure 8.6

Screws

Screws are available in all sizes and thread forms, the most common shapes for the head being round, countersunk or cheese-headed. Figure 8.7 gives a table for screwhead sizes. Common screwheads are slotted to accept an ordinary screwdriver blade for assembly. Patented head recesses in the form of a cross designed for use with power tools in mass production are now in common use.

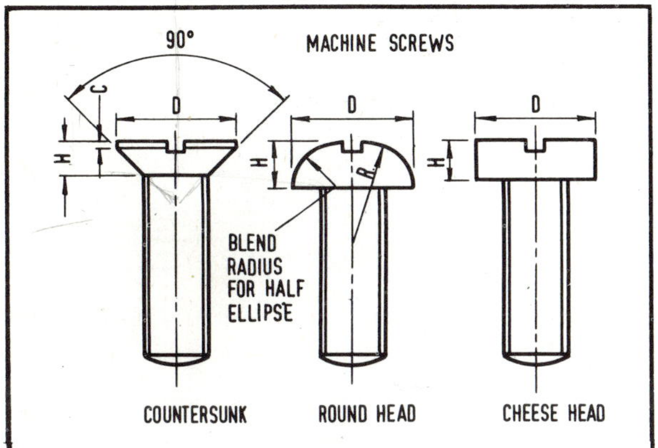

BSW and BSF (*inches*)

Size (*in*)	*Countersunk* D	H	C	*Round head* D	H	R	*Cheese head* D	H
1/4	0·438	0·133	—	0·438	0·175	0·438	0·375	0·175
5/16	0·547	0·141	—	0·547	0·219	0·547	0·469	0·219
3/8	0·656	0·169	—	0·656	0·262	0·656	0·562	0·262
7/16	0·766	0·197	—	0·766	0·306	0·766	0·656	0·306
1/2	0·875	0·225	—	0·875	0·350	0·875	0·750	0·350
9/16	0·984	0·253	—	0·984	0·394	0·984	0·844	0·394
5/8	1·094	0·281	—	1·094	0·437	1·094	0·938	0·437
3/4	1·312	0·338	—	1·312	0·525	1·312	1·125	0·525

BA (*inches*)

Size	*Countersunk* D	H	C	*Round head* D	H	R	*Cheese head* D	H
0	0·413	0·099	0·010	0·413	0·167	0·413	0·413	0·167
2	0·319	0·077	0·010	0·319	0·130	0·319	0·319	0·130
4	0·252	0·065	0·010	0·252	0·101	0·252	0·252	0·101
6	0·194	0·051	0·009	0·194	0·078	0·194	0·194	0·078
8	0·157	0·043	0·008	0·157	0·063	0·157	0·157	0·063
10	0·112	0·030	0·007	0·112	0·045	0·112	0·112	0·045

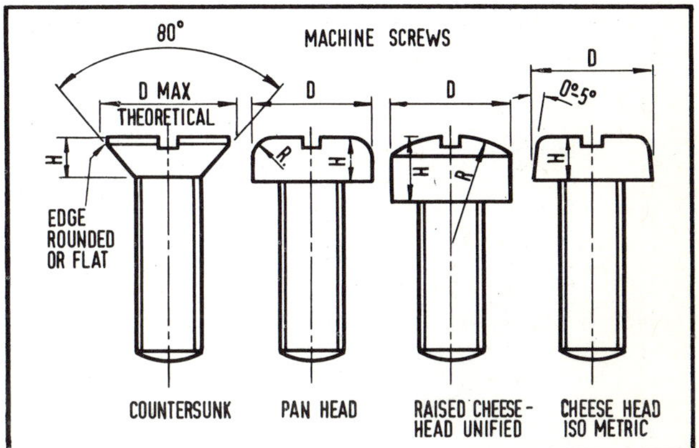

UNIFIED (*inches*)

Size	*Countersunk* D	H	C	*Pan head* D	H	R	*Raised cheese head* D	H	R
4	0·211	0·067	—	0·219	0·068	0·042	0·183	0·107	0·144
6	0·260	0·083	—	0·270	0·082	0·046	0·226	0·132	0·182
8	0·310	0·100	—	0·322	0·096	0·052	0·270	0·156	0·220
10	0·359	0·116	—	0·373	0·110	0·061	0·313	0·180	0·258
1/4	0·473	0·153	—	0·492	0·144	0·087	0·414	0·237	0·347
5/16	0·593	0·191	—	0·615	0·178	0·099	0·518	0·295	0·438
3/8	0·712	0·230	—	0·740	0·212	0·143	0·623	0·355	0·528
7/16	0·753	0·223	—	0·863	0·247	0·148	0·625	0·368	0·624
1/2	0·808	0·223	—	0·987	0·281	0·169	0·750	0·412	0·693
5/8	1·041	0·298	—	1·125	0·350	0·210	0·875	0·521	0·756
3/4	1·275	0·372	—	1·250	0·419	0·251	1·000	0·612	0·848

ISO METRIC (*millimetres*)

Size	*Countersunk* D	H	C	*Pan head* D	H	R	*Cheese head* D	H
M1	2·00	0·50	—	—	—	—	2·0	0·70
M1·2	2·40	0·60	—	—	—	—	2·3	0·80
M1·6	3·20	0·80	—	—	—	—	3·0	1·00
M2·0	4·00	1·00	—	—	—	—	3·8	1·30
M2·5	5·00	1·25	—	5·00	1·50	1·00	4·5	1·60
M3	6·00	1·50	—	6·00	1·80	1·20	5·5	2·00
M4	8·00	2·00	—	8·00	2·40	1·60	7·0	2·60
M5	10·00	2·50	—	10·00	3·00	2·00	8·5	3·30
M6	12·00	3·00	—	12·00	3·60	2·50	10 0	3·90
M8	16·00	4·00	—	16·00	4·80	3·20	13·0	5·00
M10	20·00	5·00	—	20·00	6·00	4·00	16·0	6·00
M12	24·00	6·00	—	—	—	—	18·0	7·00
M16	32·00	8·00	—	—	—	—	24·0	9·00
M20	40·00	10·00	—	—	—	—	30·0	11·00

Figure 8.7

Socket-head cap screws and hollow set screws

Hexagon socket-head cap screws are used in large quantities, particularly in tool making. The head of a cap screw can be placed below the surface of the plate in a counterbored hole and firmly secured by means of a right-angled hexagon wrench, giving a more positive fixing than a slotted screw.
Hollow set screws are tightened in the same manner.

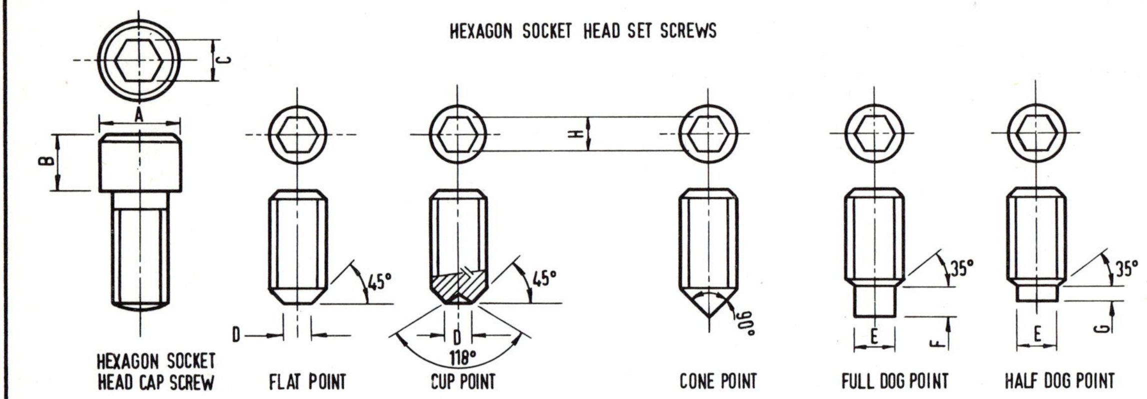

BSW and BSF (*inches*)

Size	A	B	C	D	E	F	G	H
1/4	0·375	0·250	3/16	1/8	0·156	1/8	1/16	1/16
5/16	0·437	0·312	7/32	11/64	0·203	5/32	5/64	5/64
3/8	0·562	0·375	5/16	13/64	0·250	3/16	3/32	3/32
7/16	0·625	0·437	5/16	15/64	0·297	7/32	7/64	7/64
1/2	0·750	0·500	3/8	9/32	0·344	1/4	1/8	1/8
5/8	0·875	0·625	1/2	23/64	0·469	5/16	5/32	5/32
3/4	1·000	0·750	9/16	7/16	0·562	3/8	3/16	3/16
7/8	1·125	0·875	9/16	33/64	0·656	7/16	7/32	7/32
1	1·312	1·000	5/8	19/32	0·750	1/2	1/4	1/4

UNIFIED (UNF and UNC) (*inches*)

Size	A	B	C	D	E	F	G	H
1/4	0·375	0·250	3/16	1/8	0·156	1/8	1/16	1/16
5/16	0·437	0·312	7/32	11/64	0·203	5/32	5/64	5/64
3/8	0·562	0·375	5/16	13/64	0·250	3/16	3/32	3/32
7/16	0·625	0·437	5/16	15/64	0·297	7/32	7/64	7/64
1/2	0·750	0·500	3/8	9/32	0·344	1/4	1/8	1/8
5/8	0·875	0·625	1/2	23/64	0·469	5/16	5/32	5/32
3/4	1·000	0·750	9/16	7/16	0·562	3/8	3/16	3/16
7/8	1·125	0·875	9/16	33/64	0·656	7/16	7/32	7/32
1	1·312	1·000	5/8	19/32	0·750	1/2	1/4	1/4

BA (*inches*)

Size	A	B	C	D	E	F	G	H
6	0·187	0·110	5/64	0·055	0·073	1/16	1/32	1/32
4	0·219	0·142	3/32	1/16	0·093	5/64	1/32	1/32
2	0·312	0·187	5/32	3/32	0·125	3/32	3/64	3/64

ISO METRIC (*millimetres*)

Size	A	B	C	D	E	F	G	H
M3	5·50	3·00	2·50	2·00	2·00	2·40		1·50
M4	7·00	4·00	3·00	2·60	2·50	2·80		2·00
M5	8·50	5·00	4·00	3·40	3·50	2·80		2·50
M6	10·00	6·00	5·00	4·00	4·50	3·35		3·00
M8	13·00	8·00	6·00	5·50	6·00	4·70		4·00
M10	16·00	10·00	8·00	7·00	7·00	4·90		5·00
M12	18·00	12·00	10·00	8·50	9·00	6·40		6·00
M16	24·00	16·00	14·00	12·00	12·00	8·15		8·00
M20	30·00	20·00	17·00	15·00	15·00	7·85		10·00
M24	36·00	24·00	19·00	18·00	18·00	9·65		12·00

Figure 8.8

POWER THREADS

There are three common forms of thread for transmitting power between machine elements:

1. Square thread (Figure 8.9a).
2. Acme thread, BS 1104 (Figure 8.9b).
3. Buttress thread, BS 1657 (Figure 8.9c).

Square thread

As its name implies, this thread is of square form and was the earliest type of power thread used in machine tools and power-transmitting devices. As the sides of the thread are parallel to the axis, all forces transmitted are parallel to the axis; thus no bursting forces are set up in the nut as in V-form threads. Square threads are difficult to machine because of the parallel sides of the thread; backlash increases with wear because adjustment is impossible. The square thread is used in lifting jacks, in cross slides in lathes and similar machine-tool applications, and in operating spindles on valves; it is often found on presses and vices made from wood.

For most power-transmitting devices the square thread has been superseded by the Acme thread.

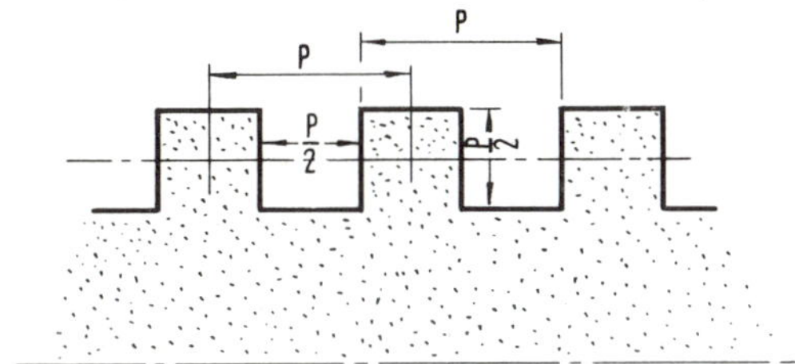

Figure 8.9*a* Square Thread

Acme thread

The acme screw thread was introduced in the late 1890s to replace the square threads used for producing traversing motions in machine tools. It is now used extensively for a variety of power-transmitting purposes. The basic design is a symmetrical V form with a 29° included angle between flanks; the thread depth is half the pitch.

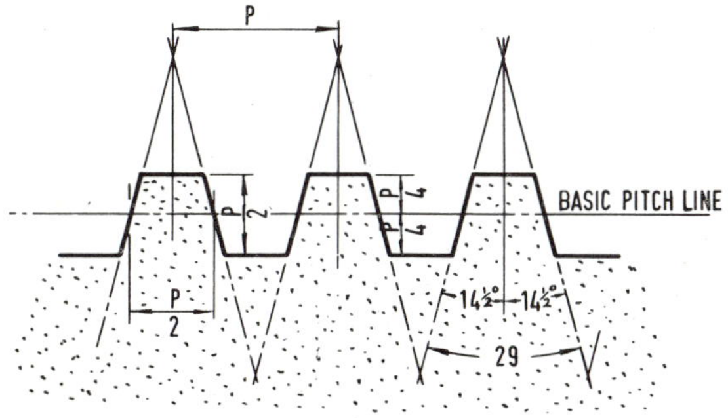

Figure 8.9*b* Acme Thread

Buttress thread, BS 1657 (Figure 8.9c)

The buttress thread is designed to resist heavy loads in one direction only, as in screw vices and presses.
BS 1657 covers two forms of the thread. The *standard* form has a pressure-resisting flank angle of 7°, with an included angle of 52° between pressure and trailing flanks. The *special* form has the pressure flank 'normal' to the axis, with an included angle of 52° between pressure and trailing flanks.

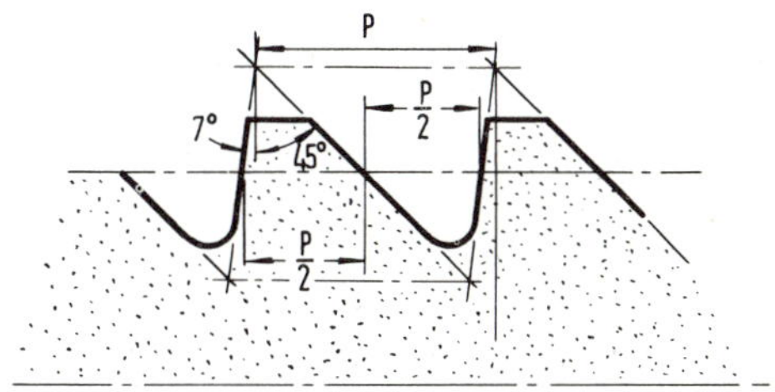

Figure 8.9*c* Buttress Thread

TABLE OF ACME SCREW THREADS

Nominal size	*Threads per inch*	*Pitch*	*Major diameter*
¼	16	0·062 50	0·250 0
5/16	14	0·071 43	0·312 5
⅜	10	0·100 00	0·375 0
7/16	10	0·100 00	0·437 5
½	10	0·100 00	0·500 0
⅝	8	0·125 00	0·625 0
¾	6	0·166 67	0·750 0
⅞	6	0·166 67	0·875 0
1	6	0·166 67	1·000 0
15	4	0·250 00	1·125 0
1¼	4	0·250 00	1·250 0
1⅜	4	0·250 00	1·375 0
1½	4	0·250 00	1·500 0
1¾	4	0·250 00	1·750 0
2	4	0·250 00	2·000 0
2¼	3	0·333 33	2·250 0
2½	3	0·333 33	2·500 0
2¾	3	0·333 33	2·750 0
3	2	0·500 00	3·000 0
3½	2	0·500 00	3·500 0
4	2	0·500 00	4·000 0
4½	2	0·500 00	4·500 0
5	2	0·500 00	5·000 0

LOCKING DEVICES FOR SCREW THREADS

Locking devices for screw threads are numerous; the following is a selection of the more common forms in general use.

Lock nut (*Figure 8.10*)

This device of using two nuts is a frictional method whereby two nuts are firmly screwed home one on top of the other. Then, with a spanner on each nut, the nuts are rotated slightly in opposite directions, wedging the nut threads on to the opposite flanks of the bolt threads. Figure 10 shows a pair of lock nuts in different positions. In practice the thick nut is sometimes fitted underneath, sometimes on top. As the top nut takes all the tensile load in the bolt the thicker nut should, however, be fitted on the top. Sometimes two thin nuts are used for locking nuts.

Figure 8.10

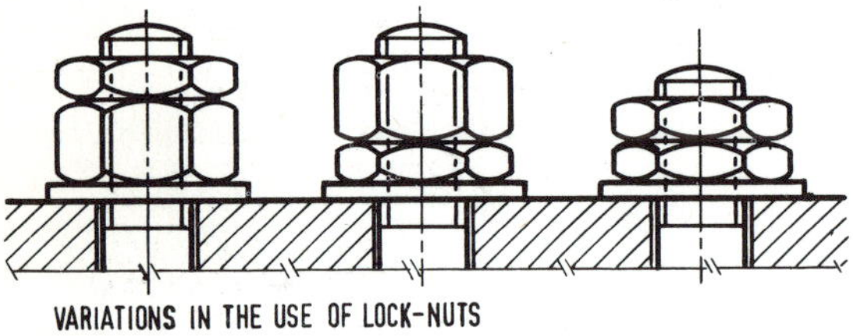

Spring washers (*Figure 8.11*)

These are made in single- and double-coil types. On tightening between the nut and the component, the spring action forces the nut threads against the bolt threads, forming a frictional lock. Spring washers often have sharp corners at the split, and these dig into the nut to produce an extra wedging action. To prevent damage to the machined surface of the component it is usual to fit a plain washer under the spring washer.

The spring washer device is now supplied in a number of forms as continuous, crinkled and serrated washers which provide spring and wedging action similar to that of the orthodox split spring washers. They are used mainly for the smaller bolt sizes and are often stamped out of thin-gauge spring steel.

Figure 8.11

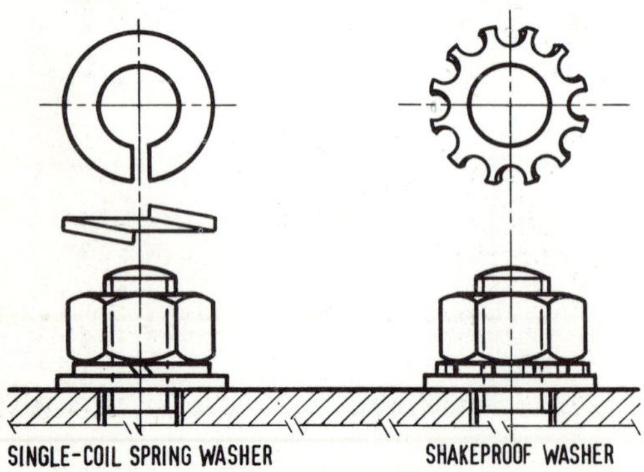

Slotted nut (*Figure 8.12*)

This is a hexagon nut with slots cut across each pair of flats. After tightening, a hole is drilled through the bolt to correspond with a pair of slots. A split pin is then inserted through the slots and hole, and the legs of the split pin are bent round the bolt and nut to secure it in position. This gives a positive locking device.

A more expensive form of the slotted nut is the castle nut which is used when it is necessary for the nut to resist high tensile forces. A shoulder is built up above the nut thickness to accept the split pin, so that locking is achieved without reducing the number of full threads in engagement with the bolt.

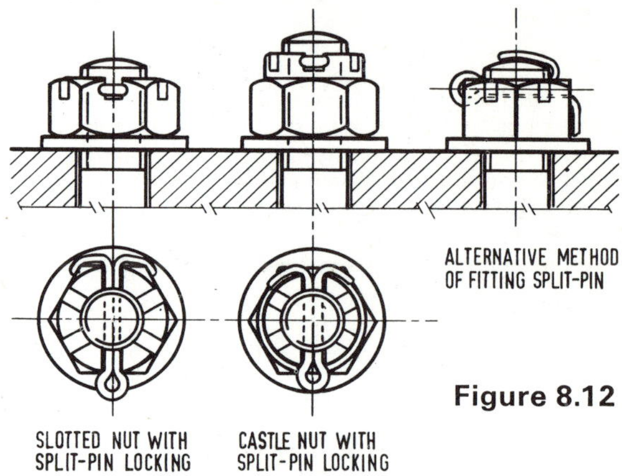

Figure 8.12

Tab washers (*Figure 8.13*)

These come in a variety of shapes and are often specially designed for a particular purpose, some locking a single nut and others a pair of nuts.

The tab washer is inserted between the nut and component. After tightening the nut, one tab is bent to locate the washer and the other tab is bent tightly against a flat of the nut to prevent rotation due to vibration. With the double-tab washer, location of the washer is inherent in the design; after tightening the nuts, only bending the tabs is necessary for locking.

Figure 8.13

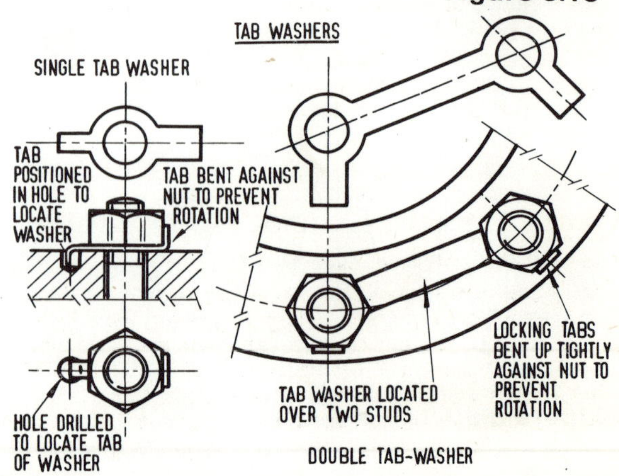

Wire locking (*Figure 8.14*)

This method of locking set bolts is widely used where there is excessive vibration and failure of the fixing due to loosening would cause structural or mechanical failure. It is used for nuts which are periodically removed for inspection, or for union nuts on tube joints where the use of split pins is impossible.

Soft wire, either steel or copper, is passed through holes drilled in the bolt heads or across the corners of nuts and pulled tight. The ends are twisted together with pliers. The wire must be fitted in a direction to prevent slackening of the nuts or bolts.

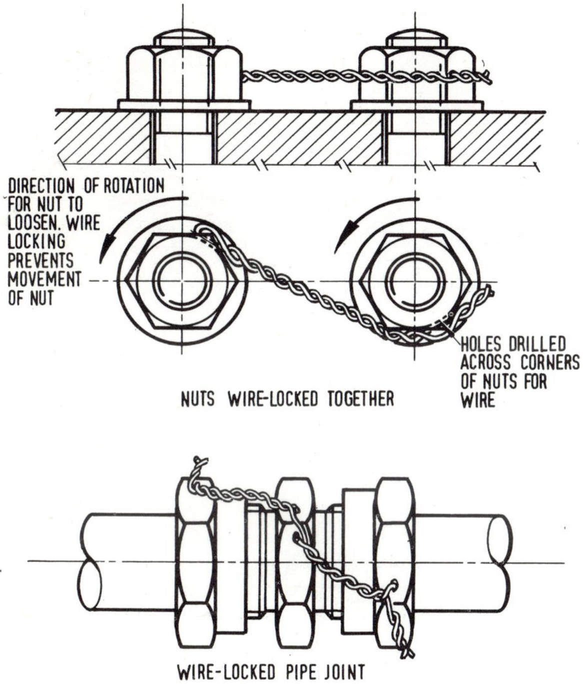

Figure 8.14

Self-locking nuts (*Figure 8.15*)

The figure shows examples of proprietary nuts having devices which produce a frictional gripping action due to fibre or plastic inserts or to mechanical de-pitching of the final threads. These nuts have the advantage of not vibrating loose after the bedding or settling down which occurs in all machinery where there is vibration.

Set bolts with plastic inserts (*Figure 8.16*)

Patented bolts of all sizes from the very smallest to several inches in diameter are manufactured with a slug or key of plastic inserted in the first few threads to prevent loosening due to vibration. This is a device similar to that used in self-locking nuts.

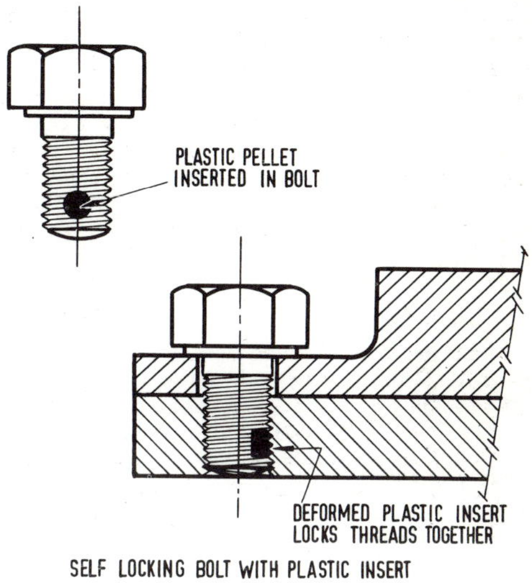

Figure 8.16

Figure 8.15

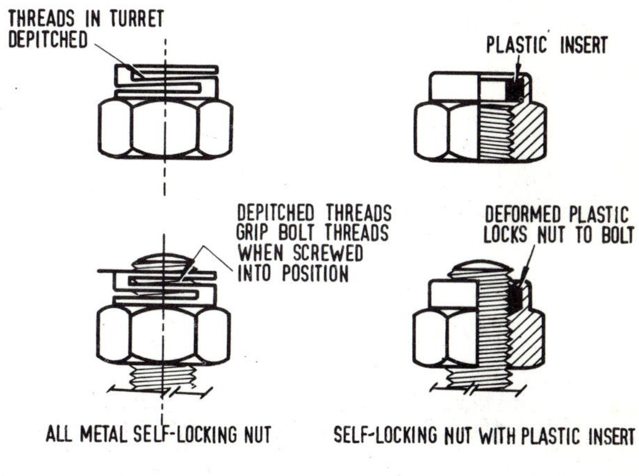

Keys and keyways (*BS 46: Part 1*)

A key is a machine element which is inserted between a shaft and hub to lock the two parts together. Most keys are made of steel to resist shearing as the drive is transmitted from one part to the other, and also to resist crushing and deformation when being fitted. From Figure 8.17 it can be seen that there are basically two types—*sunk keys* which are fitted in keyways cut in both shaft and hub, and *saddle keys* which require a keyway in the hub only.

Square and rectangular keys (*Figure 8.17*)

These keys are manufactured with opposite sides parallel, and are sunk for half their depth (measured at the side) into each of the mating components (shaft and hub). These keys can also be obtained with a slight taper of 1 : 100 on their thickness; they can also be fitted with gib heads to facilitate easy withdrawal. Keys must be a good tight fit in both parts, or fretting will take place; this will deform the key and ultimately result in failure.

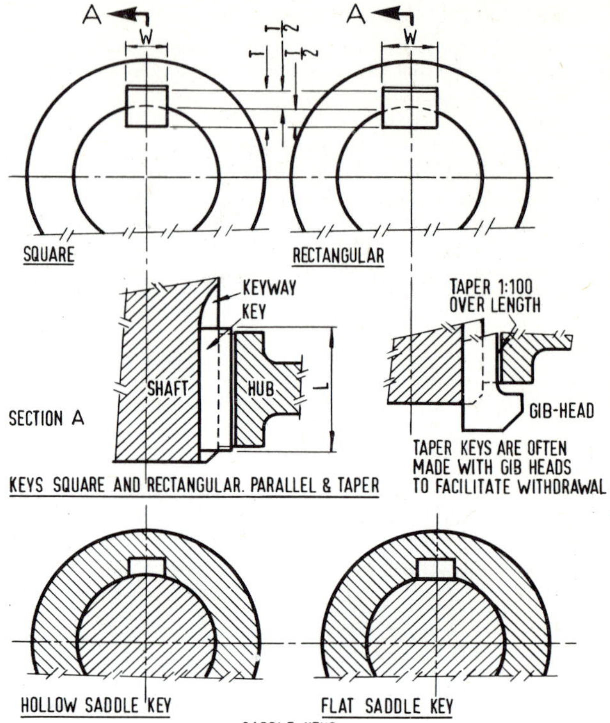

Figure 8.17

Feather keys (*Figure 8.18*)

Sometimes relative longitudinal movement is required between components, as in a gear box when a gear is slid along the shaft but must still be driven by it. In arrangements such as this feather keys are used; they are fixed firmly in the shaft and the keyway in the hub is made a sliding fit.

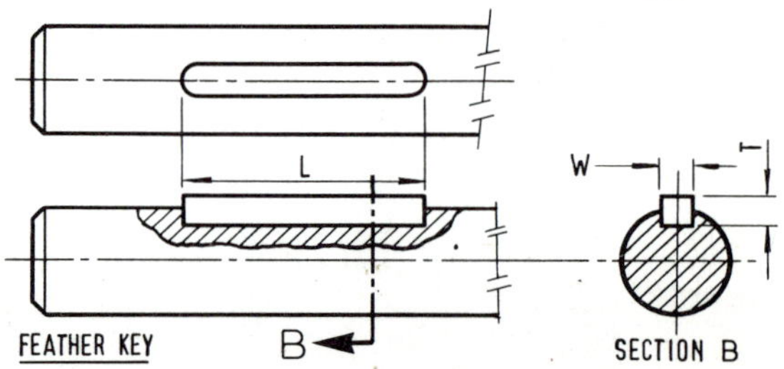

Figure 8.18

Woodruff key (*Figure 8.19*)

This is a sunk key used in conjunction with tapered shafts, the part entering the shaft being shaped as a regular arc to allow the flat side to line up with the taper on the hub on fitting. The Woodruff key is used mainly as a location device; the drive between parts is transmitted through the tapered fit.

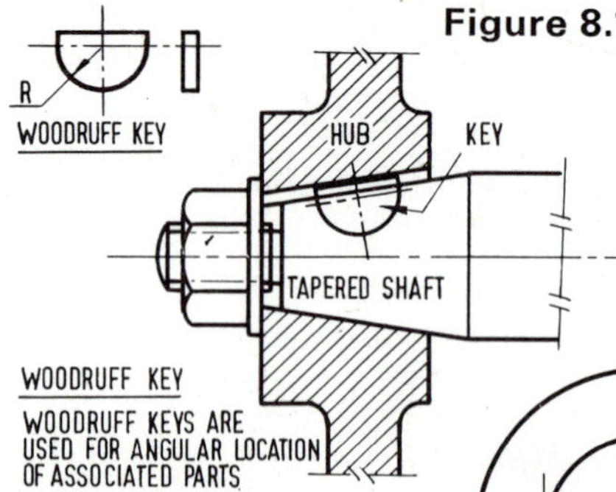

Figure 8.19

Round keys (*Figure 8.20*)

This is a simple keying device which does not require the cutting of accurate keyways. The two parts (shaft and hub) are fitted together and a hole is drilled with its centre on the contact line of the mating parts. A tight round pin is then driven into the hole to produce the locking effect. The hole is sometimes tapped and a threaded pin used; this is necessary when the pin has to be removable.

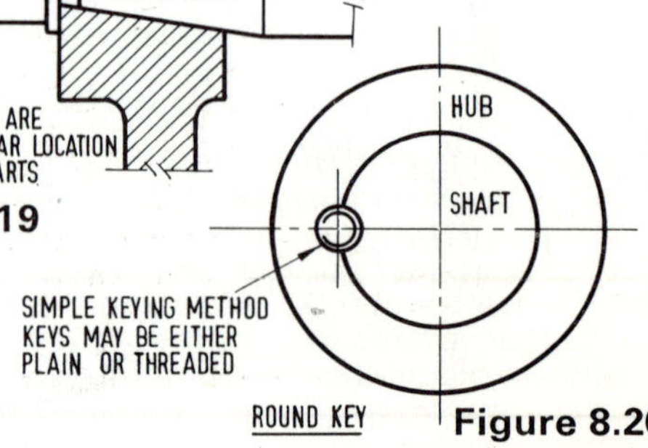

Figure 8.20

RIVETING

Riveting is a method of permanently fastening parts together. Dismantling a riveted joint usually results in the destruction of the parts.

Figure 8.21

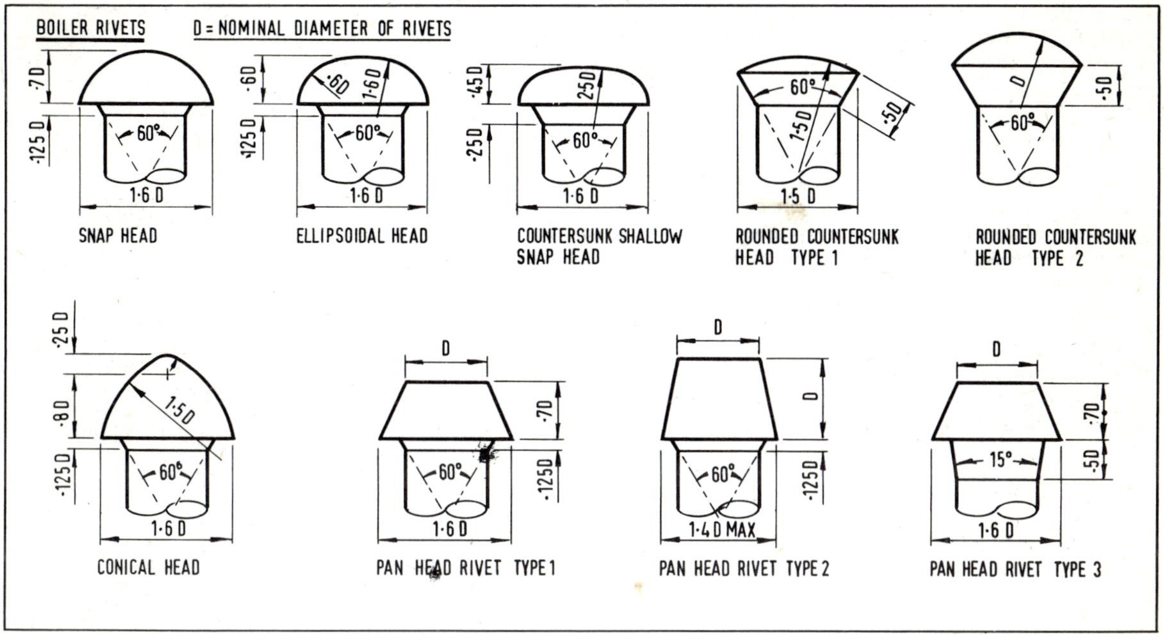

Riveted joints can be roughly divided into two classes: those required in heavy engineering such as structural boiler-making and shipbuilding, and those for light engineering where light-gauge plates and small rivets are used, as in aeronautical work. Figures 8.21 and 8.22 show some of the types of rivet available for light and heavy engineering.

These large steel rivets are produced in a range of shank diameters and are heated to red heat before fitting in structural joints.

Figure 8.22

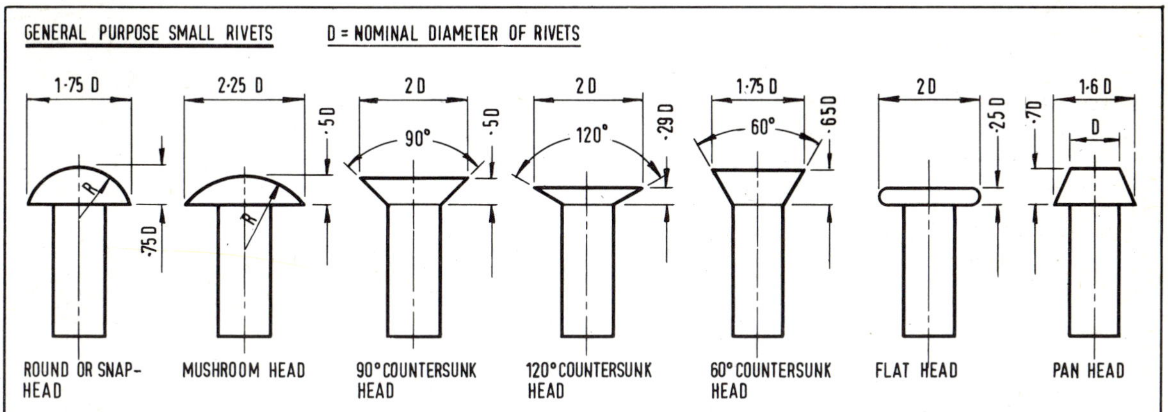

These small rivets are produced in a range of shank diameters. They are manufactured in mild steel, aluminium, aluminium alloys, copper and brass. Millions of the small-diameter rivets are used daily in industry, and are usually set cold.

In heavy engineering, joints are made with rivets that have been heated until red hot and inserted in holes which have been either drilled or punched and deburred; the shank of the rivet is quickly formed into a second head by blows from a mechanical riveting hammer. By this method the rivet is swelled into the hole and on cooling contracts and draws the plates tightly together.

In light engineering, the small riveted joints are made cold with cold rivets by upsetting the rivet shanks with a mechanical riveting hammer to form the second head, or a proprietary type of rivet such as the Tucker 'Pop' rivet shown in Figure 8.23 is used. With this a special tool applies a slow constant pressure to form the head through the medium of an expendable mandrel, which breaks off when the forming of the head is complete. The advantage of this type of riveting is that it can be carried out from one side of the work and requires only one person instead of the usual two. It is also the only method of riveting when one side of the work is inaccessible, as in closed shell structures manufactured during aircraft construction.

Holes for riveting, especially in highly stressed aircraft structures, are accurately made to rigid specifications to prevent unnecessary weakening or giving rise to high-stress concentration round the hole, and so causing failure in service.

Figure 8.23

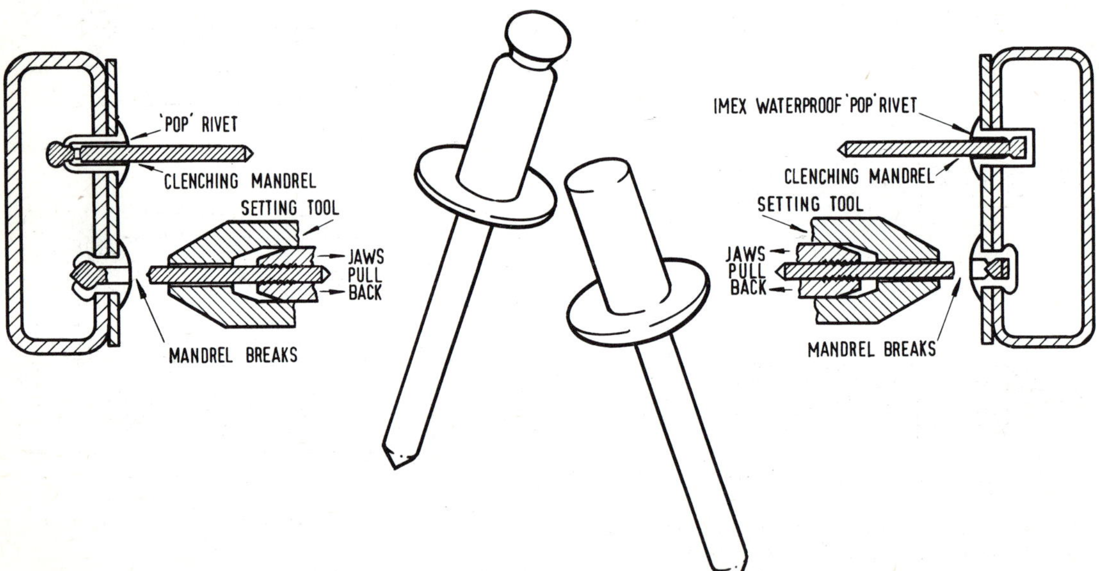

Blind riveting requires only one operator, the rivet being 'snapped up' from one side only by the use of the headed mandrel or steel pin which is supplied fixed to the rivet. The setting of the rivet is controlled by the mandrel, which can only break when the joint is tight and the head fully formed. Blind riveting is a fast and economical system; for closed shell structures it is the only satisfactory method of riveting that can be used.

Figure 8.24 shows methods of making riveted plate joints. Plates can be simply lapped by placing one plate on top of the other and riveting them together, this method being satisfactory where joints are not highly stressed. Where a smooth external surface is required, and the plates are riveted to internal structural frames, butt joints with a single strap riveted with countersunk rivets are satisfactory. Plates resisting heavy tensile loading should be butt jointed with a double butt strap; this places the rivets in double shear and, if the joint is well made, no stresses due to bending should be induced in the rivets.

Remember that riveted joints should be designed to place rivets in shear and not in tension.

Some riveting patterns are shown in Figure 8.24. The size, type, number of rows and pitch are dependent on the loading and the conditions to which the joint will be subjected in service.

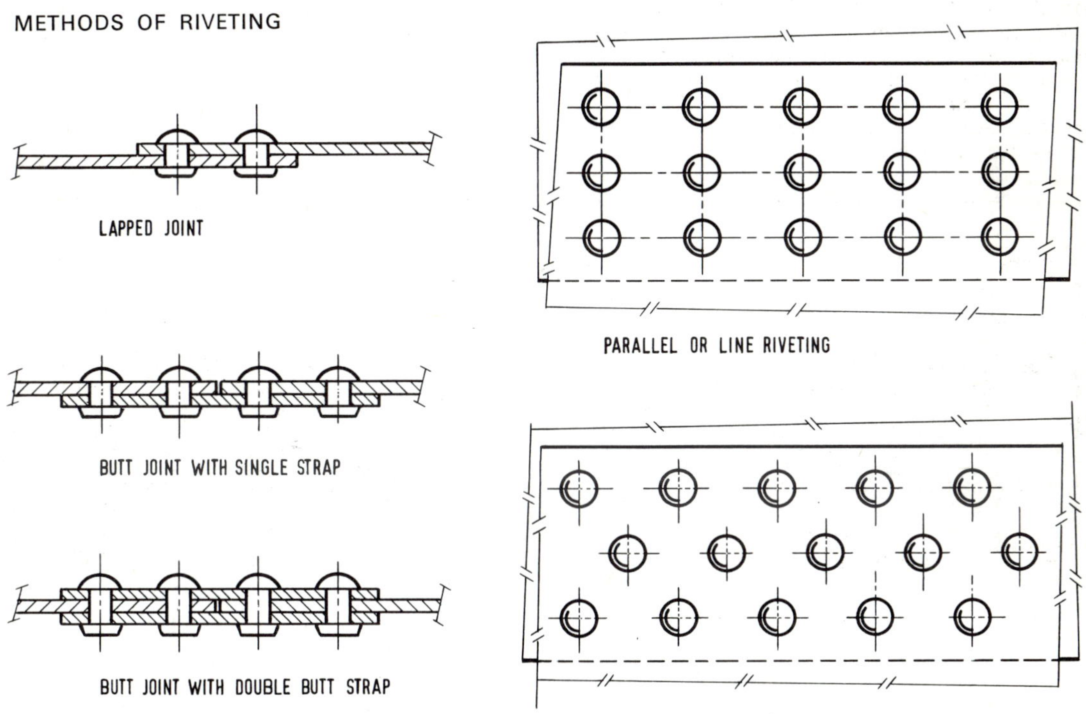

Riveted joints may have either single or multiple rows of rivets, depending on the requirement for joint strength and on the application.

Figure 8.24

INTERPENETRATION

Where solid geometric shapes meet (or appear to pass through each other) there is a contact line of intersection. This line is known as the *interpenetration line.* It is most important that the student should make a serious attempt to understand the basic geometric method which can be applied in all cases to give the correct line of intersection when producing orthographic views.

In Chapter 6, Figures 6.19 and 6.20 clearly show the method, which can be seen to be only an application of projecting points.

The interpenetration line has two distinct applications, one being approximate and the other extremely accurate.

The approximate interpenetration line often appears in orthographic views of forgings, castings and stampings at the intersection of solid geometric shapes. It also occurs where solid geometric shapes are cut by machined surfaces. It is only necessary for a draughtsman to plot a few points to enable an approximate interpenetration line to be drawn. The line conveys no information for manufacture, it is only pictorial.

The accurate interpenetration line is required in the orthographic views of sheet-metal work. It supplies information to enable plotting of the correct shape to be cut from flat sheet-metal stock. The following examples, Figures 9.1 to 9.9, show the method of plotting curves of interpenetration. Figures 9.1 and 9.2 describe a method of scribing lines along the smaller pipe, and plotting where they contact the larger. The remainder of the examples show how points are established on the interpenetration curve by drawing a nest of sections. On closer examination it will be seen that the first examples can also be described as drawing a nest of sections in the plan view.

If this basic fact of cutting the object into a number of parallel slices, more correctly termed 'drawing a nest of sections', is clearly understood, no curve of interpenetration should give the student any difficulty, providing a number of examples are carefully and methodically worked through. It is important to stress, even to those students with some experience of drawing, that it is essential to get into the habit of numbering all points for reference. The problem does not have to be very difficult before one can get lost in a mass of unnumbered projection lines, sections and points.

PROBLEM 9.1

The two views show a tube of 125 mm diameter and another of 75 mm diameter intersecting at right angles. It is required to complete the elevation with the curve of interpenetration. Note that the line of interpenetration is the curve assumed by the smaller pipe on attachment.

SOLUTION

(a) To the smaller tube in both views attach a half end view, and divide it into 6 equal parts. This is to establish 12 equally spaced points around the rim of the tube (see Figure 6.19).

(b) Number the points. Corresponding points in the two views carry the same number.

(c) Projection lines drawn horizontally from the numbered points in the plan can be regarded as scribed lines on the cylinder surface. It will be seen that these lines end on contact with the larger tube. Number each point with the same number as the other end of the projection line.

(d) The same series of projection or scribed lines is drawn horizontally in the elevation, and the lengths are established by projection from the plan.

(e) The points plotted in the elevation are points on the interpenetration line. A smooth curve joining the points will produce the line of interpenetration.

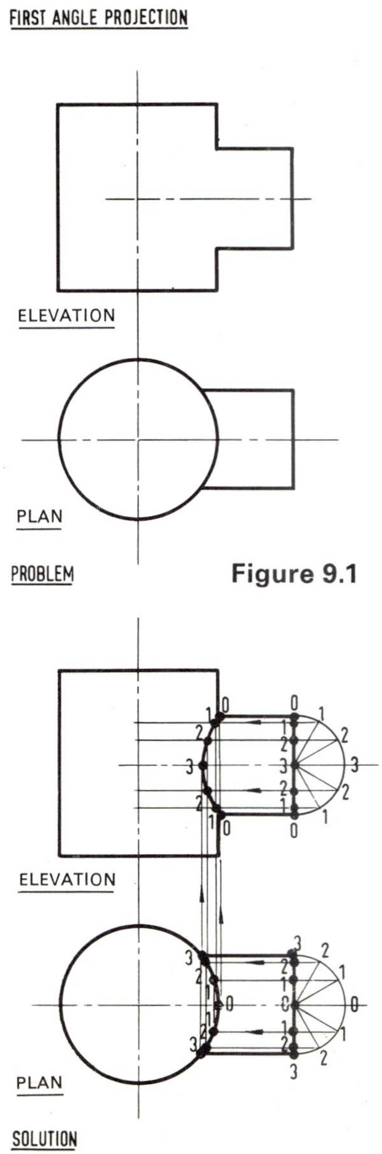

Figure 9.1

PROBLEM 9.2

The two views show a tube of 125 mm diameter and another of 75 mm diameter intersecting at 60°. It is required to complete the elevation with the curve of interpenetration.

SOLUTION

Proceed exactly as in the previous solution.

(a) To the smaller tube in both views attach a half end view, and divide it into 6 equal parts to establish 12 equally spaced points around the rim of the smaller tube.

(b) Number the points, corresponding points in the two views to carry the same number.

(c) In the plan, projection lines or scribed lines are drawn horizontally along the length of the tube until they stop at the surface of the larger cylinder. Number these endpoints with the same number as at the beginning of the line.

(d) The same series of projection lines is drawn along the length of the 75 mm tube in the elevation, and the lengths are established by projection from the plan view.

(e) The points thus plotted in the elevation are points on the interpenetration line. A smooth curve joining the points will produce the line of interpenetration.

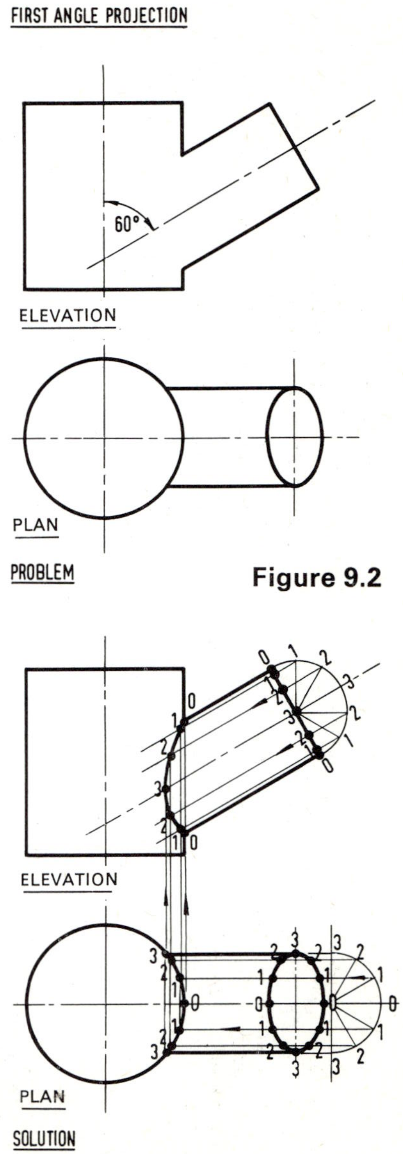

Figure 9.2

PROBLEM 9.3

A 75 mm long equilateral triangular prism, having sides 45 mm long, is intersected at right angles as shown by a cylinder of 32 mm diameter. Draw the plan and complete the elevation, as shown, with the visible line of interpenetration.

Points to note

A. Any section through an equilateral triangular prism will always be a constant-size triangle as in the plan view.

B. Any longitudinal section taken through a cylinder, parallel to the axis, will be a rectangle.

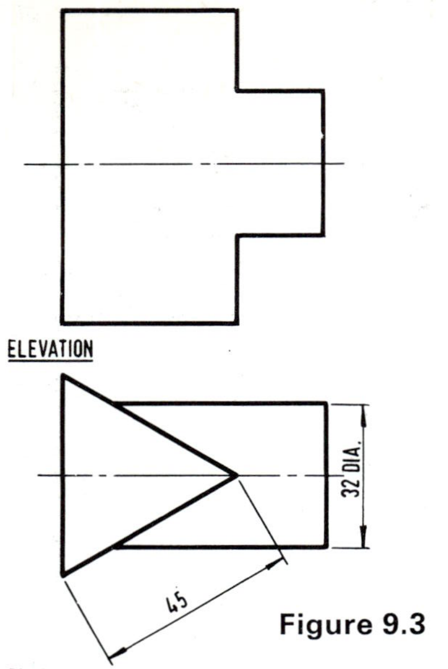

Figure 9.3

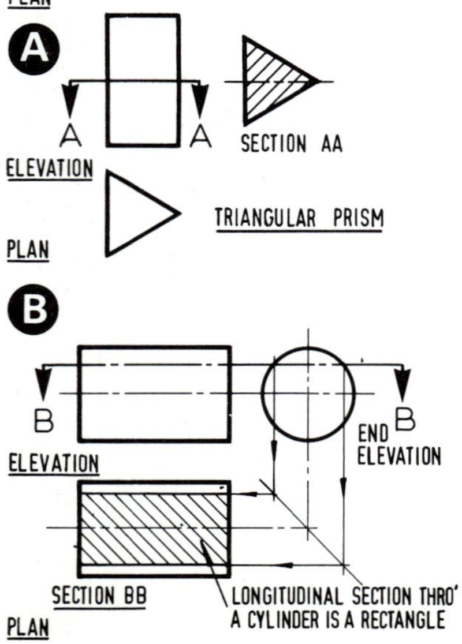

STAGE 1

To draw the plan.

1, 2. Draw centre lines.

3. Draw a construction line at right angles to the centre line in the plan view and extend it to the elevation.

4. Measure 22·5 mm (half of 45 mm) both sides of the centre line in the plan view and draw one side of the triangular prism.

5, 6. Using a 60°–30° set square, complete the plan view of the equilateral triangular prism.

7. Draw a construction line at right angles to the centre line in the plan to position the free end of the round pipe, and extend it to the elevation.

8. Measure 16 mm (half of 32 mm) both sides of the centre line and draw the end of the round pipe.

9, 10. Complete the plan view of the round pipe by joining line 8 to the equilateral triangle with horizontal lines.

STAGE 3

To establish a series of equally spaced points around the circumference of the cylinder in both plan and elevation.

20, 21. On the end view of the cylinder in the plan and elevation draw a half end elevation. Divide into 6 equal parts using a 60°–30° set square. Project the divisions on to the outline of the views to establish equally spaced points around the cylinder. Number the points clearly, as shown, corresponding points in the two views to carry the same number.

STAGE 4

To produce points along the line of interpenetration, the elevation is sectioned (or sliced horizontally) at the established points. The sections are then drawn in the plan view to identify the exact points of contact between the cylinder and the prism sections.

Consider section 1 (upper and lower sections must be identical because of the symmetry of the cylinder). This section is made up of two parts:

(a) The section *through* the triangular prism (a constant equilateral triangle at any section).

(b) The corresponding section through the cylinder.

To draw the section.

22, 23. Draw horizontal lines from points No. 1 in the plan to establish the width of the section of the cylinder at section 1. Where the lines contact the triangle, a point is established on the interpenetration line.

24. Project these contact points to the section lines in the elevation.

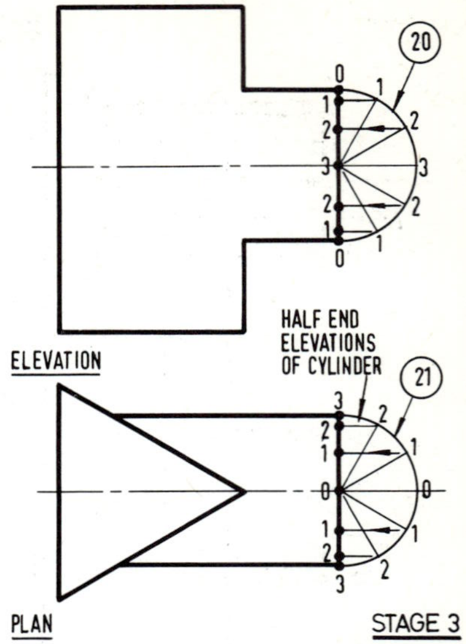

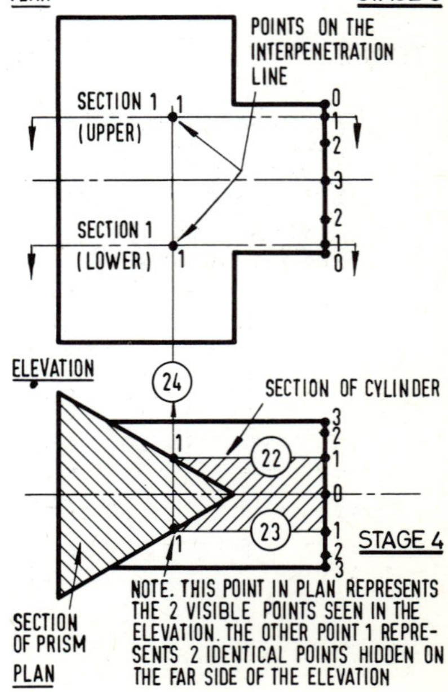

Note: To check the accuracy of your drawing, point A should be vertically above point B.

STAGE 2

To draw the outline of the elevation.

11. Project the edge of the triangular prism from point C into the elevation.

12, 13. Draw lines equally spaced about the centre line in the elevation to establish the height of the triangular prism.

14. Open compasses to dimension ED (radius of the round pipe) and transfer this dimension to both sides of the centre line in the elevation. Draw the end of the round pipe.

15, 16. Complete the outline of the round pipe by joining line 14 to the outline of the prism.

17, 18, 19. Line in the remainder of the outline of the prism.

Note: This sequence of drawing the plan and elevation accurately relates the views to each other.

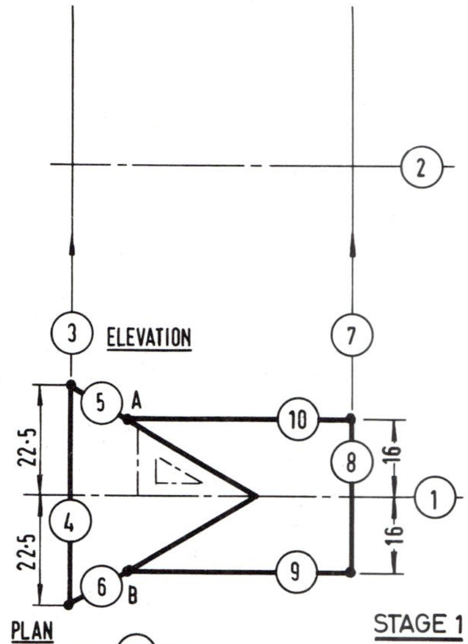

STAGE 1

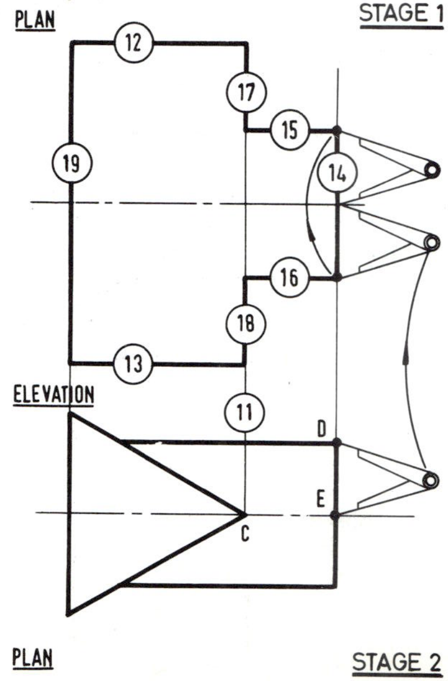

STAGE 2

STAGE 5

Draw a series (or nest) of sections 0 1 2 3 2 1 0 in the plan view to establish several points on the interpenetration line.

25. Project these points to the corresponding section lines in the elevation and, using a french curve, join the points with a smooth curve to give the line of interpenetration.

STAGE 6: SOLUTION

The student's drawing in this exercise should appear as in the illustration, with the following features clearly shown:

(a) The visible outline drawn as a thick, clean line.

(b) All points and sections suitably numbered with good clear numbers, as shown in the illustration.

(c) All construction, section and projection lines shown as sharp, clean, thin lines.

A drawing thus produced conveys clearly to the tutor or examiner the student's knowledge of the technique of orthographic projection.

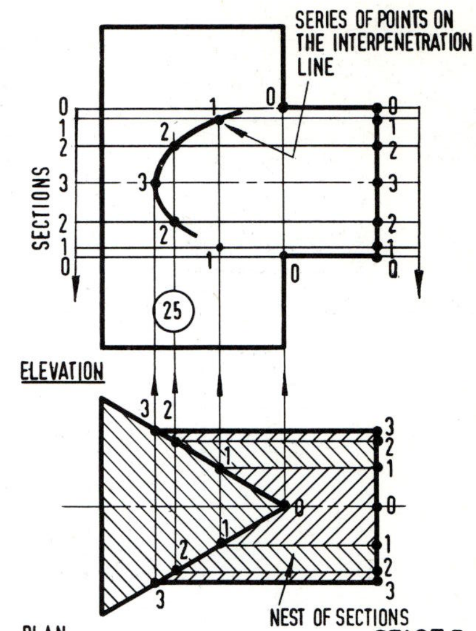

STAGE 5

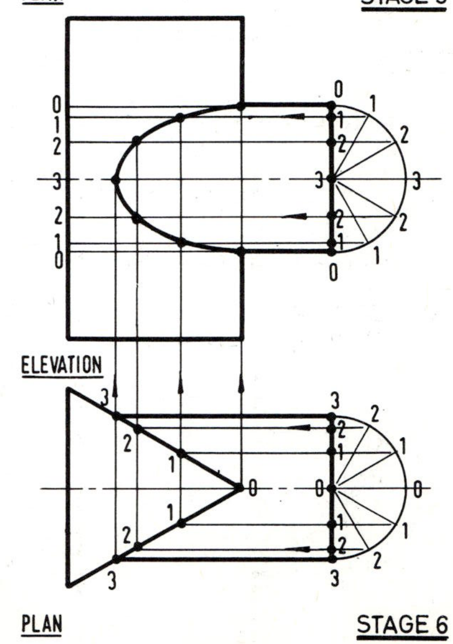

STAGE 6

SOLUTION

PROBLEM 9.4

A right cone, having a base diameter of 75 mm and a vertical height of 125 mm, intersects a right cylinder of 64 mm diameter at right angles. The intersection of the axes is at a point 50 mm from the base of the cone.

Draw the given views and complete the elevation with the line of interpenetration.

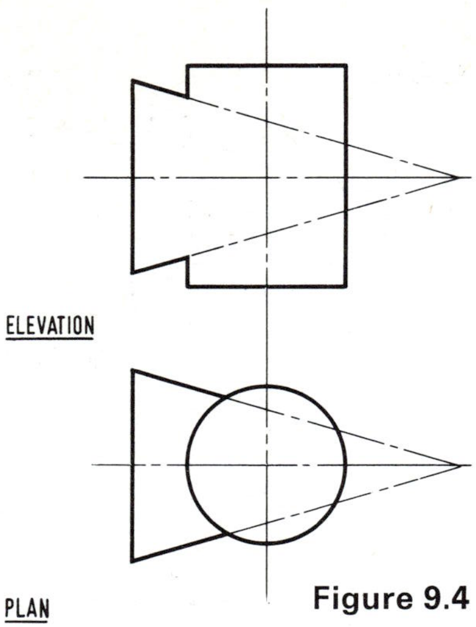

Figure 9.4

STAGE 1

To set out the plan and elevation of the cone.

1, 2. Draw horizontal centre lines with at least 1·5 times the base line between them.

3. Draw a vertical construction line to the L.H.S. extending from plan to elevation.

4, 5. Measure 37·5 mm to one side of the intersection of the plan centre line and line 3. Take this distance in the compasses and step off either side of both centre lines. Join with thick lines between the points, to produce the base lines of the cone in the elevation and the plan.

6. Draw a parallel construction line at a distance from line 3 equal to the vertical height of the cone (125 mm), intersecting the centre lines at point A.

7, 8, 9, 10. Join the base lines to point A, with construction lines to produce the configuration of the cone in both plan and elevation.

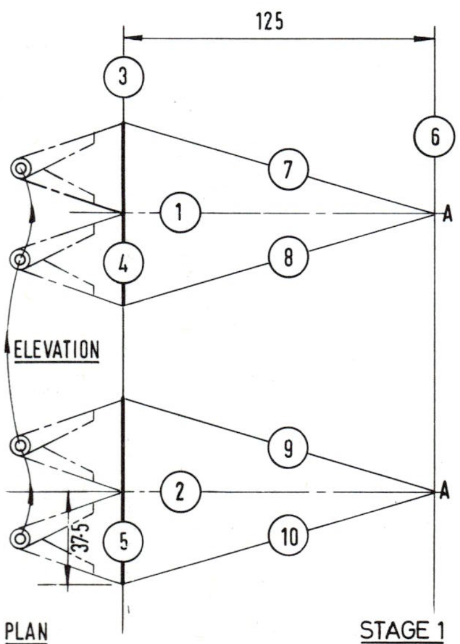

STAGE 4

To establish equally spaced points around the base of the cone in plan and elevation.

24. In the elevation, attach a half end elevation to the base, and divide in 6 equal parts with a 60°–30° set square. Project the divisions to establish equally spaced points around the base.

25. Repeat the above procedure in the plan view.

Note: Number the points as shown, the corresponding points in the two views carrying the same numbers.

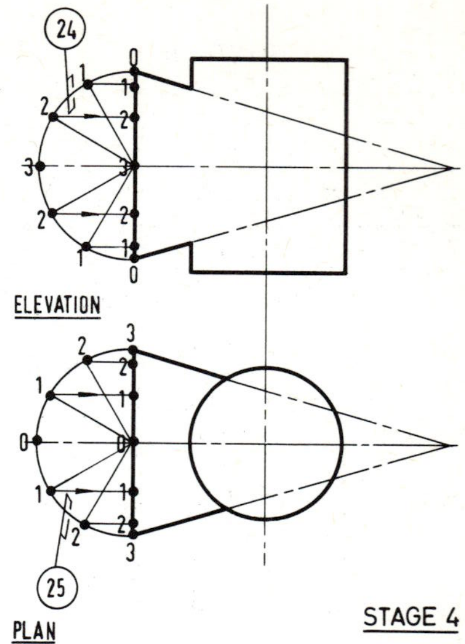

STAGE 5

To draw the construction for establishing the points of interpenetration.

26. Draw generator lines from equally spaced points around the base of the cone to apex A. Somewhere along these generator lines contact is made with the cylinder. These contact points are points on the line of interpenetration.

27. Draw generator lines from base to apex in the plan. Where the generator lines cut the circle, these must be points of contact. Number these points with the same number as the generator lines.

28. Point No. 2 is shown as an example of the projection method. Project contact point No. 2 in the plan to the corresponding generator lines in the elevation, to establish two points on the interpenetration line.

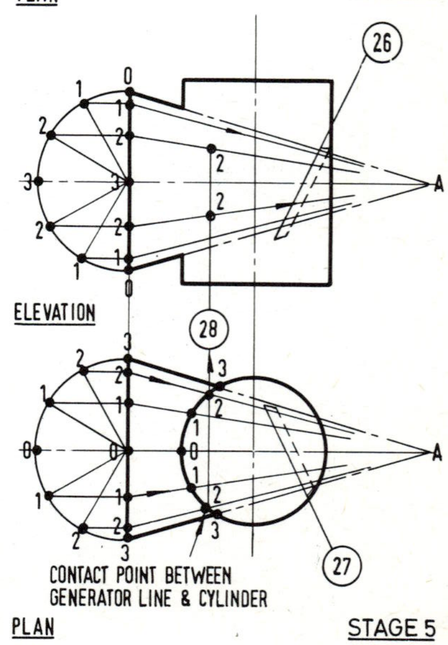

STAGE 2

To draw the cylinder in the plan and elevation.

11. Measure 50 mm from the base of the cone along the centre line in the plan, and draw a perpendicular line to represent the axis of the cylinder.

12. In the plan, from the axis of the cylinder along the centre line, measure 32 mm. Take this distance in the compasses and draw a circle (thick line) to represent the cylinder in the plan view.

13, 14. Project the diameter of the cylinder in the plan to the elevation. 15, 16. At a suitable distance (45 mm) either side of the centre line of the cone in the elevation, draw the top and bottom of the cylinder (thick lines).

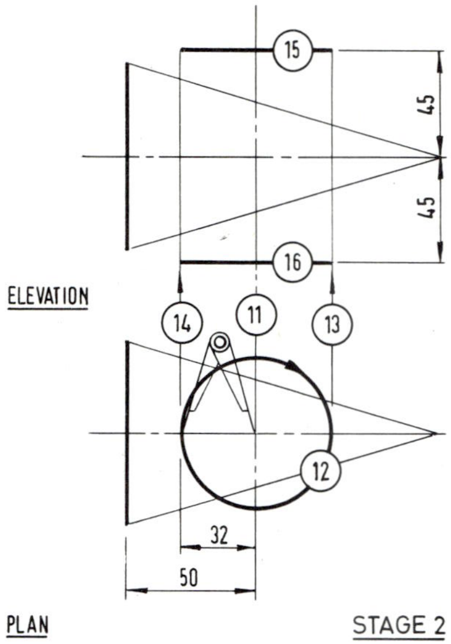

STAGE 3

To complete the outline of the plan and elevation prior to plotting the curve of interpenetration.

17, 18. Line in the slant sides of the cone as shown (thick lines).

19, 20, 21, 22, 23. Line in the outline of the cylinder and cone as shown (thick lines).

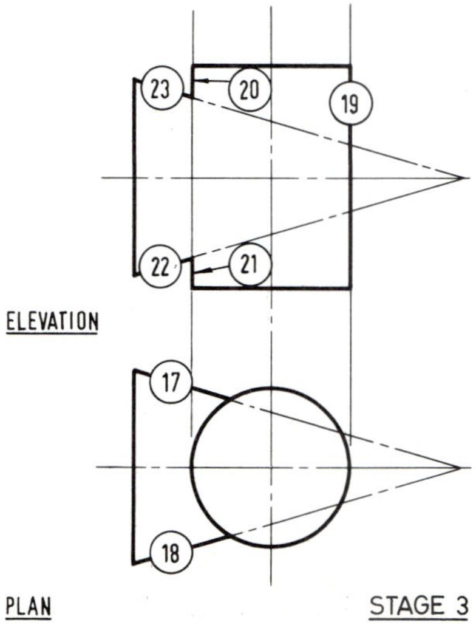

STAGE 6

To plot a series of points on the interpenetration line in the elevation.

29. Project the contact points on the cylinder in the plan to the corresponding generator lines in the elevation. Number the points to identify them. Through this series of plotted points draw a smooth curve. This is the line of interpenetration between the cone and the cylinder.

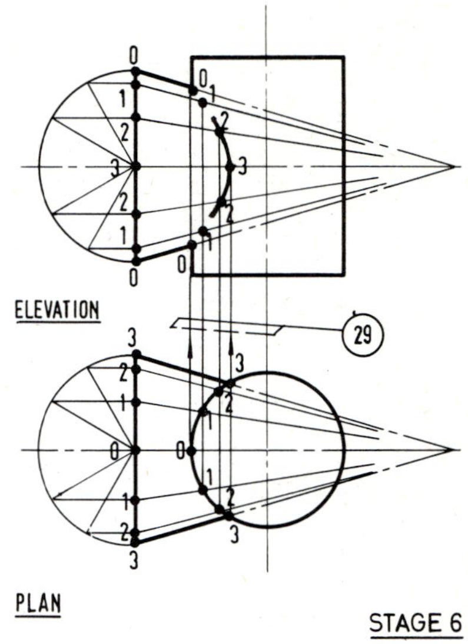

STAGE 7: SOLUTION

The student's drawing in this exercise should appear as in the illustration with the following features clearly shown:

(a) The visible outline drawn as a thick, clean line.

(b) All points and sections numbered with good clear numbers, as shown in the illustration.

(c) All construction, section and projection lines shown as sharp, clean, thin lines.

A drawing thus produced conveys clearly to the tutor or examiner the student's knowledge of the technique of orthographic projection.

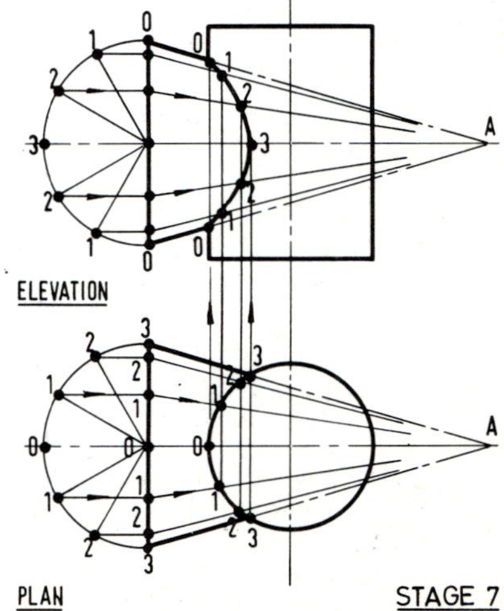

SOLUTION

PROBLEM 9.5

A pipe 65 mm in diameter and 115 mm long is intersected at 60° by a square pipe having sides of 40 mm as shown in the accompanying views. Draw the plan and complete the elevation as shown, with the visible line of interpenetration.

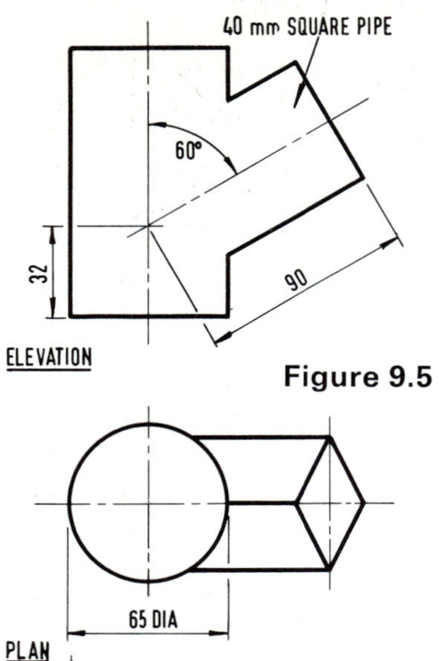

Figure 9.5

STAGE 1

To lay out the plan and elevation.

1, 2. Draw the centre lines intersecting at right angles.

3. Measure 32·5 mm (half of 65 mm diameter) from the intersection of the centre lines. Take the distance in the compasses and draw a circle to represent the plan view of the pipe of 65 mm diameter.

4, 5. Project the diameter of the drawn circle to the elevation.

6. At a convenient distance from the plan draw the horizontal base of the round pipe in the elevation.

7. Measure 115 mm above line 6 and draw the horizontal top of the round pipe.

8. Line in the L.H.S. of the round pipe.

9. Measure 32 mm along the centre line of the round pipe and draw the centre line of the square pipe, using a 60°–30° set square.

10. Measure 30 mm along the square pipe from the intersection of the centre lines to point A, and draw a construction line at right angles to establish the length of the square pipe.

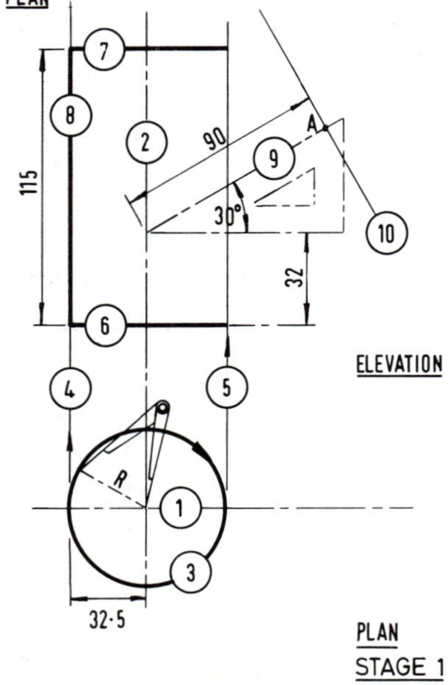

23. Project point A and point A_1 (which is directly behind point A) in the elevation to the plan.

Join points A, E, A_1 and D in sequence, and join to the plan view of the round pipe as seen at Stage 4.

STAGE 4

To establish a set of equally spaced points around the end of the square pipe in the elevation and plan.

24. Divide the sides in the end elevation into an equal number of parts and project them to the end of the square pipe in the elevation. Number them 0 1 2 3 2 1 0 as shown.

25. Project the points in the elevation to the plan.

Note: The corresponding points in the plan and elevation carry the same number.

STAGE 5

To draw projection (or scribed) lines along the square pipe to establish points along the line of intersection.

26. In the elevation, from the established points 0 1 2 3 etc., draw parallel lines along the square pipe as far as the centre of the round pipe. These lines must end where they strike the round pipe.

27. In the plan view draw the same set of parallel lines to ascertain where the scribed lines terminate as they contact the round pipe. Give these points the same number.

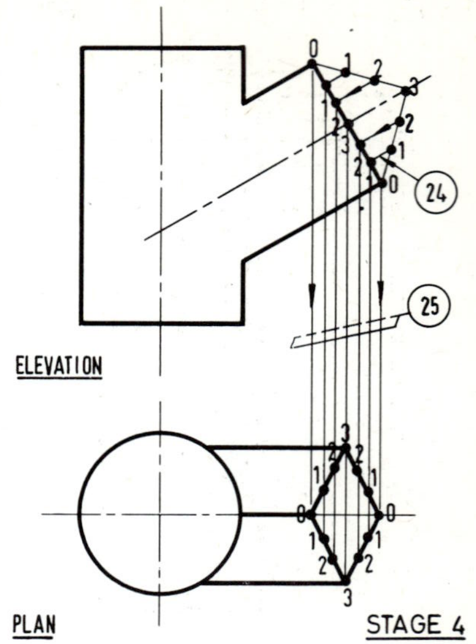

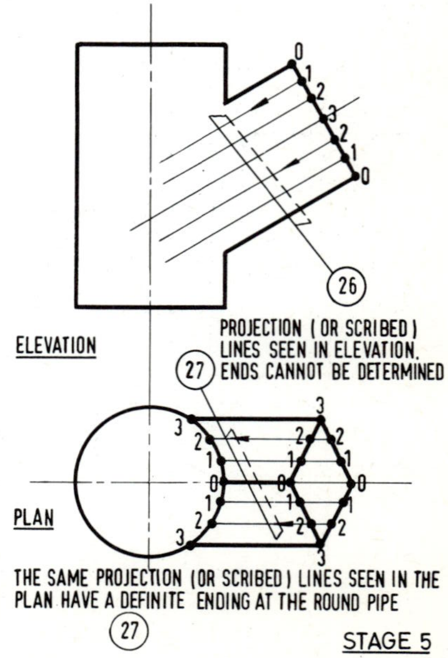

STAGE 2

Geometry required to draw the half end elevation of the square pipe to enable views of the square pipe to be drawn.

11. Draw a construction line at 45° to the centre line of the square pipe at point A, using a 60°–30° set square and a 45° set square as shown in position 1.

12. Measure 20 mm (half of the 40 mm side length of the square pipe) from point A to point B along line 11. Using a 60°–30° set square as shown in position 2, draw the line CD through point A.

13. Using a 60°–30° set square and a 45° set square as shown in position 1, draw line CE. Lines 12 and 13 form the half elevation of the square pipe.

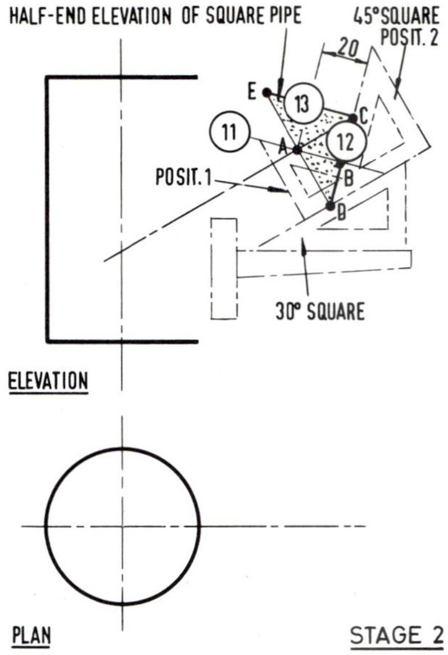

STAGE 3

To complete the construction of the square pipe in the elevation and plan.

14. Line in the end of the square pipe between points E and D.

15, 16. Draw the outline of the square pipe, using a 60°–30° set square.

17, 18. Complete the visible outline of the round pipe.

19, 20. Project points E and D from the elevation to the plan.

21, 22. To draw the outline of the square in the plan, take the half width AC in the elevation in the compasses and transfer it to the plan either side of the centre line. Draw construction lines 21 and 22.

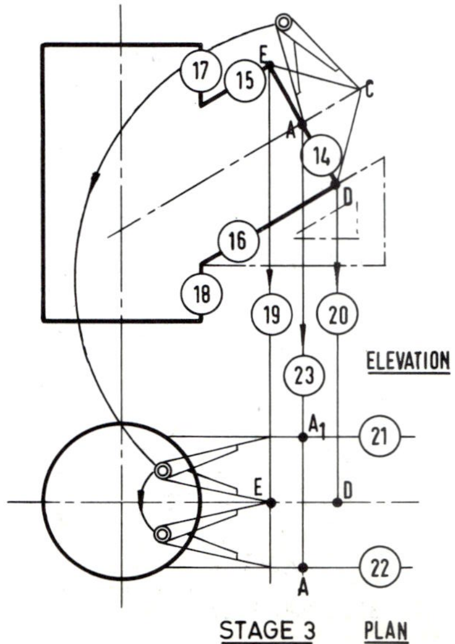

STAGE 6

To finally plot points on the interpenetration line in the elevation.

28. Transfer by projection the end points of the scribed lines in the plan to the elevation—point 3 to scribe line 3, point 2 to scribe line 2 etc.—to build up a series of points on the interpenetration line. Using a french curve, draw a smooth line through the points to complete the elevation.

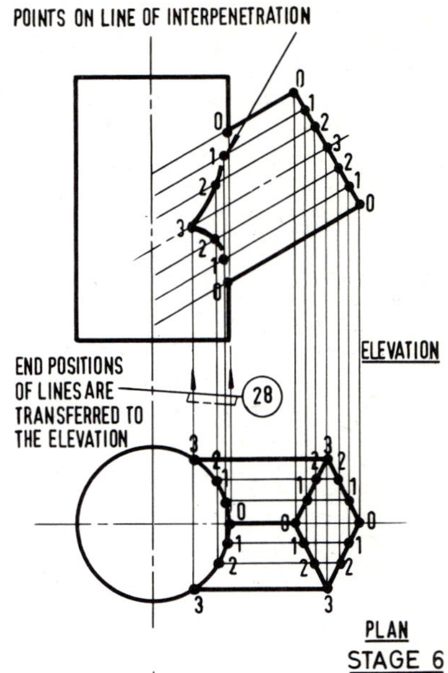

STAGE 7: SOLUTION

The student's drawing in this exercise should appear as in the illustration, with the following features clearly shown:

(a) The visible outline drawn as a thick, clean line.

(b) All points and sections numbered with good clear numbers, as shown in the illustration.

(c) All construction, section and projection lines shown as sharp, clean lines.

A drawing thus produced conveys clearly to the tutor or examiner the student's knowledge of the technique of orthographic projection.

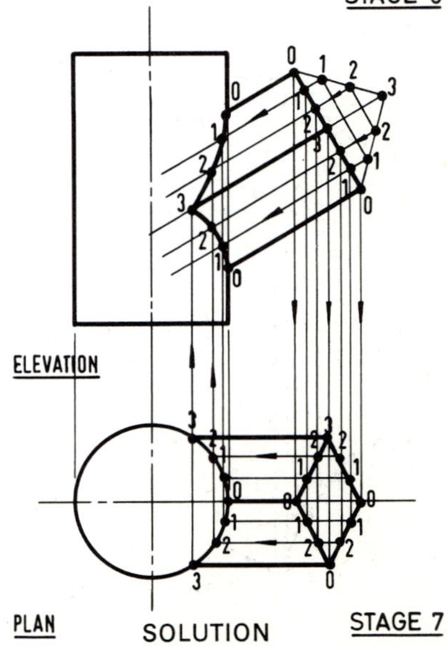

PROBLEM 9.6

The two views show a lathe-turned end to a special strut. It was turned from 75 mm diameter ground stock. The shank is reduced to 38 mm diameter with a blend radius of 65 mm and the end finished hemispherically. A flat location face is milled on the end parallel to the axis as shown. Draw the plan and elevation as shown, completing the elevation with the line of interpenetration.

Points to remember

A. Objects can often be broken down into elements each of which is a simple geometric solid about which fundamental information is known.

B. Any section at right angles to the axis of any object turned on a lathe is a circle.

C. Any longitudinal section through a cylinder is a rectangle.

D. Any section through a hemisphere at right angles to the base is a semicircle.

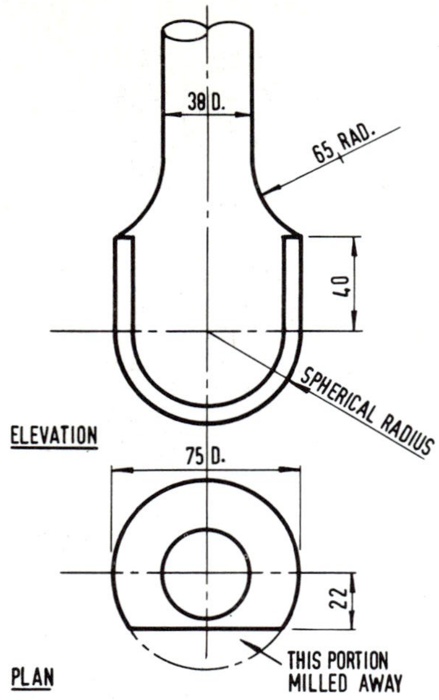

Figure 9.6

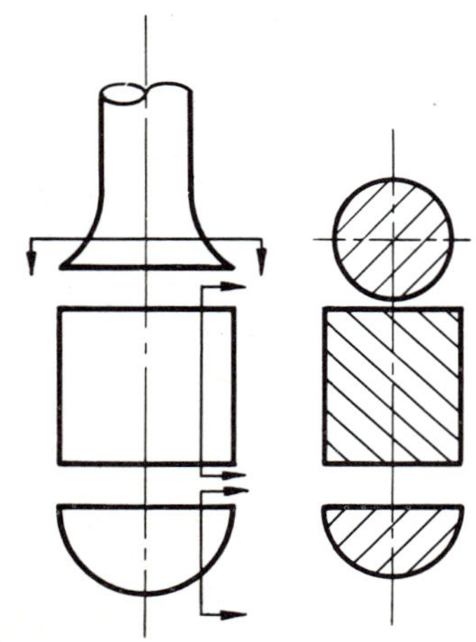

STAGE 1

To draw the plan view.

1, 2. Set up centre lines at right angles.

3. At 22 mm below the horizontal centre line draw a thin construction line to establish the position of the milled surface.

4. Measure 37·5 mm along the horizontal centre line. Take this distance in the compasses and draw the thick visible outline of the plan view.

5. Measure 19 mm along the horizontal centre. Take this distance in the compasses and draw a circle

(Do not set the compasses from the ruler as a dimensional error could be introduced into your construction.) and, with centre R, strike an arc cutting construction line 18 at S.

20. Draw a horizontal construction line through point S, cutting line 17 at V to establish the point of tangency.

21. With centre S, and radius SR in the compasses, strike arc RV as a thick line.

22, 23, 24, 25, 26. Line in the remainder of the outline.

STAGE 4

To establish the height of the interpenetration curve, which is the point where the milling cutter leaves the surface of the turned end.

27. With radius OW in the compasses, and centre O, draw a construction circle. This circle represents the largest remaining uncut circular section.

28. Project the radius of this section to cut the blend radius in the elevation to establish the position of section T.

29. Draw a horizontal section line at the intersection on the blend radius. Point W in the elevation is now established on the centre line, and is the top point of the interpenetration curve.

30. Draw a construction line at the top of the parallel portion. The interpenetration curve must lie symmetrically between points BB and W.

STAGE 5

To plot a series of points on the curve of intersection.

31. In the elevation establish a

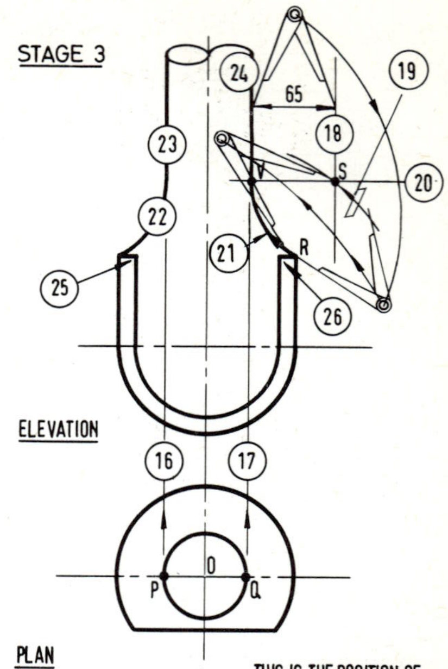

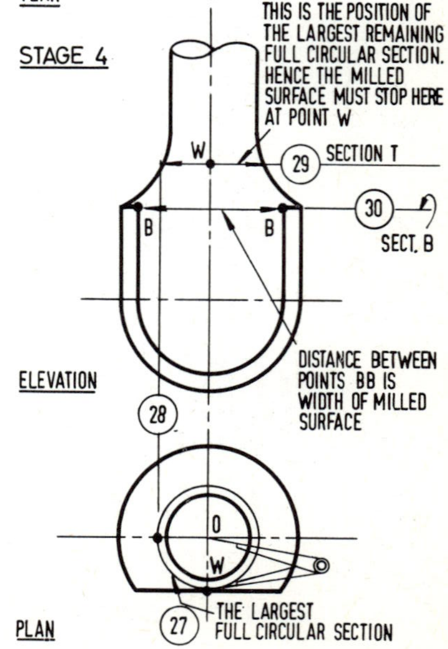

to represent the visible outline of the shank.

6. Line in between the ends of the circular arc with a thick line.
The plane view is now complete.

STAGE 2

To draw the elevation.

7, 8. Establish another horizontal centre line at a convenient distance above the existing horizontal centre line of the plan and on to it project points M and N.

9, 10. Taking distances ON and OM in the compasses, draw semi-circles to represent the boundary of the milled face and the outline of the hemispherical end, respectively. Draw as thick visible outlines.

11. Draw a faint construction line above the centre line to establish the straight portion of the end.

12, 13, 14, 15. Draw thick perpendicular lines from the ends of the circular arcs (9 and 10) to the construction line (11) to continue the boundary of the milled face and the outline of the end.

STAGE 3

To continue to draw the elevation, realising that a simple tangency problem arises when drawing the blend radius.

16, 17. From the plan draw projection lines from points P and Q to establish the shank diameter in the elevation.

18. Measure horizontally 65 mm from the visible outline of the shank and draw a thin vertical construction line.

19. Take the original distance of 65 mm between shank and line 18.

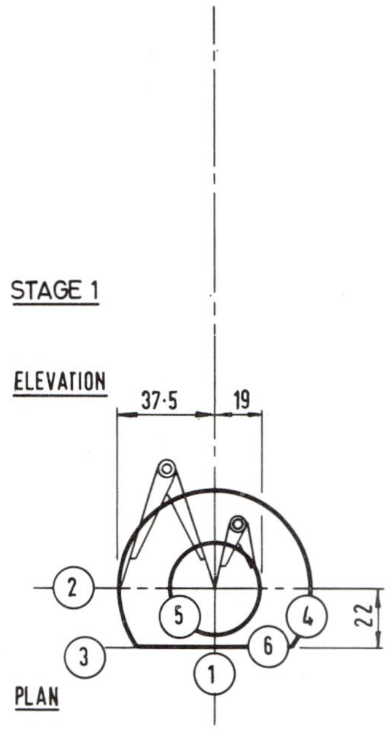

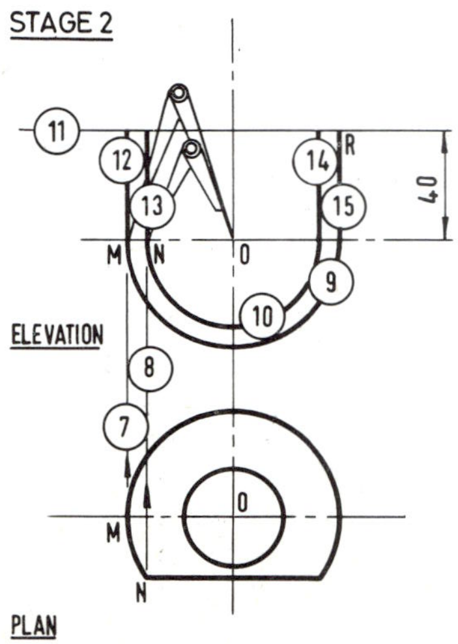

number of horizontal sections between Section T and Section B. They need not be equally spaced. Number them 1, 2, 3 etc.

32. Project the radii of these sections to the horizontal centre line in the plan on the L.H.S. only.

33. Using the projected radii, draw one of the sections in the plan view. It can now be seen where the circular section to the milled flat points 1, 2, 3, 4 etc.

34. Project these points to the corresponding sections in the elevation on the R.H.S. only, to establish a series of points on the curve of intersection. The points on the L.H.S. are transferred across the centre line by using a compass. This gives a mirror-image set of points and enables a symmetrical curve to be produced. Using a french curve, draw the line of intersection by joining the points with a smooth curve.

STAGE 6: SOLUTION

The student's drawing in this exercise should appear as in the illustration, with the following features clearly shown:

(a) The visible outline drawn as a thick, clean line.

(b) All points and sections suitably numbered with good clear numbers, as shown in the illustration.

(c) All construction, section and projection lines shown as sharp, clean, thin lines.

A drawing thus produced conveys clearly to the tutor or examiner the student's knowledge of the technique of orthographic projection.

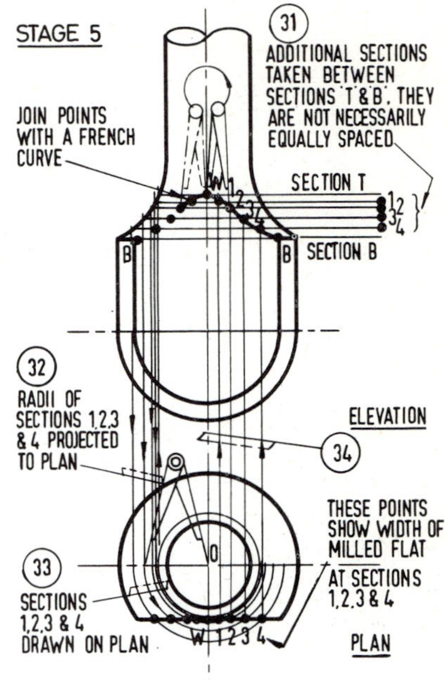

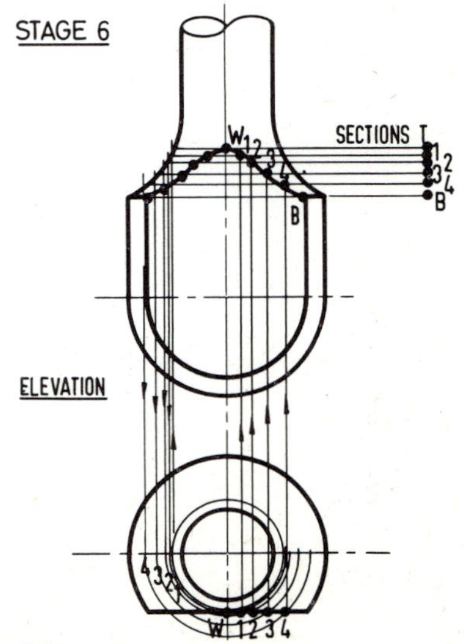

SOLUTION

PROBLEM 9.7

A solid 32 mm square prism intersects with a solid hemisphere of 65 mm diameter. The axis of the square prism is offset from the axis of the hemisphere by 5 mm in two directions at 90°. Draw the two views and complete the elevation with the visible curve of interpenetration. Do not show hidden detail.

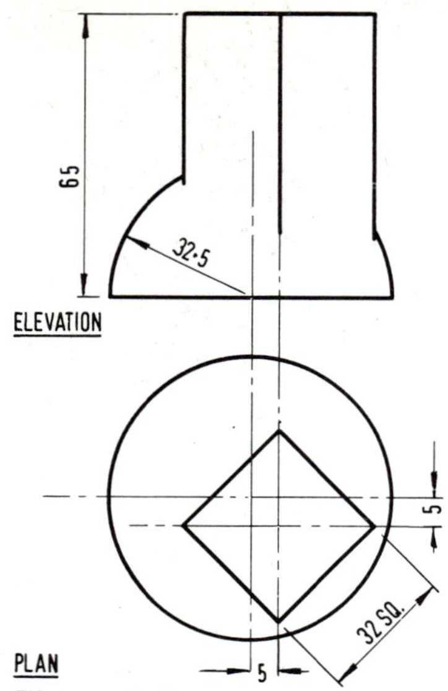

Figure 9.7

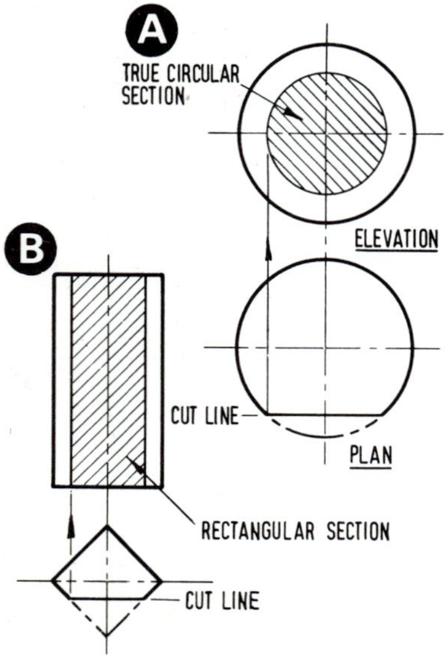

Points to remember

A. Any straight cut through a sphere produces a cut face which is a true circle.

B. Any straight cut along the length of a square prism produces a cut face which is rectangular.

13. Project the sides of the square prism from the plan view to the elevation.

14. Line in the top of the prism.

STAGE 3

To establish points on the interpenetration line, the plan view is sectioned, and the sections are drawn on top of each other in the elevation.

In the plan divide the sides of the square prism into equal parts and number as shown. Consider the section cut by section line 1; it is made up of two parts:

(a) The section through the hemisphere (a semicircle).

(b) The corresponding section through the square prism.

To draw the section

15. Project length CD from the plan to the elevation. This is the radius for drawing the semicircular section of the hemisphere.

16. Project the points numbered 1 from the plan to the elevation to establish the width of the section of the square prism. Where the projection lines cut the section of the hemisphere, points on the interpenetration are established.

STAGE 4

To draw a nest of sections of the hemisphere (sections 0, 1, 2, 3, 4).

17. Project, from the plan to the elvation, radii for drawing sections. Draw the semicircular sections. Number the sections in the elevation as shown, to identify them at all times.

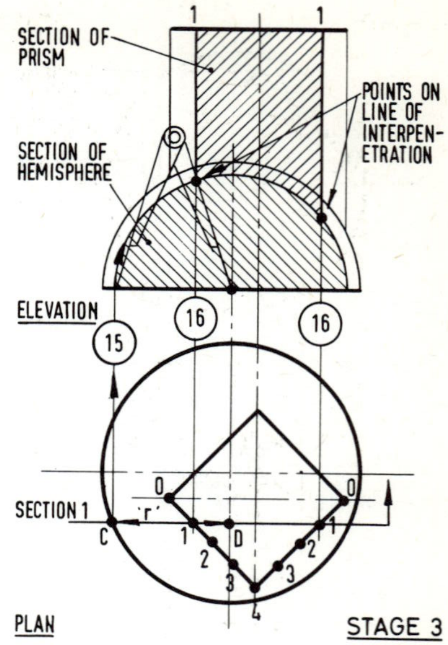

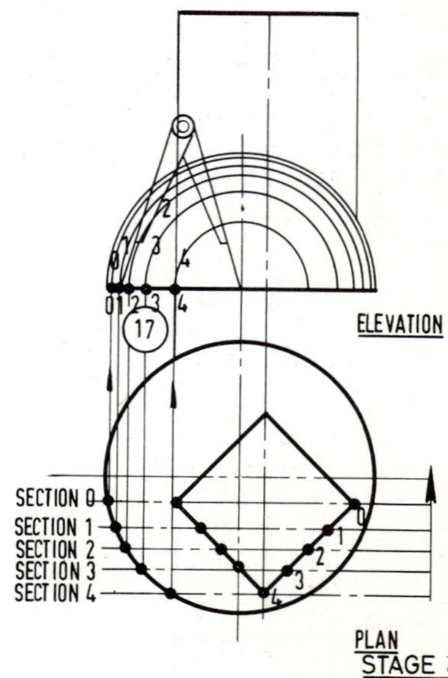

STAGE 1

1, 2. Draw centre lines.

3. Measure off 32·5 mm radius along the centre line, open the compasses to this distance, and draw the plan circle of the hemisphere.

4. Draw the base line for the elevation at a suitable distance from the plan circle.

5. Project the width (diameter of the hemisphere) vertically upwards from the plan to the base line.

6. Open the compasses to distance OA (radius of hemisphere) and draw a semicircle. Line in the base of the semicircle.

Note: By this sequence, a plan and elevation of the hemisphere have been drawn within the two views accurately related to each other.

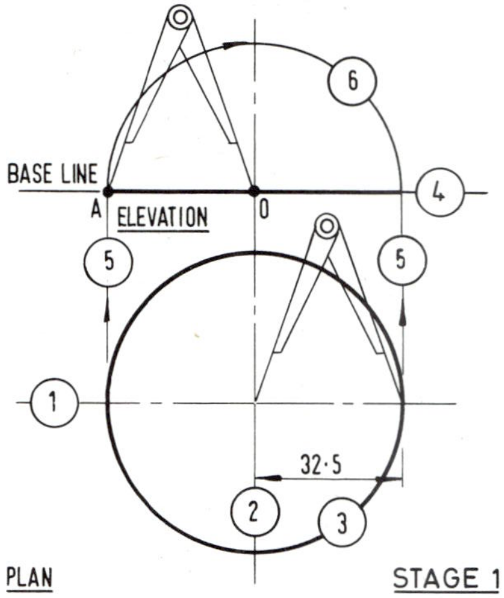

STAGE 2

To draw the square prism in the correct position.

7, 8. Measure 5 mm horizontally and 5 mm vertically from point O in the plan view, and draw centre lines for the square prism, intersecting at point B.

9. Draw a construction line 45° through point B.

10. On each side of B, measure along the 45° construction line half the width of the square prism (16 mm) and draw, at right angles, two sides of the square prism. The sides finish on the centre line at C and D.

11. Using a 45° set square, complete the square.

12. Establish the height of the prism in the elevation.

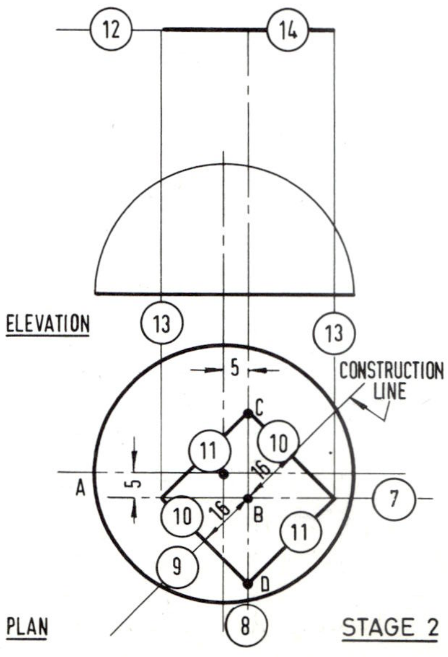

STAGE 5

To draw corresponding sections of the square prism (sections 0 0, 1 1, 2 2, 3 3).

18. Project the numbered points from the plan to the top of the square prism in the elevation. Project down from each of these points to the corresponding numbered section of the hemisphere, to locate the points on the curve of interpenetration. Join the points with a smooth curve.

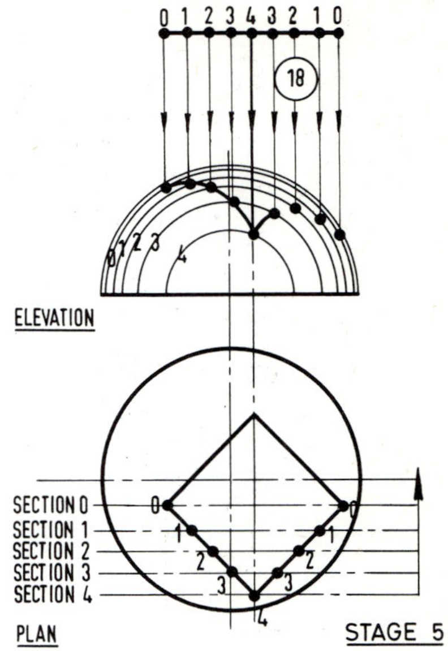

STAGE 6: SOLUTION

The student's drawing in this exercise should appear as in the illustration, with the following features clearly shown:

(a) The visible outline drawn as a thick, clean line.

(b) All points and sections suitably numbered with good clear numbers, as shown in the illustration.

(c) All construction, section and projection lines shown as sharp, clean, thin lines.

A drawing thus produced conveys clearly to the tutor or examiner the student's knowledge of the technique of orthographic projection.

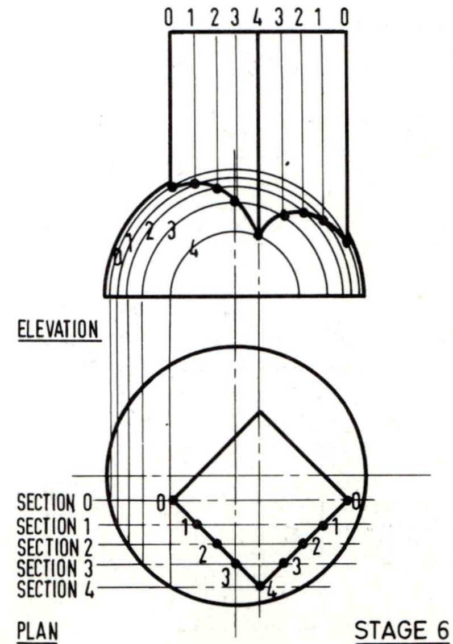

SOLUTION

PROBLEM 9.8

A 90° bend in a pipe of 70 mm diameter has a centre-line radius of 125 mm. A junction pipe of 50 mm diameter enters the bend with its centre line 120 mm from and parallel to one end. Draw the views as shown, completing both the plan and elevation with the lines of interpenetration.

Points to remember

A *torus* is a solid circular ring. It is the shape of an inflated inner tube of a bicycle tyre. Any section (or slice) taken longitudinally through the torus is, as shown, a flat circular ring of constant width.

STAGE 1

Using the given information, lay out the plan and elevation.

1, 2. Draw the centre lines at right angles in the plan.

3. At a distance of 125 mm from the intersection of the centre lines, draw a vertical construction line through the plan and elevation to establish the end of the pipe bend.

4. At a suitable distance from the plan draw a horizontal construction line as the base line for the elevation.

5. Measure 35 mm from the intersection of lines 1 and 2. Take this distance in the compasses and draw a full circle, lining in the R.H.S. (thick line), with the L.H.S. as hidden detail.

6, 7. Draw the sides of the pipe attached to the circle.

8. Complete the plan view of the pipe bend.

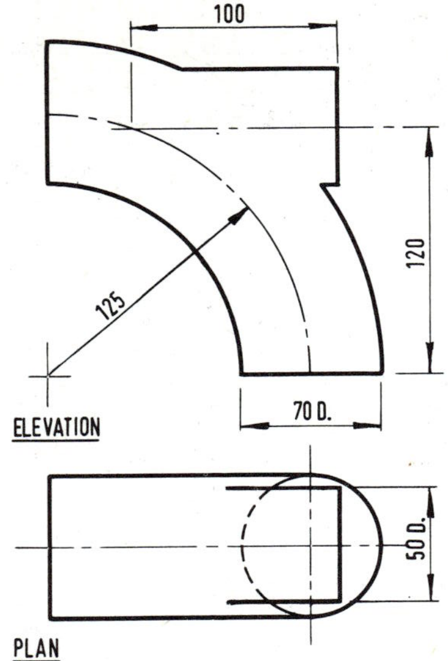

Figure 9.8

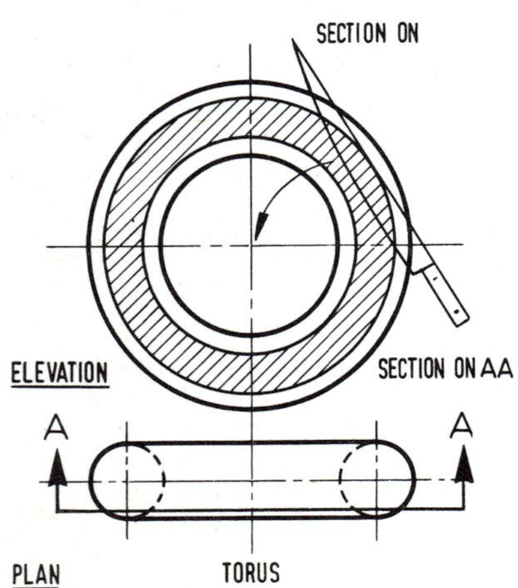

using a 60°–30° set square. Project the divisions to the end of the pipe and number them as shown.

28. In the plan, repeat as above but remove the half elevation slightly to the right to prevent confusion during later projection. Give the same numbers to corresponding points in plan and elevation.

STAGE 4

To find points on the interpenetration line by taking a series of sections through the composite figure, at points established on the junction pipe in the plan view. Consider section 2 only.

29. In the plan draw a horizontal section line as a construction line through point 2 (an identical section lies at the rear).

30, 31. Project the intersection points of the section line and the circumference of the 90° bend pipe (points 2A) into the elevation.

32, 33. Draw arcs from the points 2A as shown. The hatched portion between these lines is section 2 in the elevation.

34, 35. Draw section 2 of the junction pipe. Where the sections meet are contact points on the line of interpenetration.

STAGE 5

To produce a series of points on the interpenetration lines, using the method shown at Stage 4 and using several sections.

36. Draw the section lines 3, 2, 1, 0, 1, 2, 3 on the plan view.

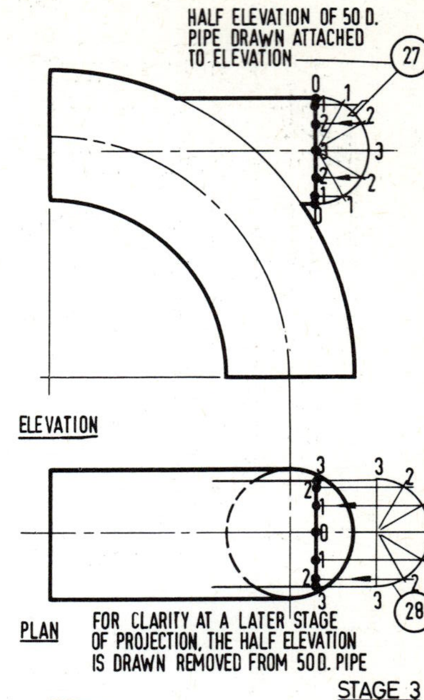

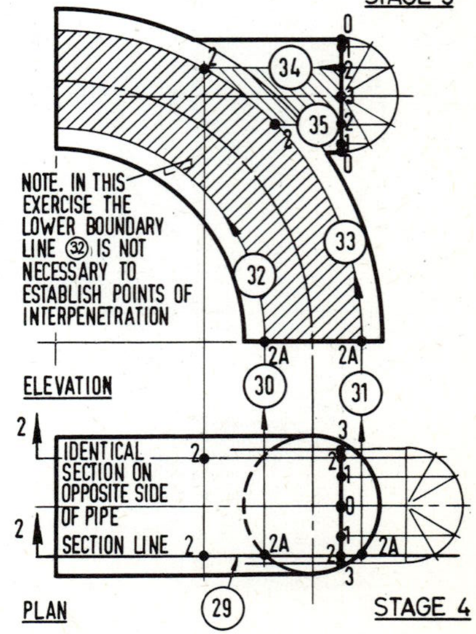

9, 10, 11. Project the diameter of the pipe bend and the centre line to the elevation and establish points A, B and C on line 4.

12. Line in AC, the lower end of the pipe in the elevation.

13, 14, 15. With the compasses set at OA, OB and OC, draw 90° arcs to represent the pipe bend.

16. Complete the pipe bend by drawing the end between the outside arcs. Line in 13 and 16.

STAGE 2

To position the intersecting junction pipe.

17. At a distance of 120 mm from the base of the elevation, position the centre line of the junction pipe.

18. At a distance of 100 mm along the centre line of the junction pipe, draw a vertical construction line to fix the end of the junction pipe.

19. In the elevation, measure 25 mm on one side of the centre line of the junction pipe, and transfer this dimension to either side of the centre line in the plane and elevation. Line in the end of the pipe between the points.

21, 22, 23, 24. Line in the remainder of the outline in the elevation.

25, 26. Draw the sides of the pipe in the plan view as construction lines.

STAGE 3

To establish a set of equally spaced points around the end of the junction pipe.

27. Draw a half elevation attached to the junction pipe in the elevation and divide it into 6 equal parts,

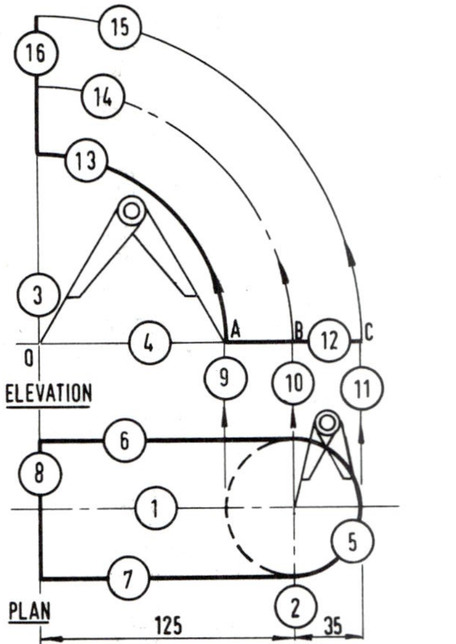

STAGE 1

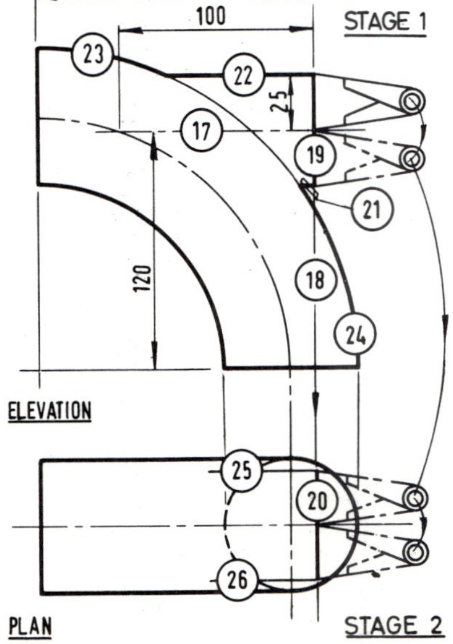

STAGE 2

37. Project the intersection points of the section line and the 90° bend pipe to the elevation.

38. Draw the upper boundary lines of the section of the 90° pipe bend.

39. Draw the corresponding sections of the junction pipe. Where the corresponding sections of the two pipes meet must produce a series of contact points on the interpenetration line in the elevation. Number the points accordingly.

40. Project these points to the corresponding section lines in the plan view.

41. Join the points in the elevation with a smooth curve.

42. Join the visible points in the plan view with a smooth curve.

43. Join the hidden points on the underside with a dotted hidden-detail line.

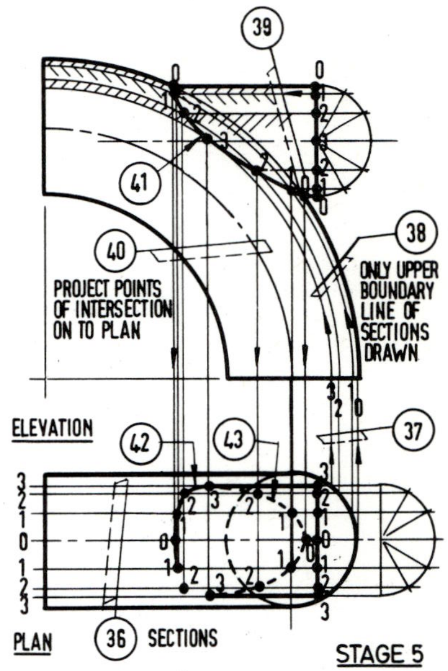

STAGE 5

STAGE 6: SOLUTION

The student's drawing in this exercise should appear as the illustration, with the following clearly shown:

(a) The visible outline drawn as a thick, clean line.
(b) All points and sections numbered with good, clear numbers, as shown in the illustration.
(c) All construction, section and projection lines shown as sharp, clean, thin lines.

A drawing thus produced conveys clearly to the tutor or examiner the student's knowledge of the technique of orthographic projection.

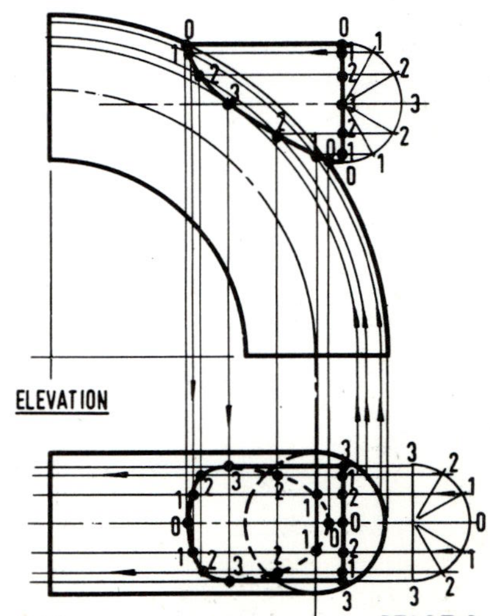

SOLUTION

STAGE 6

PROBLEM 9.9

The drawing shows two views of a regular tetrahedron having sides 125 mm long, with one side horizontal. It is intersected by a cylinder of 50 mm diameter, the centre line of which is positioned horizontally 40 mm from the base. Draw the views as shown and complete both the plan and elevation with the line of intersection.

Points to remember

A. A tetrahedron is a solid four-sided figure. Each side is a triangle.

B. A regular tetrahedron is a solid four-sided figure with each side of identical shape. Each side is an equilateral triangle.

C. Any section (or slice) taken parallel to a side is always a triangle.

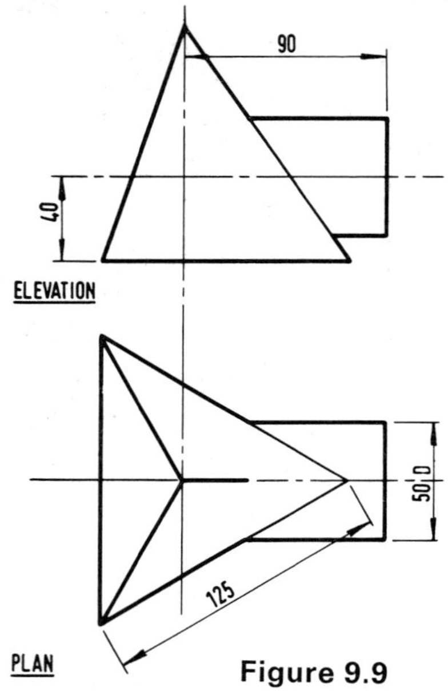

Figure 9.9

STAGE 1

To construct the tetrahedron in the plan and elevation.

1. Set out a horizontal line on the plan.

2. Draw a vertical construction line to L.H.S. of the centre line and measure off 62·5 mm to one side of the intersection, transferring this dimension to the other side with compasses.

3. Line in one side of the figure between these points.

4, 5. Take length AB in the compasses and, with centre A, draw an arc cutting the centre line to establish point C. Draw AC and BC as thin lines.

6. Find centre O by construction (Bisect angle BAC). Draw the construction line.

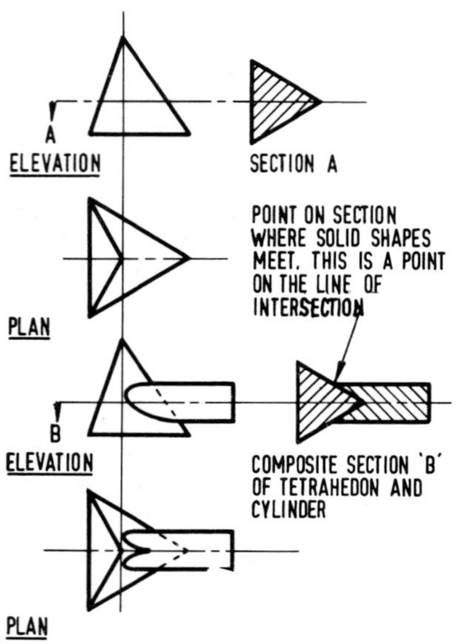

STAGE 3

To establish a series of equally spaced points around the end of the cylinder

27. In the elevation attach a half end elevation to the cylinder and divide into six equal parts, using a 60°–30° set square. Project through points to the end of the pipe in the elevation and number as shown.

28. Repeat the above process in the plan view and number as shown.

Note: Each numbered point in the elevation represents two points of the same number, one visible and one directly behind on the far side of the cylinder. In the plan the same points carry the same number as in the elevation but they have been rotated through 90°. Hence point 1 is directly above point 7 and point 2 is directly above point 6 etc.

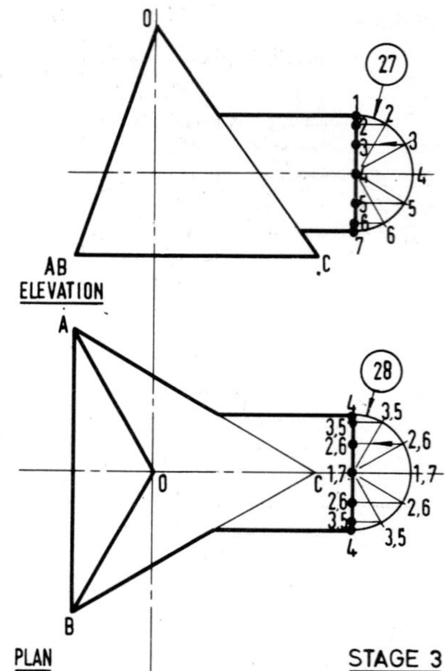

STAGE 4

To draw a nest of horizontal sections through the tetrahedron at the levels of the established equally spaced points around the cylinder.

29. Draw lines of construction thickness horizontally across the elevation from the points 1, 2, 3, 4 etc. to establish the sections being considered.

30. Consider section 3. Project the point 3A where the section line cuts line OC into the plan view.

31. Using a 60°–30° set square, draw a triangle from point 3A with sides parallel to the base. This is the shape at section line 3.

32. Repeat with the other sections from 1 to 7, to build up a nest of sections in the plan view. Number them as shown for reference.

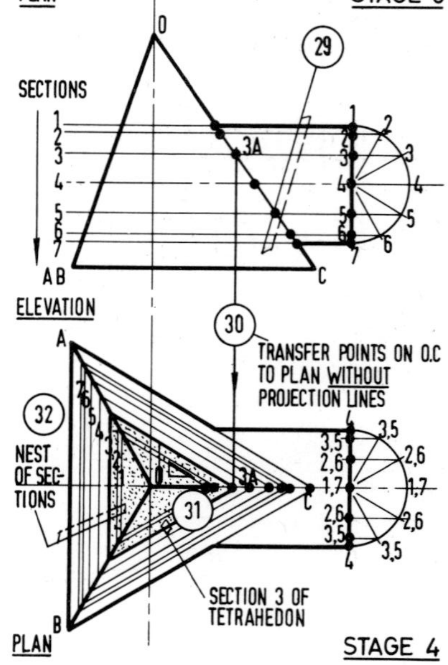

7. Project the datum line vertically upward from point O to the elevation.

8, 9. At a convenient distance from the plan, set up a horizontal base line, and establish points A, B and C on it by projection.

10. From the plan it is seen that line OC will appear in the elevation as a true length. Also OC equals AB. With centre C in the elevation, and AB as radius, strike an arc cutting datum line 7, to establish point O in the elevation.

11, 12, 13. Line in (thick) these three lines, as all points on the tetrahedron are now established.

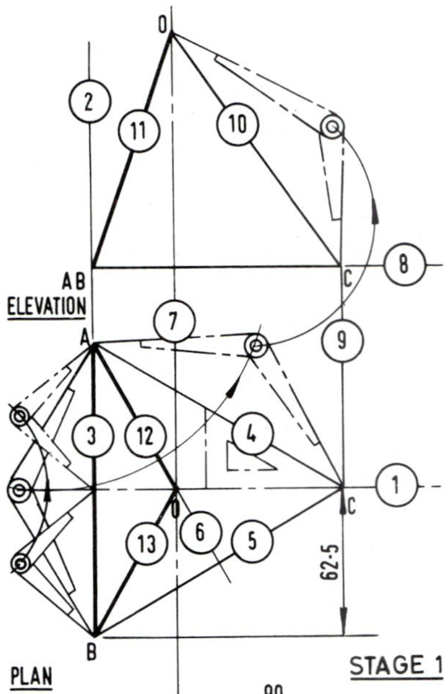

STAGE 2

To add the cylinder of 50 mm diameter in plan and elevation.

14. Draw the horizontal centre line of the cylinder 40 mm from the base.

15. Draw a vertical construction line to establish the end of the cylinder.

16, 17. In the plan, measure the half-width of the cylinder (25 mm radius) to one side of the centre line and transfer it with compasses to the three other positions in the plan and elevation, as shown. Line in between points.

18, 19, 20, 21. Line in the sides of the cylinder in the plan and elevation.

22, 23, 24, 25, 26. Line in the sides of the tetrahedron in both plan and elevation.

The two views thus drawn by construction and projection are related accurately to each other.

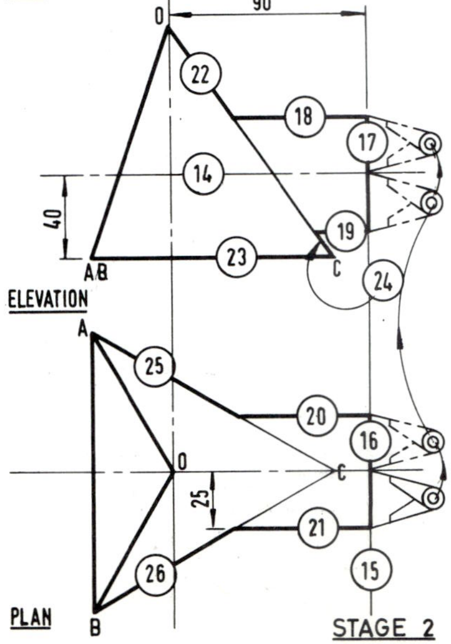

STAGE 5

To establish points on the line of intersection in both plan and elevation.

33. Draw horizontal construction lines in the plan view from the numbers around the cylinder to the corresponding numbered triangular sections and number the joining points accordingly. A series of points on the interpenetration line in the plan view has been established.

34. Project these points to the corresponding sections in the elevation and number accordingly. Join the points with a smooth curve, remembering that the points 4 to 7 in the plan are joined by a dotted or hidden-detail line.

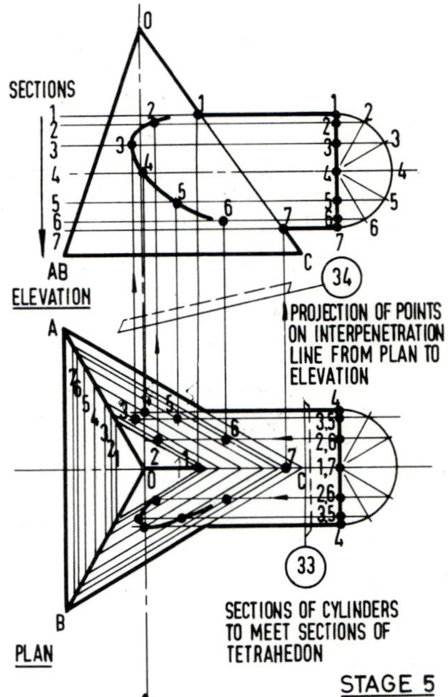

STAGE 6: SOLUTION

The student's drawing in this exercise should appear as in the illustration, with the following features clearly shown:

(a) The visible outline drawn as a thick clean line.

(b) All points and sections suitably numbered with good clear numbers, as shown in the illustration.

(c) All construction, section and projection lines shown as sharp, clean thin lines.

A drawing thus produced conveys clearly to tutor and examiner the student's knowledge of the technique of orthographic projection.

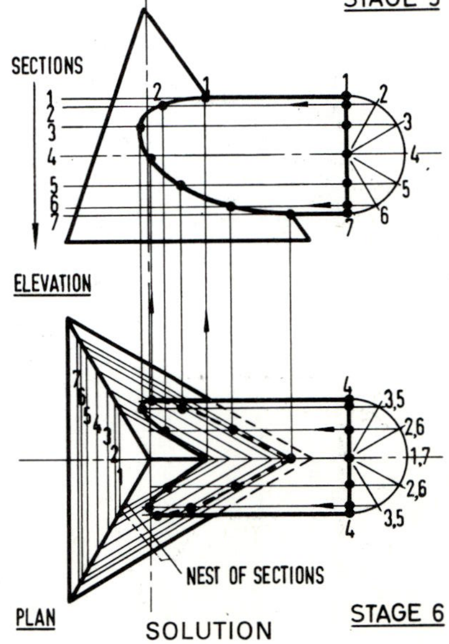

RABATMENT OR FINDING THE TRUE LENGTHS OF LINES

To enable sheet-metal patterns to be laid out on flat sheet stock, the true lengths of all lines must be known. Before considering the techniques of sheet-metal development, a method of finding the true lengths of apparent-length lines in orthographic views must be learned.

The method shown in Figure 9.10 is satisfactory for all straight lines which do not lie in the principal planes of projection, and the simple extra diagram requires the minimum of line work to establish the true length without adding numerous lines across the main orthographic views.

The technique is that of plotting the perpendicular components of a line to find its true length; it is similar to that used in mechanical science when perpendicular components of force are plotted to find the resultant force.

Figure 9.10

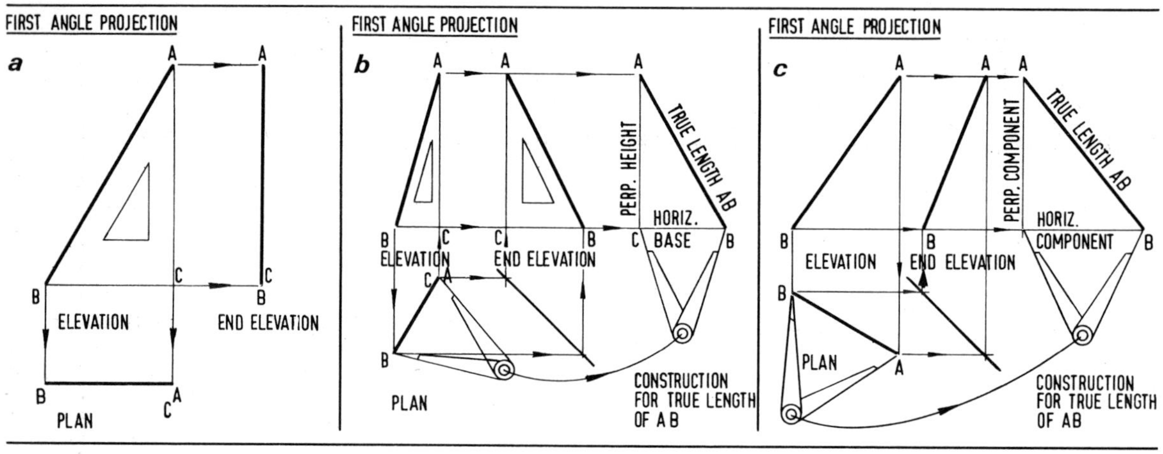

HOW TO FIND THE TRUE LENGTH OF A STRAIGHT LINE FROM ORTHOGRAPHIC VIEWS

Note that only two views are necessary.

Figure 9.10*a* shows an ordinary set square used in drawing, with the long side AB (the hypotenuse) drawn as a thick black line. Consider this side AB.
In the elevation, because it is viewed at right angles, it appears as its true length.

In the end elevation, because it slopes away from the viewer, its apparent length is equal only to the perpendicular height AC.

In the plan, because it slopes down from the viewer, its apparent length is equal only to the horizontal base length BC.

Summary

(a) The hypotenuse has perpendicular components, i.e. two lines at right angles, equal to the perpendicular height AC and the horizontal base BC.
(b) Hence, if the two perpendicular components can be drawn, completion of the triangle will establish the true length of the hypotenuse.
(c) If the line appears horizontal in the plan view, it appears as its true length in the elevation, and vice versa.
(d) If a line appears vertical in the end elevation, it appears as its true length in the front elevation, and vice versa.

Figure 9.10*b* shows three views of the same set square after it has been revolved 60° about line AC. The line AB does not lie in any of the principal planes of drawing, hence its true length is not shown in any view.

To find the true length

(a) Project AC, the perpendicular height of point A above point B, from the elevation.
(b) Transfer the horizontal base length BC from the plan, and draw it at right angles to the perpendicular height.
(c) Complete the triangle by joining A to B. This gives the true length of AB.

Figure 9.10*c* shows three views of a line which does not lie in any of the three principal planes of drawing, hence its true length is not shown in any view.

To find the true length proceed exactly as shown in Figure 9.10*b*

(a) Project the perpendicular component of AB, i.e. the vertical height of point A above point B.
(b) Transfer with dividers the horizontal component of AB, i.e. the length of line AB seen in the plan view, and draw it at right angles to the perpendicular component.
(c) Complete the triangle, and the length of the hypotenuse will be the true length of AB.

PROBLEM

The drawing shows the elevation and plan of an aircraft engine nacelle with the proposed route ABCDEF of an oil pipe. Find the length of the proposed pipe line between points A and F.

METHOD

Find the true length of the pipe between adjacent points, then add them together in a straight line for measurement.

Stage 1
From the elevation project the vertical component (VC) of the pipe length between A and B. From the plan transfer the horizontal component (HC) of the pipe. Complete the triangle to find the true length of AB.

Stages 2, 3, 4 and 5
Repeat the above method to find the true lengths between points BC, CD, DE and EF.

Stage 6
Add all five true lengths together and measure the total length.

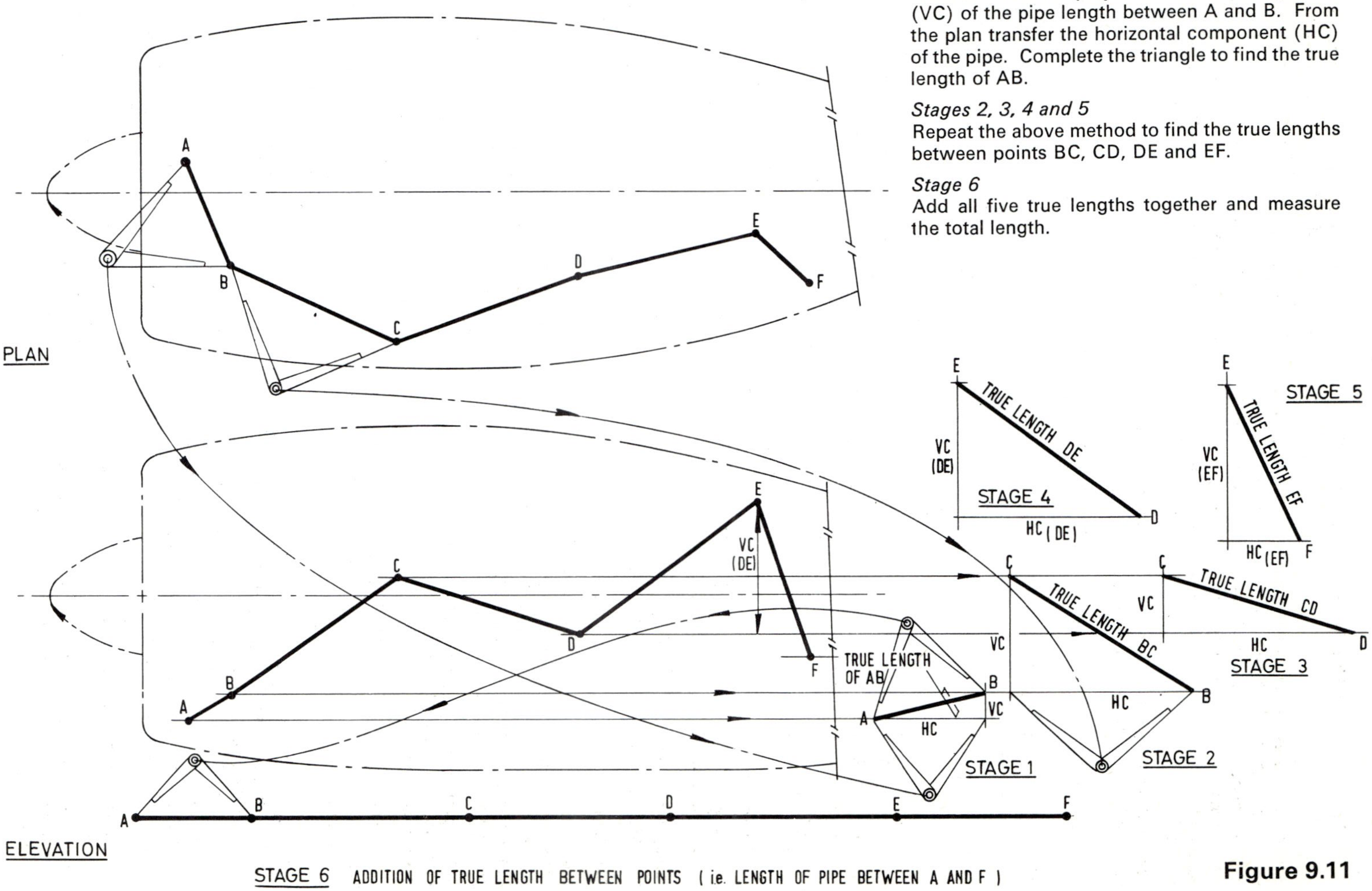

Figure 9.11

SHEET-METAL DEVELOPMENT

This is the technique of producing, from the orthographic views of a three-dimensional sheet-metal fabrication, the correct shape in flat sheet metal that, when folded up or rolled, will produce the exact three-dimensional shape required.

Figures 9.12 to 9.15 show orthographic and pictorial views of simple sheet-metal objects, together with their sheet-metal developments.

When laying out the shape, it should be arranged so that, in folding, the edges of the developed shape to be brought together for joining (or seaming) are as short as possible. The short join requires less time to make and needs less solder or weld material; as a result, less heat is needed when making the joint and so the risk of distortion is reduced.

There are three methods of sheet-metal development—parallel line, radial line, and triangulation. Figures 9.16 to 9.22 show examples of all these methods in developing a methodical approach to these problems.

When drawing a sheet-metal development, it is essential at all times to number or letter the points as plotting proceeds; unreferenced lines and points can easily lead to errors.

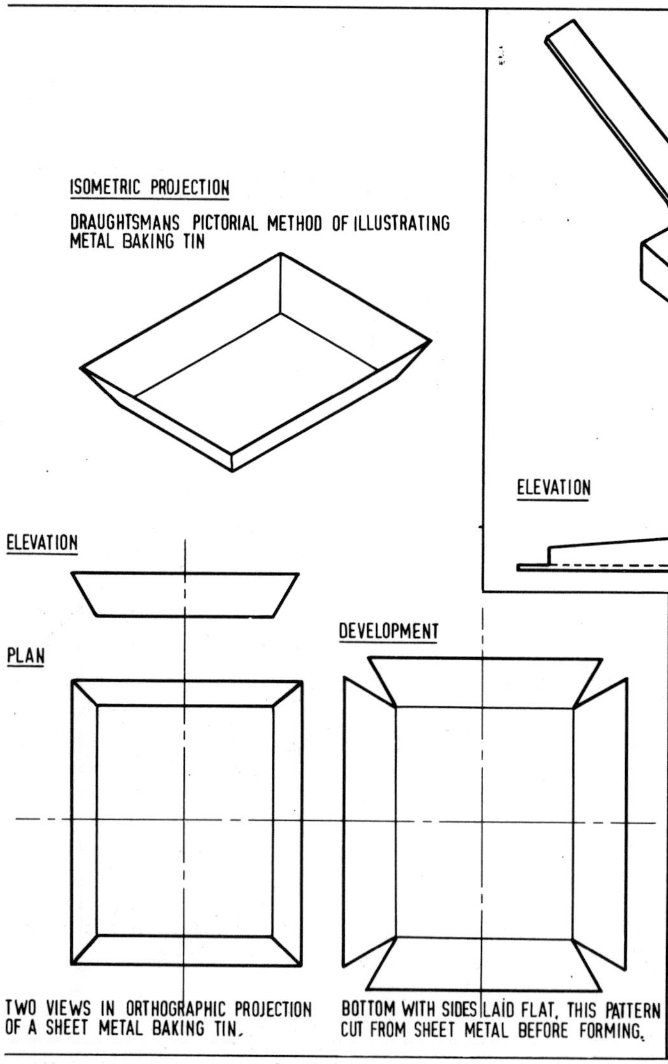

Figure 9.12

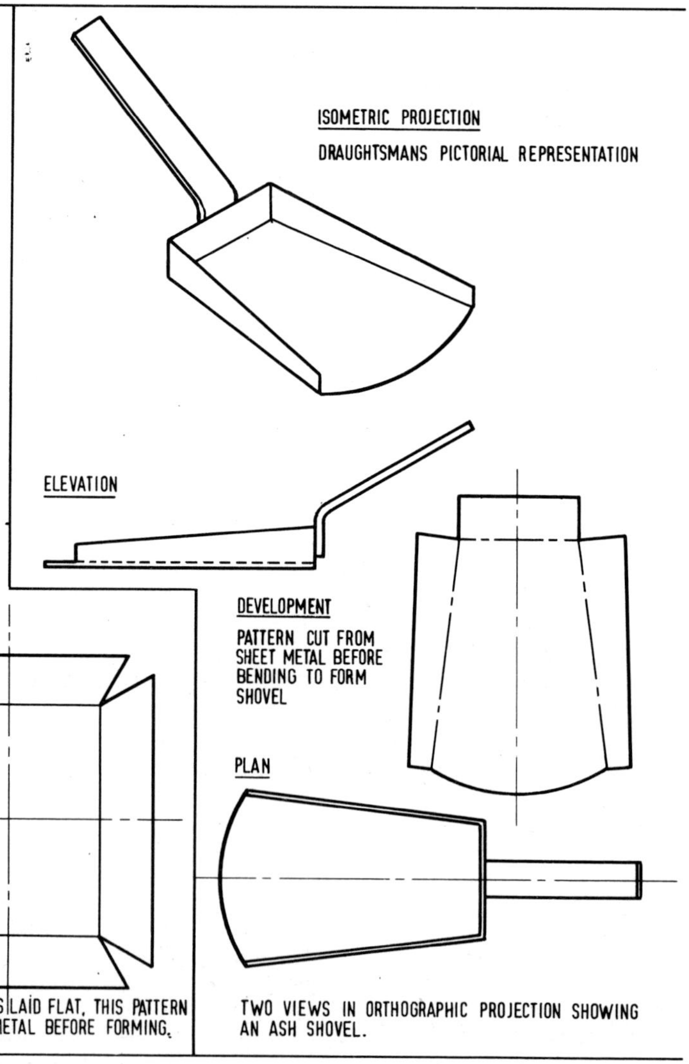

Figure 9.13

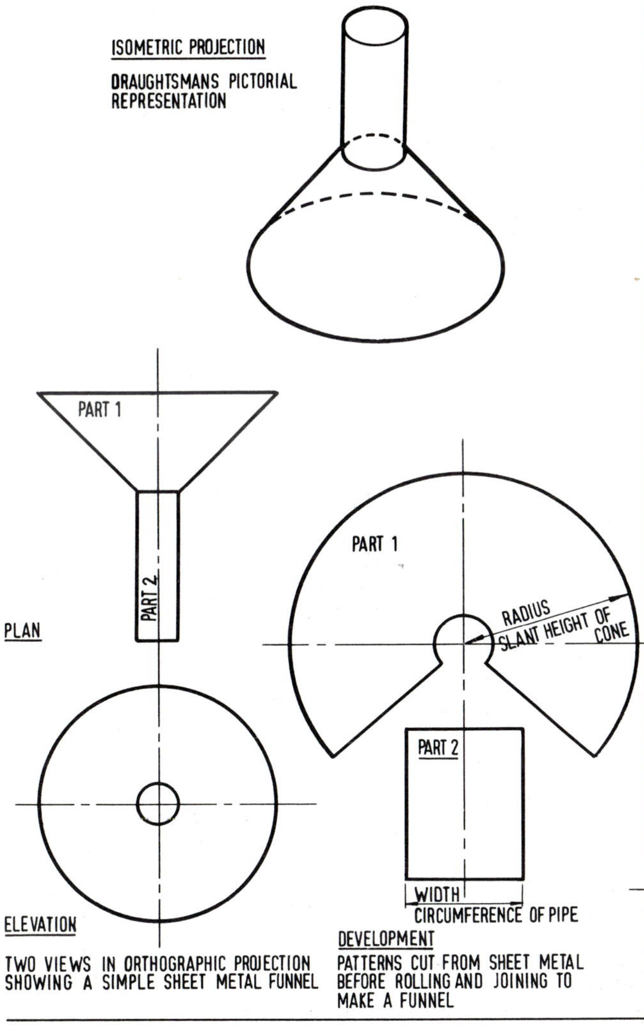

Figure 9.14

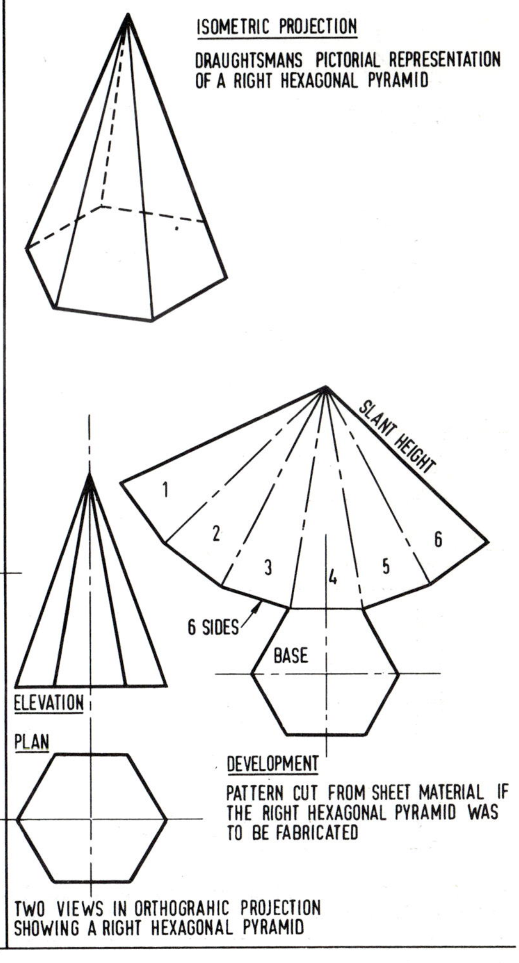

Figure 9.15

SHEET-METAL DEVELOPMENT: PARALLEL-LINE METHOD

PROBLEM 9.16

The drawing shows a plan and elevation of a sheet-metal cylindrical pipe, one end of which is cut off at 45°. Draw the two views and then draw the sheet-metal development.

Points to remember

A. A cylinder can be rolled from a flat rectangular sheet.

B. The width of the sheet is equal to the circumference of the pipe, which is calculated from the formula πd, where $\pi = 3{\cdot}142$ and d = diameter of the pipe.

STAGE 1

To draw the two views in sequence so that they correspond dimensionally.

1, 2. Draw centre lines at right angles.

3. To one side of the intersection of the centre lines, measure off half the pipe diameter (37·5 mm). Take this distance in the compasses and draw a circle to represent the plan view of pipe.

4, 5. Project the diameter of the circle in the plan view to the elevation.

6. Draw the base line of the cylinder between the projection lines.

7. Measure the height of the pipe (125 mm) above the base line and draw a horizontal construction line.

8. Draw the top of the pipe cut off at 45° and then line in the elevation as shown.

By use of projection, the two views correspond dimensionally to each other.

STAGE 2

To produce a series of equally spaced points around the base of the pipe when viewed in the elevation.

9. Divide the plan view into a number of equal parts, using a 60°–30° set square, and number the points on the circumference as shown.

10. Project vertically upwards the numbered points in the plan to the base of the pipe in the elevation.

Points in both views are given corresponding numbers.

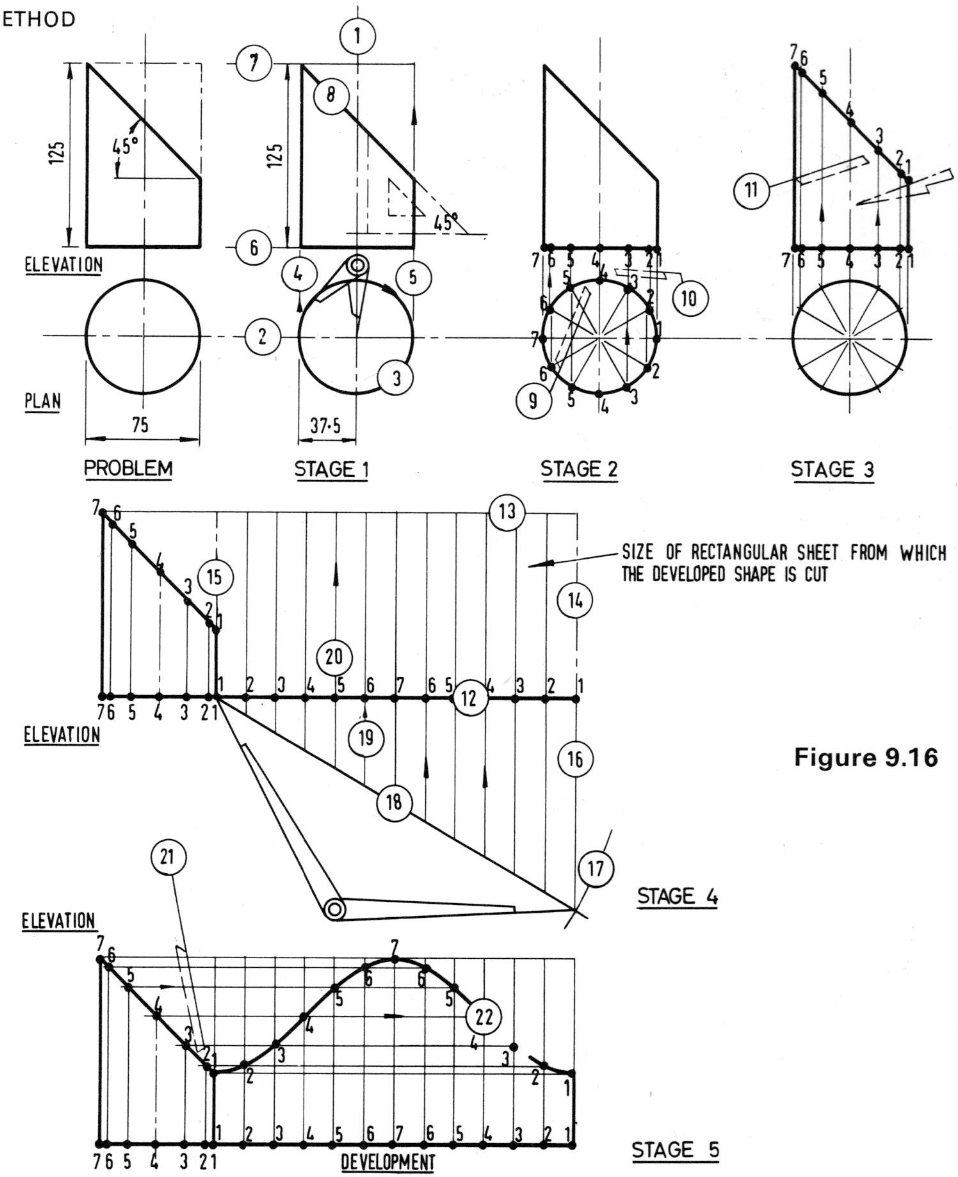

Figure 9.16

STAGE 3

To draw parallel ordinates on the pipe to enable the development to be drawn.

11. From the numbered points on the base draw vertical lines to the top edge of the pipe, numbering also the upper points of the lines.

Note: The lengths of these lines are the lengths of the pipe at equally spaced intervals around the circumference.

The pipe will be cut on the shortest length and opened out.

STAGE 4

To draw the rectangular flat sheet from which the pipe is cut and rolled.

12. Draw from the base of the elevation a line equal in length to the circumference $(\pi d) = 3{\cdot}142 \times 75 = 235{\cdot}65$ mm. Draw as accurately as possible.

13. Project a horizontal line from the top of the pipe to establish the height of the rectangular plate.

14, 15. Draw the sides of the rectangular plate in chain-dot line.

16, 17, 18, 19. Divide the base of the rectangle (line 12) into the same number of equal parts as the plan was divided into in Stage 2 (using the method shown in Figure 4.5).

20. Draw the vertical ordinates from the equally spaced points on the base and number as shown, in the same sequence as they follow around the pipe.

STAGE 5

To draw the final developed shape.

21. Project the length of the ordinates in turn to the correspondingly numbered ordinates on the rectangle, making a pencil dot, and numbering each one as plotted.

22. Join up the series of plotted points, using a french curve, not freehand, and line in the development as shown.

STAGE 6

The student's drawing in this exercise should appear as in the illustration, see stage 4 of Problem 9.17 on page 128.

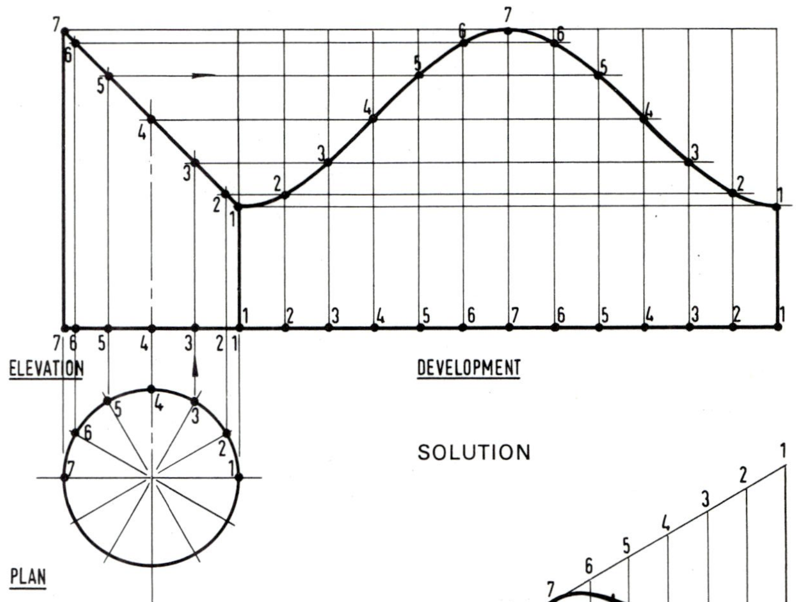

STAGE 6

SOLUTION

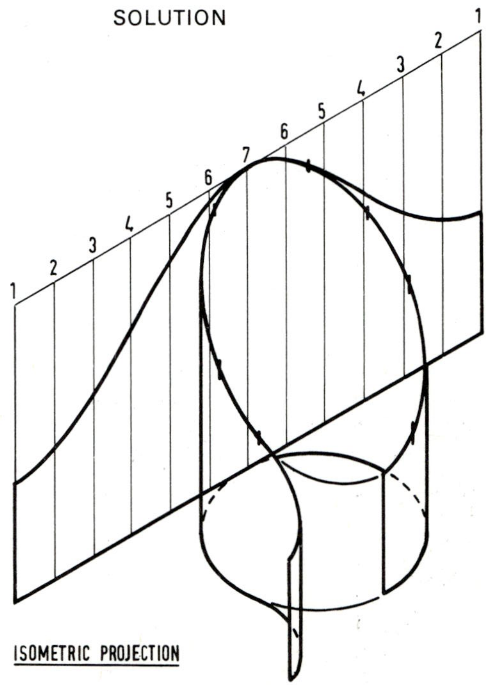

A draughtsman's pictorial representation of the pipe being unrolled against a flat rectangular sheet to produce the developed shape.

PROBLEM 9.17

A sheet metal T junction is made from a pipe of 100 mm diameter and another of 75 mm diameter as shown. Draw the given views, plot the line of interpenetration, and then draw the development of both pipes.

Points to remember

A. Arrange the shortest length for seaming.
B. Do not seam across a hole.

STAGE 1

To draw the given views and plot the line of interpenetration.

Draw the given views by projecting dimensions between views as described in previous exercises.

1, 2, 3, 4, 5. To draw the line of interpenetration refer to Figure 9. 1. Draw a smooth curve of interpenetration through the plotted points with a french curve.

STAGE 2

To draw the development of the branch pipe of 75 mm diameter.

Note: As the pipe development will be symmetrical about a centre line, it is practice to draw only half, or slightly more than half, of the development.

6. From the square end of the pipe project a constructional line at right angles.

7. Draw a base line a little longer than half the circumference of the junction pipe. Erect a perpendicular centre line at exactly the half-distance.

8. Divide the half base line into six equal parts to correspond to the numbered divisions around the pipe in the elevation, and erect ordinates. Number the ordinates, starting with the shortest as seen in the elevation, i.e. 0 0. Always add an extra division beyond the centre line to enable a more accurate curve to be drawn at the middle.

9. Project the lengths of ordinates in the elevation to the corresponding ordinates on the development, numbering them as they are plotted.

10. Join the plotted points to give the developed shape of the junction pipe, and line in as shown.

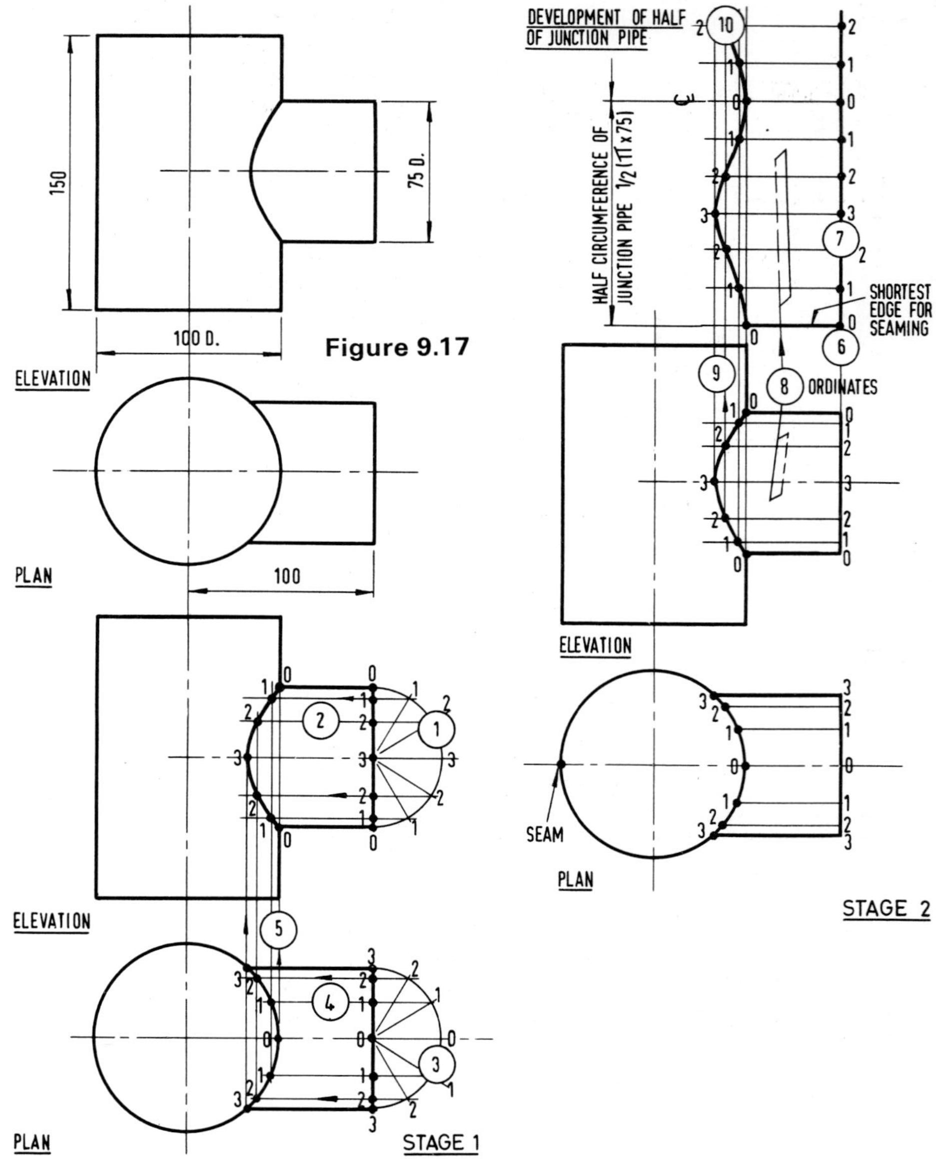

Figure 9.17

STAGE 3

To draw the development of the pipe of 100 mm diameter, including the hole for entry of the junction pipe.

Note: Draw the development so that the hole is complete and not split by the seam.

11. Draw vertical ordinates in the elevation through the numbered points on the intersection line, and give them the same numbers as the points.

12, 13. Project the length of the 100 mm diameter pipe on to the development.

14, 15, 16. Draw the rectangular sheet from which the 100 mm diameter pipe can be rolled, height between lines 12 and 13, and width by calculation of the length of the circumference ($\pi \times 100$ mm). Only a half-width plus enough to accommodate the full hole need be drawn. Erect a centre line at the half-width.

17. Erect ordinates both sides of the centre line, at positions fixed by transferring the distances between ordinates from the plan view with the compasses. Number the ordinates with the corresponding numbers as the transfer takes place.

18. By horizontal projection transfer the numbered points on the interpenetration line in the elevation to the corresponding numbered ordinates in the development.

19. Through the plotted points draw a smooth curve to give the developed shape of the required hole.

STAGE 4

The student's drawing in this exercise should appear as in the illustration, with the following features clearly shown:

(a) The visible outline drawn as a thick clean line.
(b) All points suitably numbered with good clear numbers as shown.
(c) All construction and projection lines clearly shown as sharp, clean thin lines.

A drawing thus produced conveys clearly to the tutor or examiner the student's knowledge of the technique of orthographic projection.

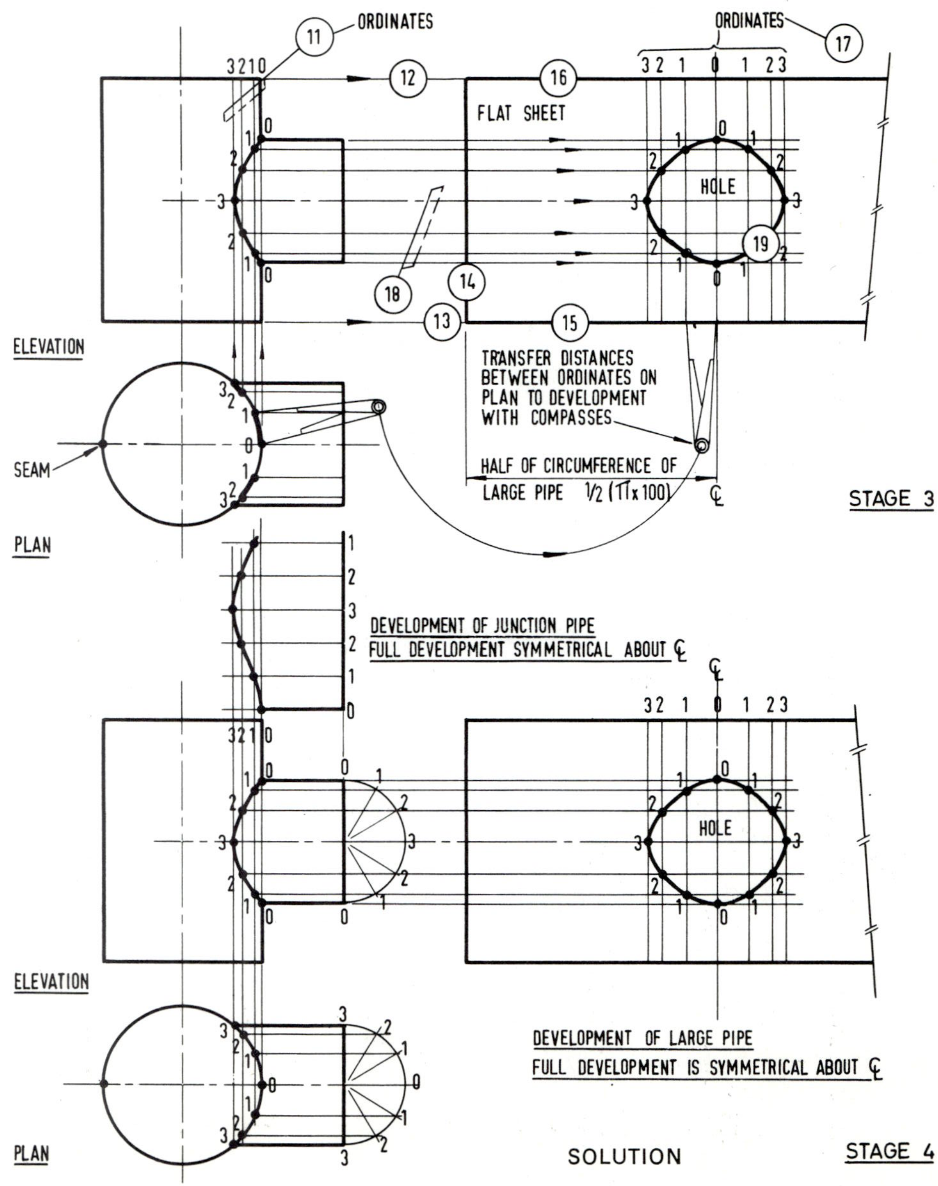

PROBLEM 9.18

A sheet-metal offset T junction-piece is made from a pipe of 100 mm diameter and another pipe of 75 mm diameter as shown. Draw the given views, plotting the line of interpenetration. Show the rear part of the interpenetration line hidden from view as a dotted hidden-detail line. Then draw the development of both pipes.

Points to remember

A. Arrange the shortest length for seaming.
B. Do not allow the seam to pass through a developed hole.

STAGE 1

To draw the given views and plot the line of interpenetration, and then draw the development of the offset junction pipe.

Draw the given views by projection of the dimensions between views as described in previous exercises.

To draw the curve of interpenetration.

1, 2. Add a half end elevation to the junction pipe, and divide it into 6 equal parts and number the points as shown. Project the points to the edge of the pipe, and produce horizontal projection lines to the centre line.

3, 4. Repeat as above, in the plan view, giving corresponding points the same number. Draw the numbered horizontal projection lines on the junction pipe until contact is made with the large pipe. Number these points as shown.

5. Project the points to the identically numbered projection lines in the elevation, numbering each point as plotted.

6. Join the plotted points with a smooth curve, the rear part being shown as hidden detail.

To plot the development.

7. Draw a line equal in length to the circumference of the junction pipe ($\pi \times 75$ mm).

8. Erect a perpendicular at one end of the line equal in length to the longest ordinate (3 3) in the elevation.

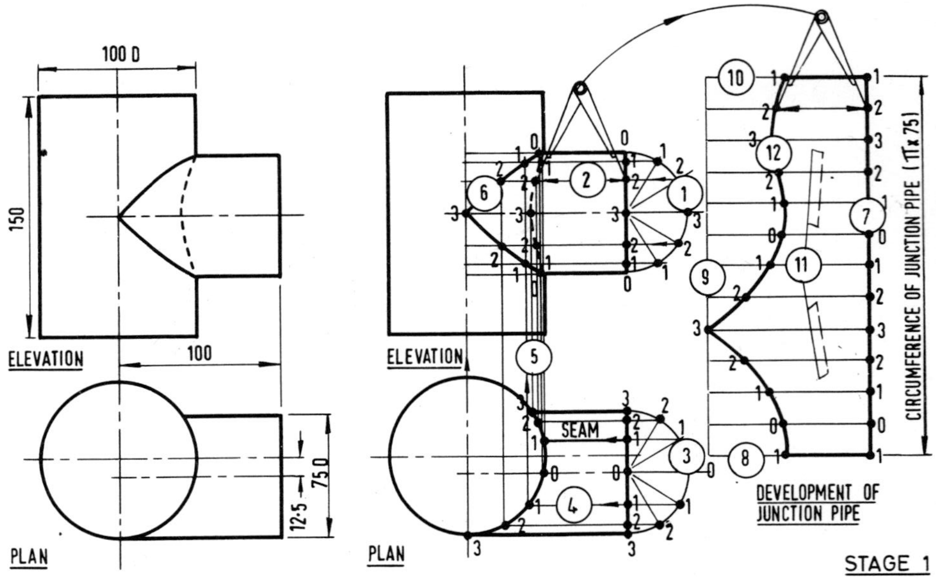

Figure 9.18

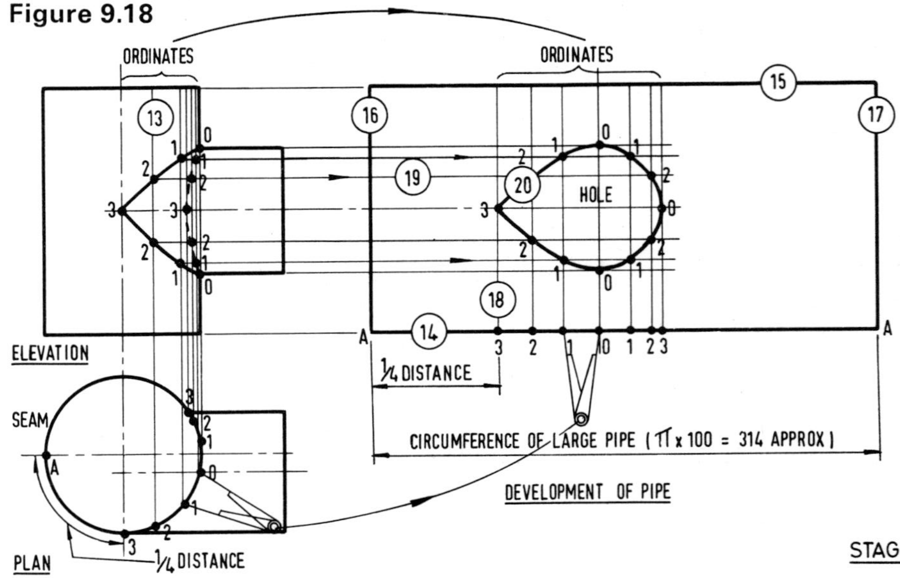

9, 10. Complete the rectangle on sides 7 and 8. This is the size of sheet from which the development is cut.

11. Erect ordinates by dividing line 7 into twelve equal parts. Select the shortest ordinate in the elevation (1 1 at rear) and transfer the length to the first ordinate of the development. Transfer the rest of the ordinate lengths to the development in order of rotation, numbering them at the same time.

12. Draw a smooth curve through the plotted points to complete the development of the junction pipe.

STAGE 2

To draw the development of the 100 mm pipe.

13. Draw ordinates through the plotted points on the interpenetration line from top to bottom of the elevation.

14, 15, 16, 17. Draw a rectangle having width equal to the length of the circumference of the pipe of 100 mm diameter ($\pi \times 100$ mm) and height projected from the elevation.

18. Erect the first ordinate (3) at a point a quarter of the length along the base of the rectangle. It can be seen from the plan view that it is situated a quarter of the way round the circumference from the seam. Transfer the chordal distances between ordinates in sequence (as seen in the plan to the development) numbering them as plotting proceeds. Erect ordinates at these numbered points.

19. Project points from the elevation to the correspondingly numbered ordinates in the development.

20. Join the plotted points with a smooth curve to give the developed shape of the hole.

STAGE 3

The student's finished drawing should appear as the illustration, with the following features clearly shown:

(a) The visible outlines drawn as thick, clean lines.
(b) All points clearly numbered as shown.
(c) All construction and projection lines shown as thin, clean lines.

A drawing thus presented conveys clearly to the lecturer or examiner the student's knowledge of the technique of orthographic projection.

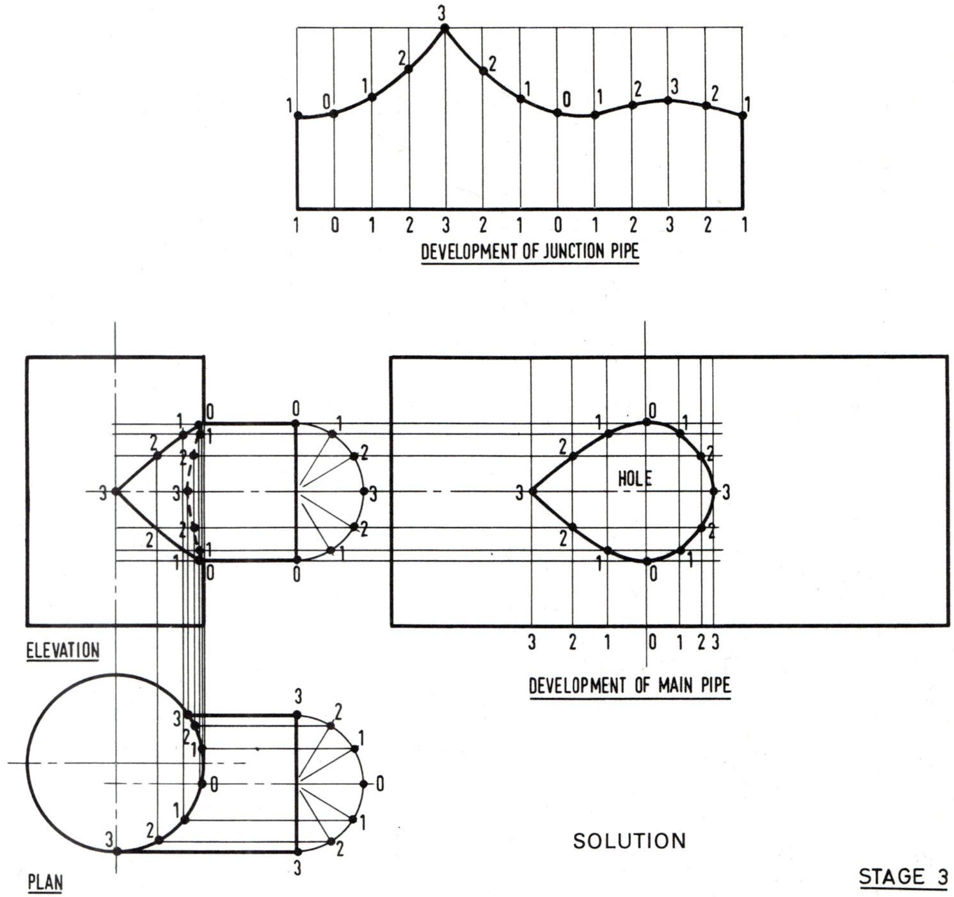

SOLUTION

STAGE 3

SHEET-METAL DEVELOPMENT:
RADIAL-LINE METHOD

PROBLEM 9.19

A plan and elevation are shown of a right cylinder obliquely truncated at 30°. Draw as much of the two views as is necessary, and draw the development.

Points to remember

A. There is no need to complete the plan view. The development could be achieved by drawing only the elevation and a half-plan attached, for plotting equispaced points around the base as seen in the completed exercise.

B. It is necessary to find true lengths of cut generator lines. The only true-length lines seen in the elevation are the sides. Lines drawn elsewhere from apex to base are apparent lengths only. If the cone were revolved about the axis, the lines, on reaching the side position, would appear as true lengths. True length of line is found, when drawing, by projecting the upper point of the lines to the slant sides (see Stage 3).

C. If the cone were laid on its side and revolved with the apex held stationary, a circle of slant-side radius will be traced out by the base. The sector of the circle traced out by one revolution of the cone base will be the development of the full cone. The angle of the sector (see Stage 2) can be calculated.

STAGE 1

To draw the plan and elevation of the cone so that the views will correspond dimensionally with each other.

1, 2. Draw centre lines intersecting at right angles.

3. Measure half the diameter (25 mm) to one side of the intersection. Take this distance in the compasses and draw the circle representing the base of the cone.

4. At a suitable distance from the plan, draw a base line for the elevation, of construction-line thickness.

5. Measure off the vertical cone height and draw another horizontal construction line cutting the vertical centre line at A.

6, 7. Project the exact diameter of the plan to the elevation base line.

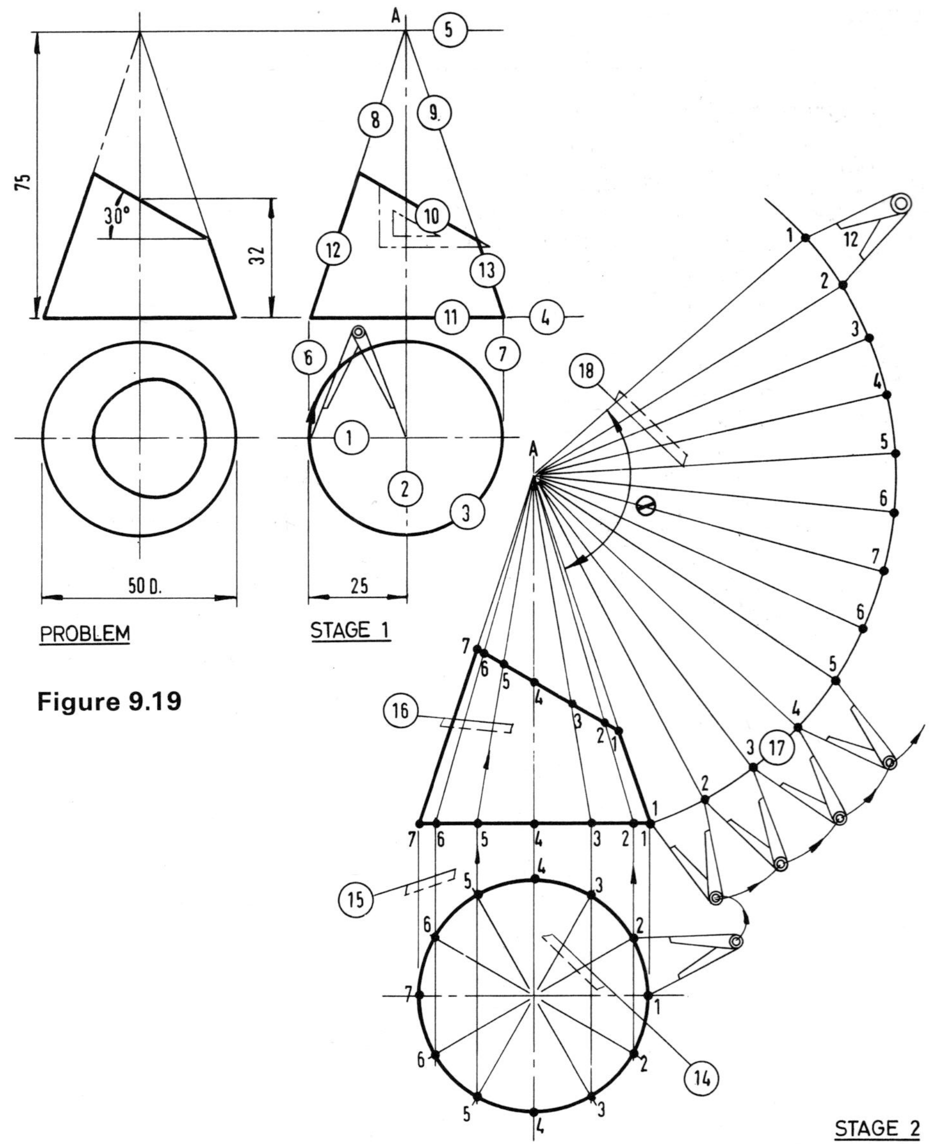

Figure 9.19

8, 9. Construct the elevation of the full cone between points on the base and the apex A.

10. Measure 32 mm up the centre line from the base and, using a 60°–30° set square, draw the top line of the figure.

11, 12, 13. Line in the remainder of the truncated cone in the elevation.

STAGE 2

To lay out the development of the full cone.

14. Divide the plan into twelve equal parts, using a 60°–30° set square, and number the points as shown.

15. Transfer the numbered points by projection to the base of the cone in the elevation. Number the points as in the plan view.

16. Join the numbered points on the base to the apex, producing a series of generator lines all equal in length and equally spaced around the surface of the cone. Number the points where they cross the truncated surface as shown.

17. With the slant height as radius, draw an arc attached to the side of the elevation. This represents the path traced out by the rolling cone.

18. Using the compasses, take from the plan view the base chordal distance between generators and step off 12 divisions along the circular arc, numbering them in the rotational sequence as numbered in the plan view. Join the apex A to the numbered points. A full development of the cone has been laid out, complete with the generator lines.

Note: The angle θ can be calculated from the following formula, and used as an alternative method of laying out the development:

$$\theta = \frac{360^\circ \times \text{radius of cone}}{\text{slant height of cone}}$$

STAGE 3

To ascertain the true lengths of the generator lines around the obliquely truncated cone.

19. The true lengths of these generator lines can only be seen when viewed at right angles. This can be

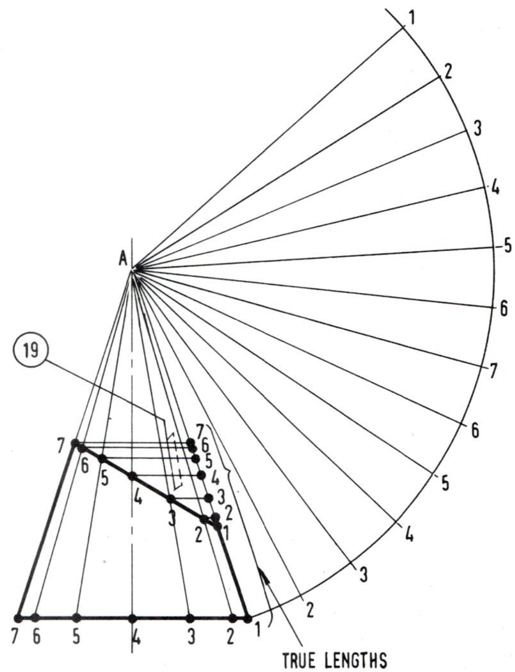

STAGE 3

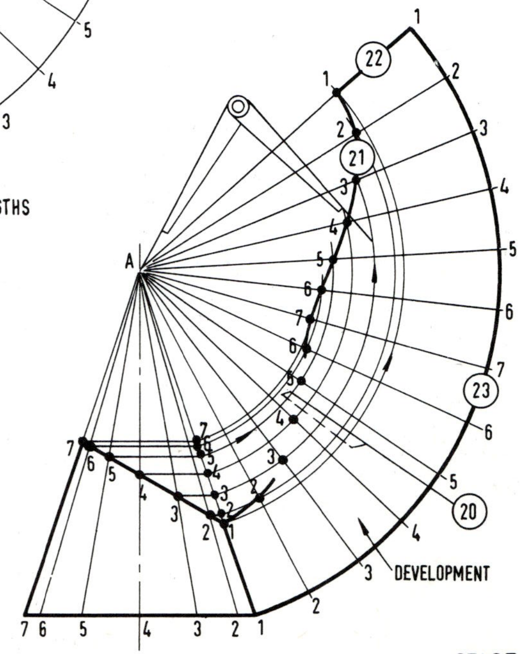

STAGE 4

achieved by revolving the cone about its axis to bring them to the side position.

As rotation is impossible on a drawing, the same result can be obtained by projecting the points horizontally along their rotational paths to the slant side.

The true lengths are thus projected on the slant side of the cone and numbered correspondingly.

STAGE 4

To draw the development of the truncated cone by plotting the lengths of the generators on the development of the full cone.

20. With centre A, and compasses set to radii A7, A6, A5 . . . A1, draw arcs transferring points 1 to 7 to the generators of the same number, thus transferring the generator lengths in sequence.

21. Using a french curve, join the plotted points with a smooth curve.

STAGF 5

The illustration shows how a student's drawing should appear, using the minimum of lines, with the following features clearly shown:

(a) The visible outline as a bold clean line.
(b) All points suitably numbered with good clear numbers.
(c) All projection and construction lines shown as sharp, clean, thin lines.

A drawing thus produced conveys clearly to the tutor or examiner the student's grasp of the technique of projection and development.

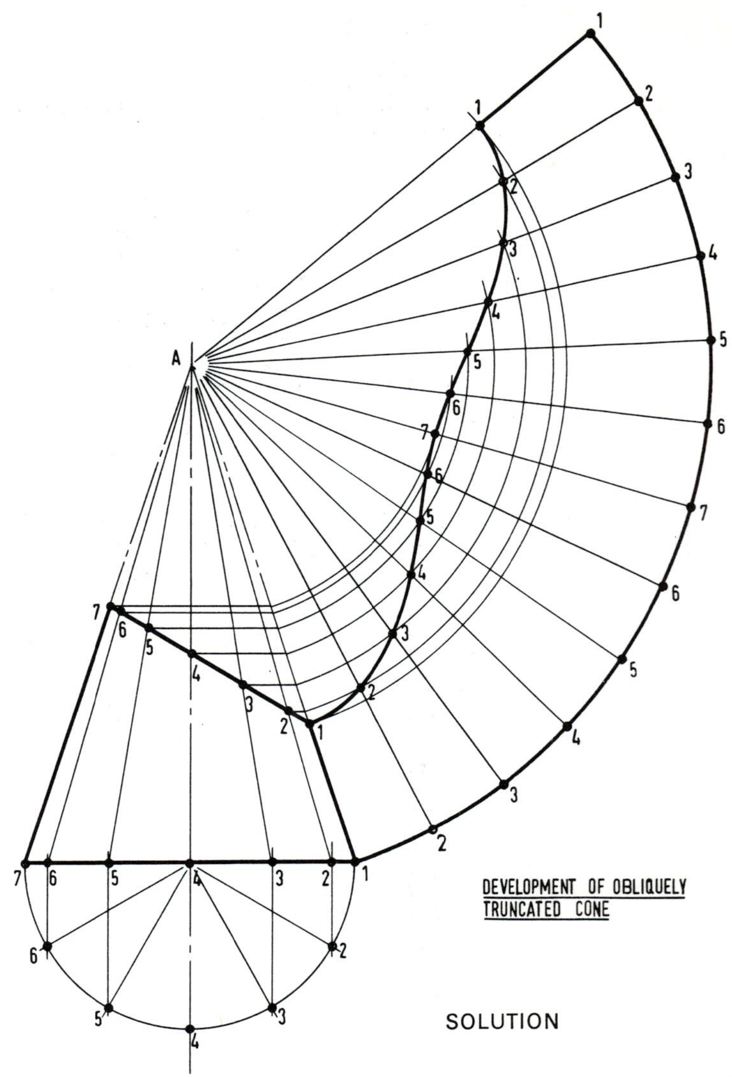

SOLUTION

STAGE 5

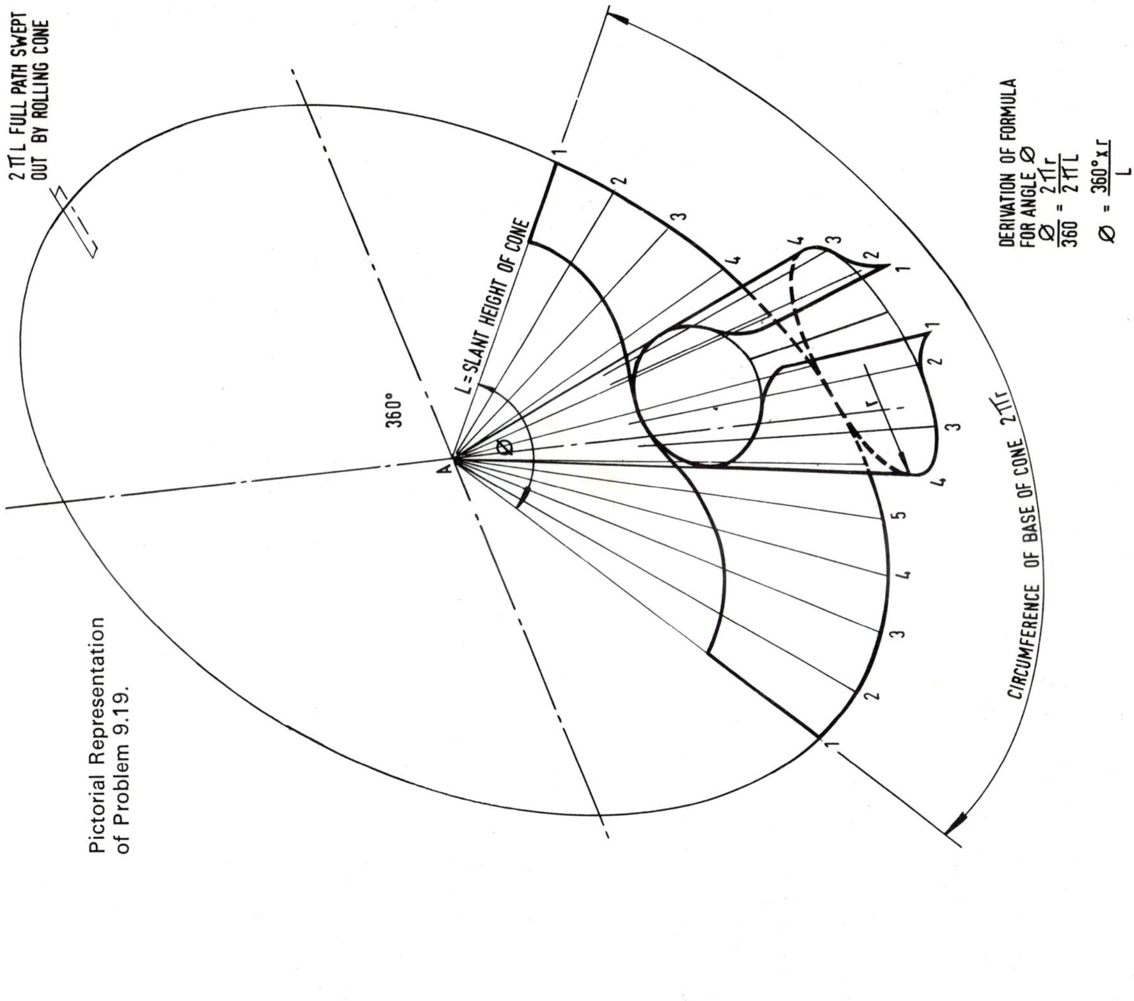

Pictorial Representation of Problem 9.19.

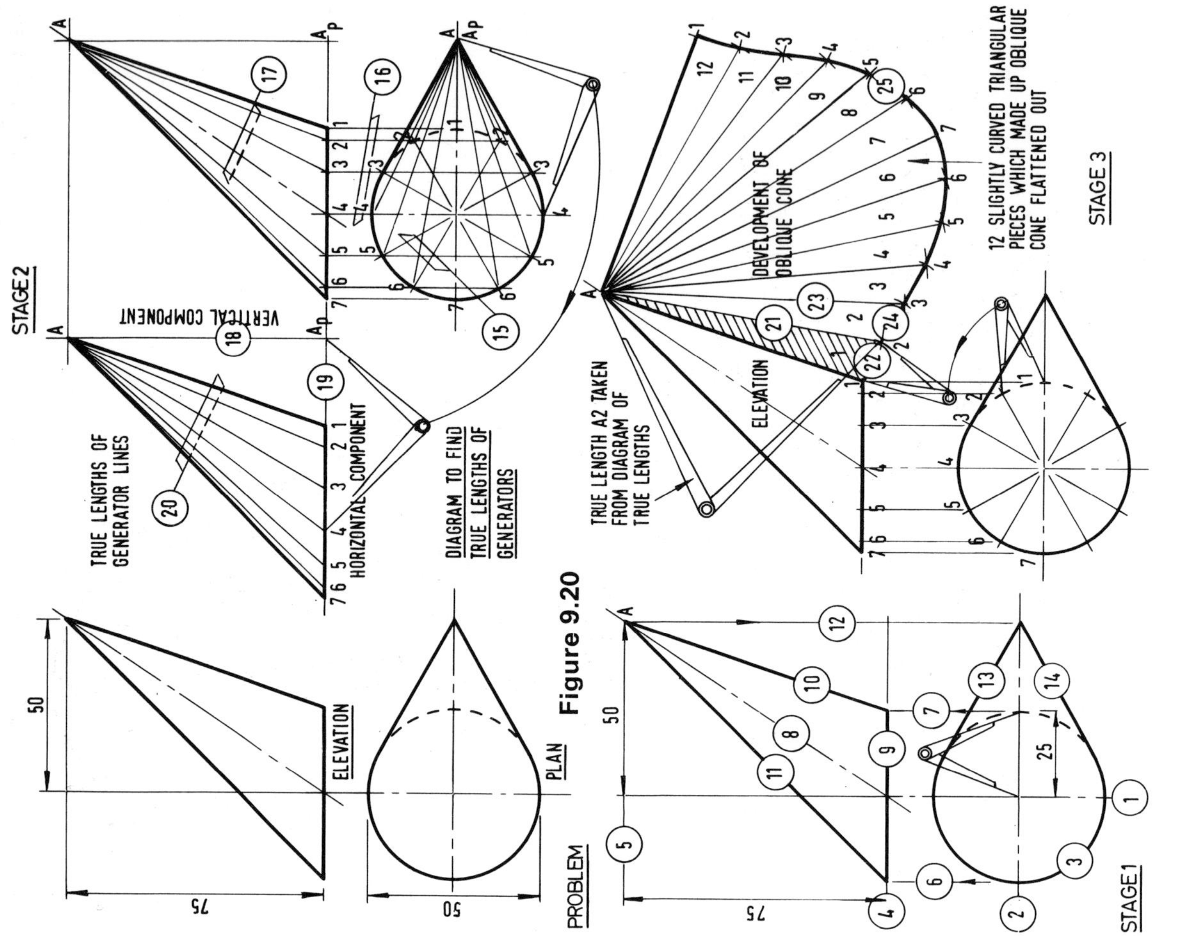

Figure 9.20

PROBLEM 9.20
A sheet-metal oblique cone is of the dimensions shown. Draw the elevation and plan, and then draw the sheet-metal development.

Points to remember

An oblique cone is *not* an ordinary cone with the bottom cut at an angle. The base is circular and all sections parallel to the base are circular.

STAGE 1

To draw the plan and elevation so that the two views correspond dimensionally.

1, 2. Set up horizontal and vertical intersecting centre lines.

3. Measure half the diameter of the base of the cone from the intersection and, taking this as radius, draw a circle to represent the plan view of the base of the cone.

4. At a convenient distance from the plan, set up the base line for the elevation.

5. Draw a parallel line above the base line at a distance equal to the vertical height.

6, 7. Project precisely the diameter of the base of the cone in the plan to the base line in the elevation.

8. Measure the offset of the apex A from the vertical in the elevation and draw the axis of the cone.

9, 10, 11. Line in the outline of the cone.

12. Project the apex A to the plan view.

13, 14. Join the apex tangentially to the base circle in the plan view to complete the two views.

STAGE 2

To draw generator lines equally spaced around the cone in the plan and elevation, and to ascertain their true length.

15. Divide the circular base of the cone in the plan into 12 equal parts, numbering the circumference points as shown.

16. Project these points to the elevation and number them identically.

17. In both plan and elevation join the numbered base points to apex A.

To ascertain the true lengths of the generators before transfer to development. Refer to the method of ascertaining the true length of a line (*Figure 9.10*).

18, 19. Set up a diagram by drawing a right angle with the vertical leg equal to the perpendicular height of cone A Ap. With the compasses, transfer the lengths of the generators as seen in the plan to the horizontal leg of the right angle, numbering them as transfer proceeds. The drawing shows the transfer of Ap 4.

20. Join point A in the diagram to the numbered points to give the true length of the generators.

STAGE 3

To draw the development of the surface of the oblique cone.

It is practice to attach the development to the shortest side of the oblique cone in the elevation. The method is to construct a series of adjacent triangles, using the chordal distance between the generators, taken from the plan, and the true lengths of the generators.

21, 22. Starting with the true-length generator A1 as base, construct the first triangle, using true-length generator A2 and the chordal distance between points 1 and 2 in the plan as the other two sides.

23, 24. Using generator A2 as base, construct No. 2 triangle, using the true-length generator A3 and the chordal distance between points in the plan for the other two sides. Repeat this process until the 12 triangular sections of the cone have been constructed.

25. Using a french curve, draw a smooth curve through the numbered points of each triangle to produce the developed shape.

STAGE 4: SOLUTION

The illustration shows how a student's finished drawing should appear, with the following features clearly shown:

(a) The visible outline as bold, clean lines.
(b) All points suitably numbered with good, clean numbers.
(c) All projection and construction lines shown as sharp, clean, thin lines.

A drawing thus produced conveys clearly to the tutor or examiner the student's knowledge of the technique of the development.

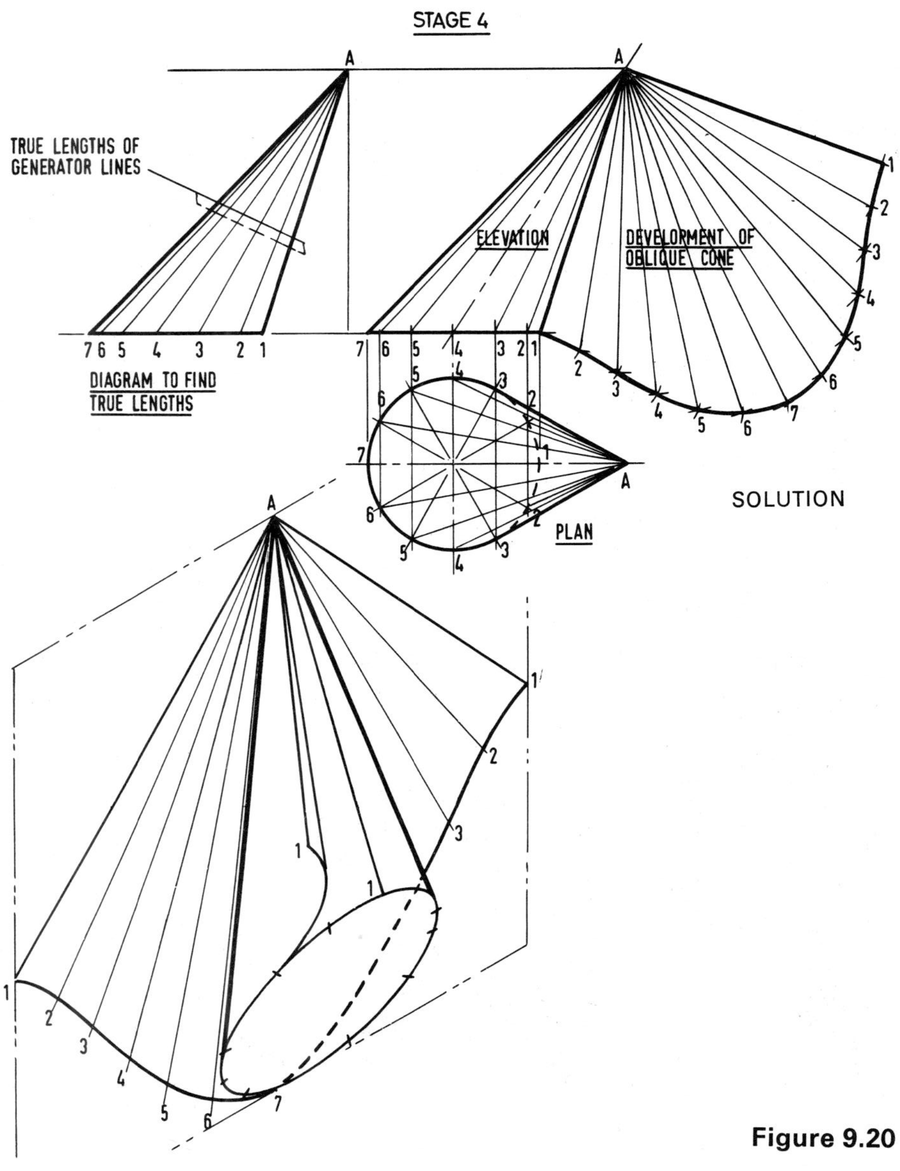

Figure 9.20

PROBLEM 9.21

To join round sheet-metal piping to ducting of square section, a transformation piece is necessary. The drawing shows a plan and elevation of this sheet-metal component. Draw the two views as shown and then plot the sheet-metal development. Use scale 1 :2.

Points to remember

(a) The transformation piece is made up of four identical triangular flat sides (M, N, O and P) each having one of the sides of the square end as its base.
(b) The corners (1, 2, 3 and 4) between these triangular side sections must be curved to give the circular entry for the round pipe.
(c) The seam is down the centre of one triangular side (O), where it is easiest to make and also shortest in length.

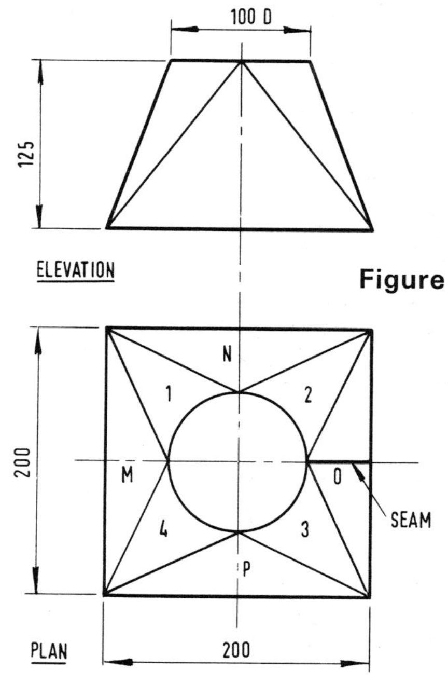

Figure 9.21

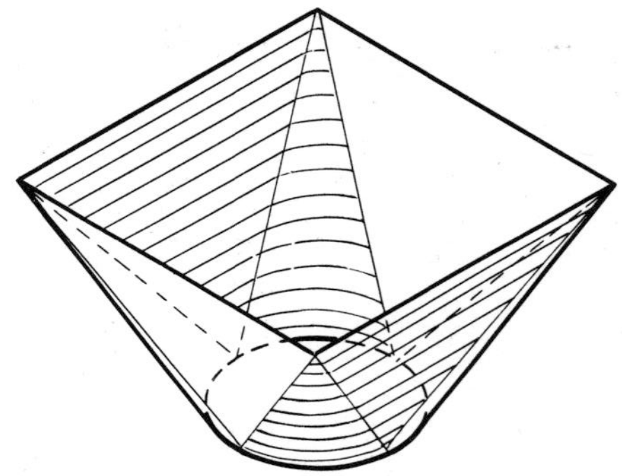

Pictorial Illustration of the Transformation Piece in Problem 9.21

To produce the sheet-metal development of the transformation piece.

The method is to divide the surface area into a number of triangular pieces, and to plot their true shapes adjacent to each other. Note that in this development, due to the symmetry of the transformation piece, only two true-length sides have to be found. The lengths of AO, A3, BO, B3, CO, C3, DO and D3 are all equal, and the lengths of A1, A2, B1, B2, C1, C2, D1 and D2 are also all equal.

1. Divide the circular end to the transformation piece into 12 equal parts, numbering the points as shown.

2. Join the points A, B, C and D to the points 1 and 2 in the respective corner sections. The surface is now divided into suitable triangular areas for plotting the development.

3, 4, 5. Draw a diagram for finding the true lengths of the lines by erecting a right angle with the vertical leg equal to the perpendicular height of AO and A1, and a horizontal leg to which the lengths of OA and A1, taken from the plan, are transferred by compasses.

6. By completing the triangle on the vertical and horizontal components of lines OA and A1, the true lengths of the lines are obtained.

To lay out the development

7, 8, 9. *To draw the true shape of the triangle M,* draw AB (the true length is obtained from the plan) and letter the ends. Complete the triangle by construction, using the true lengths of AO and BO.

10. Attach triangles Q, R and S in sequence, using for construction the rèlevant true length line and the chordal distance between points on the circle in the plan view. Continue plotting the remainder of the development by attaching the other triangular sections in the order in which they appear on the plan view. Number all points as the layout of the development proceeds. The numbered points 0 to 3 for the corner sections are joined with a smooth curve.

11, 12. The final triangular sections which are brought together for seaming are attached to the lines CO and DO in the development. To construct these triangles, the true length of line EO is taken directly from the elevation, and the true lengths of lines CE and DE are taken directly from the plan view.

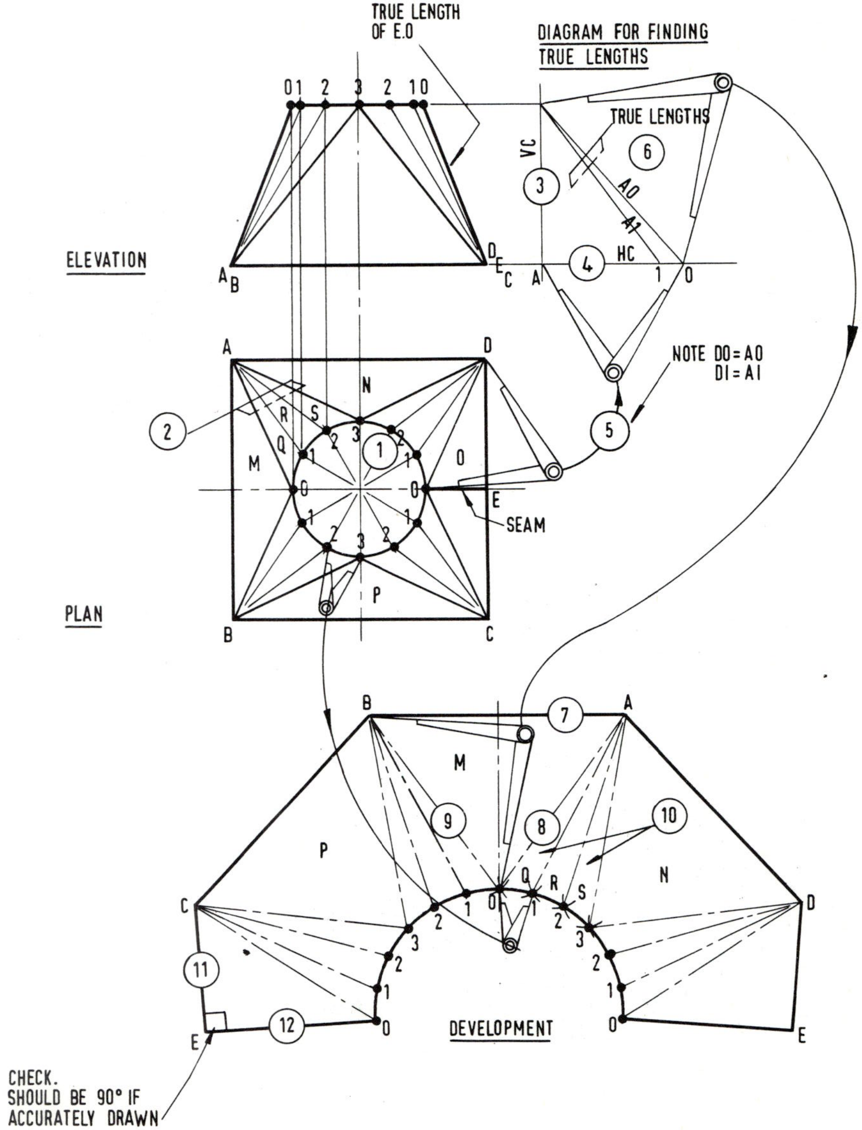

PROBLEM 9.22

The drawing shows two views of a simple sheet-metal hopper with one side vertical. Draw the two views as shown and an extra view in the direction of arrow A.

Then plot a sheet-metal development with the seam in the position shown. Use scale 1 : 2.

Points to remember

A. The view in the direction of arrow A will give the true shape of that side.

B. The side BCON is a trapezium symmetrical about the centre line. Knowing the length of the lines BC and NO, and the true distance between them, obtained from the elevation, there is no difficulty in plotting this side.

C. The two remaining sides are identical, unsymmetrical trapeziums. Although the lengths of the four edges are quickly found, any number of shapes can be drawn using these dimensions, as shown.

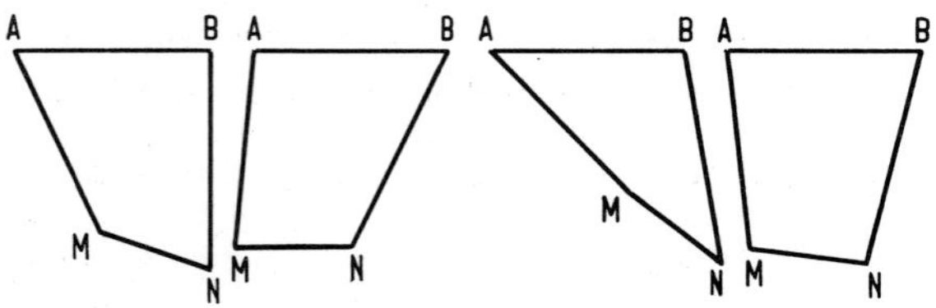

The method adopted to draw these sides is to insert on the plan view a diagonal, dividing the side into two triangles. The triangles are then accurately plotted adjoining each other to give the required shape.

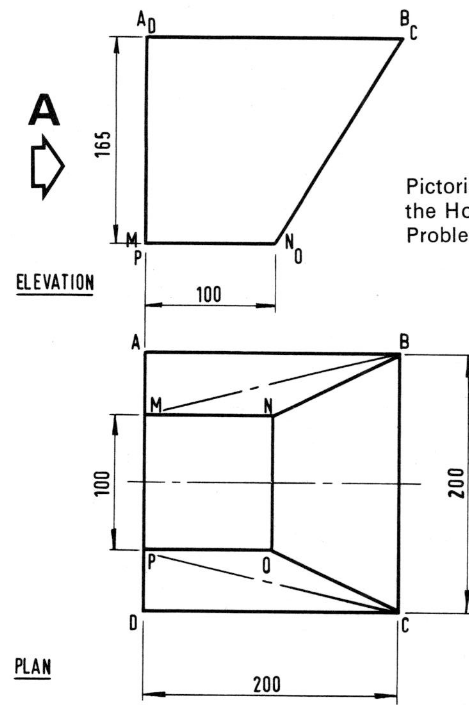

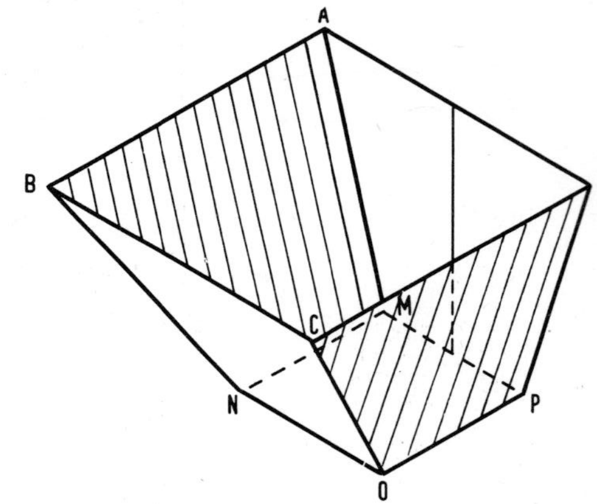

Pictorial Illustration of the Hopper in Problem 9.22

Figure 9.22

To produce the sheet-metal development of the hopper

Accurately draw the three views by projection, and then insert, as chain-dotted lines, the diagonals as shown in the end elevation and plan. The development is symmetrical about a centre line. Letter all points as the development is constructed.

1. Construct the trapezium BCON symmetrically about a centre line, using the true lengths for BC and ON (taken from the plan) and the true height between BC and ON (taken from the elevation by compasses).

2. Construct triangle 2, attached to the above figure on side BN. Length MN is a true length taken from the plan and BM is found from the construction of the true-length diagram.

3. Add to triangle 2 the triangle 3 by construction. Length AB is a true length taken from the plan and the length of AM is taken from the end elevation where it is seen as a true length.

4. Add triangle 4 on side AM. The lengths of AY and AM are transferred by compasses from the end elevation, which shows the true lengths.

5. Complete one half of the development up to the seam by adding triangle 5 to line AY. The lengths AX and XY are also taken from the end elevation.

A check on accuracy of construction can now be made. The angles at X and Y should be right angles.

The other half of the development is a mirror image of the half constructed above.

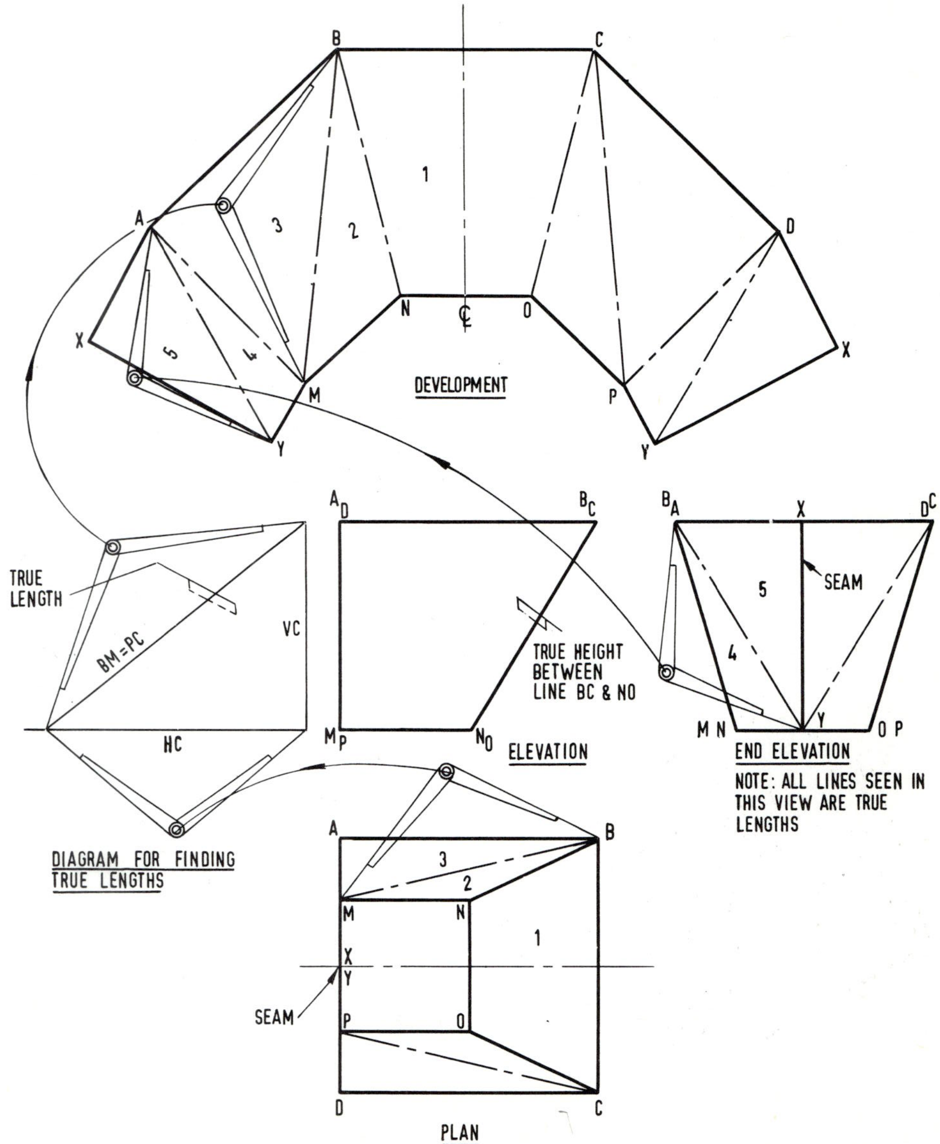

Nearly all the engineering drawing for manufacture in industry is produced in orthographic projection. This method of drawing views projected from each other in the principal planes of projection, and where necessary in supplementary auxiliary planes, is a precise method; it allows solid shapes and structures to be proved on paper before expensive production takes place.

However, the draughtsman requires at times to be able to produce a pictorial view of an assembly or detail part. This may be necessary for clarification on assembly or erection where there are many vertical and horizontal members close together, or for piping runs etc. (Figure 10.1).

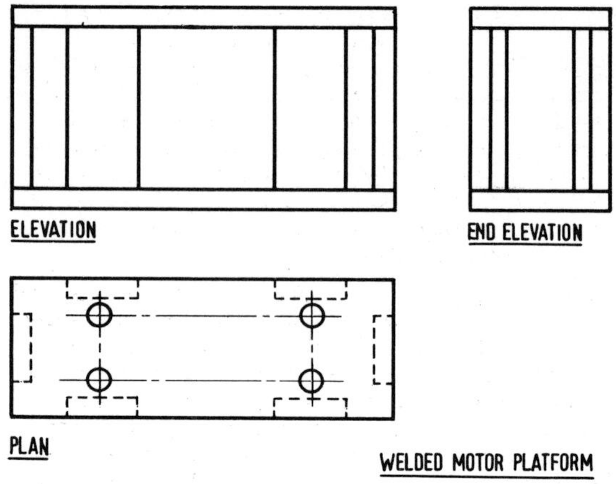

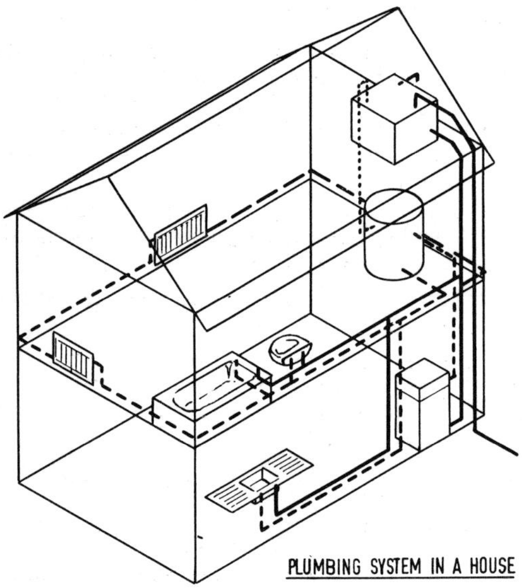

Figure 10.1

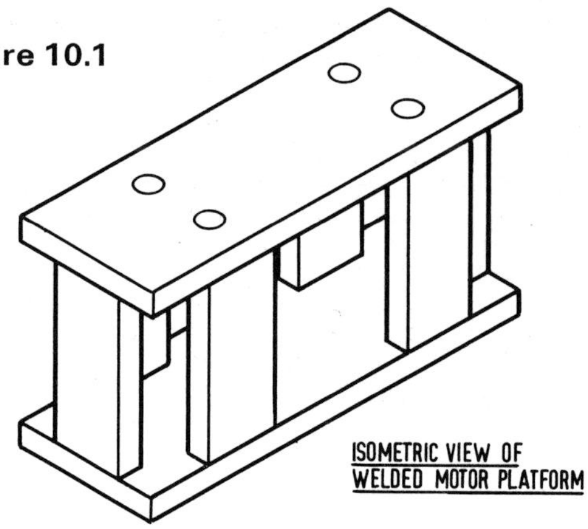

Often it is necessary to illustrate to non-technical administrative staff the shape of the proposed design, for assessment before permission is granted for quantity production.

Some industries use the method of orthographic projection for drawing individual details for manufacture, but assembly drawings are made as isometric to enable semi-skilled labour to visualise and assemble the parts. These pictorial assembly drawings are also used later to illustrate instruction manuals and spare-part schedules for users and maintenance staff. This method has a drawback when major design changes are required; assembly drawings covering the area for redesign have then to be produced orthographically before redesign can begin.

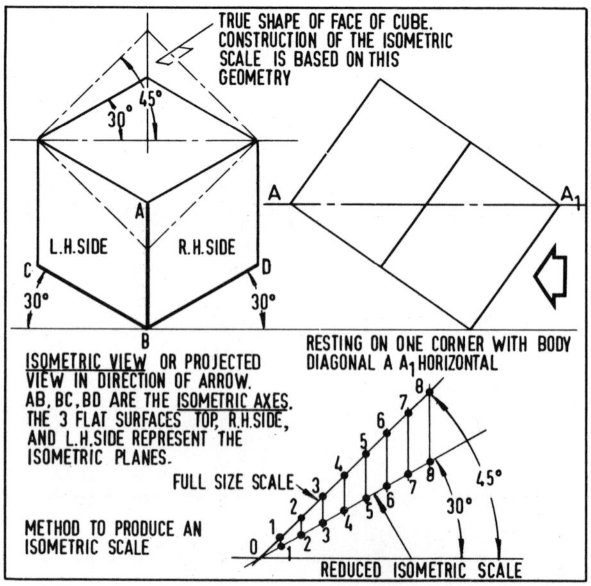

Figure 10.2

Most draughtsmen are practical engineers and have not the artist's ability to sketch an impression of an object quickly. The artist's drawing, on the other hand, is invariably lacking in precise detail and is therefore of limited use in engineering; it is sometimes used to convey simple rough ideas of detail parts between engineers and draughtsmen during initial design stages, but even then the freehand sketching of ideas is usually carried out in three orthographic views rather than the artistic perspective sketch.

Isometric and oblique drawing must not be confused with the perspective drawing of the technical illustrator. Isometric and oblique techniques are the draughtsman's method of producing a pictorial view of an object, using his instruments to guide all the drawn lines; free-hand work is not necessary. No account is taken of perspective and the object has a distorted look, the furthermost end appearing to be too large. However, isometric and oblique drawings have the advantage that parallel lines on the object remain parallel on the pictorial view.

It is usual to produce pictorial drawings from previously drawn orthographic drawings, and in so doing the lengths of lines on the isometric are taken direct from the orthographic views. The pictorial view then appears to be slightly larger than actual size, but this does not matter because the shape is more important than the size. The construction of a reduced scale, or isometric scale, to produce a more realistic size is shown in Figure 10.2, but in practice little use is made of this scale.

ISOMETRIC DRAWING

An isometric drawing is a pictorial view which shows three faces of the object simultaneously.
In producing such a drawing, it is advisable first to draw up a rectangular 'crate' which represents the smallest rectangular block from which the object could be cut. In doing this the isometric axes and planes on which the drawing is built up are established.

To establish the isometric planes and axes, consider the cube shown in Figure 10.2. The cube rests on one corner and, with the body diagonal AA horizontal, the cube is then viewed in the direction of the arrow. In this view three faces can be seen, the top, the right-hand side and the left-hand side.

The lines BC and DB make angles of 30° with the horizontal, AB being perpendicular. These three lines are the isometric axes about which all isometric drawings are made. The three flat surfaces—top, left-hand side and right-hand side—are called the isometric planes.

With the cube, the length of each side is equal, and it can be seen that sides that are parallel on the object remain parallel in the isometric view.

There are four main elements in the technique of isometric drawing:

1. Drawing lines which are isometric.
2. Drawing lines which are non-isometric.
3. Drawing circles in isometric projection.
4. Drawing curved profiles.

Drawing lines which are isometric (Figure 10.3)

These lines are easily drawn once the isometric 'crate' is established. The lengths of these lines are taken directly from the orthographic view and they are drawn on, or parallel to, the isometric axes, with the aid of a 60°–30° set square.

Figure 10.3

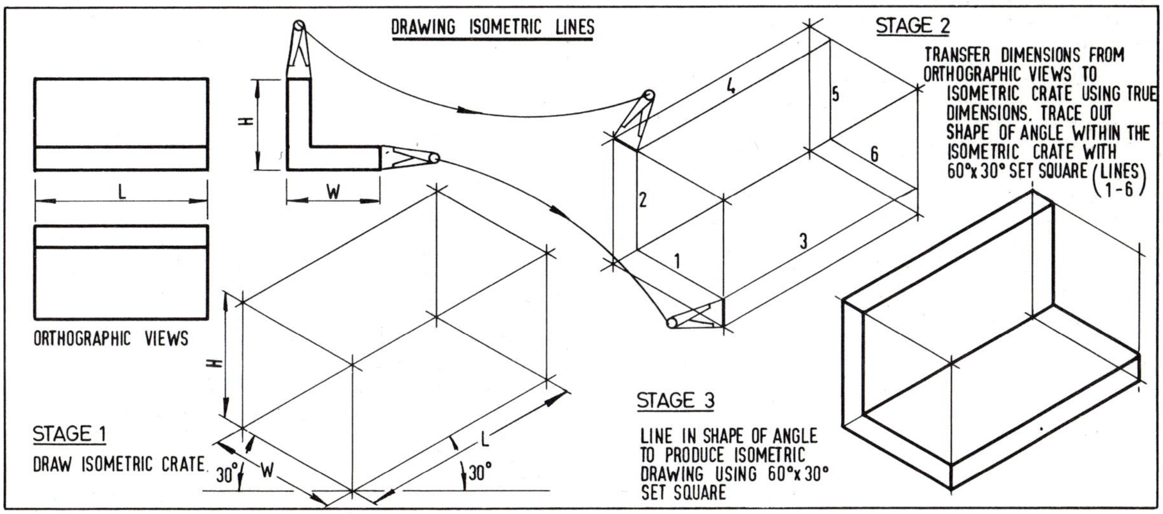

Drawing lines which are non-isometric (*Figure 10.4*)

Non-isometric lines are lines which are not parallel to the isometric axes. The method of drawing these is to establish the end points of a line within the isometric crate by measurements taken directly from the orthographic view. The points are then joined with a straight line.

Figure 10.4

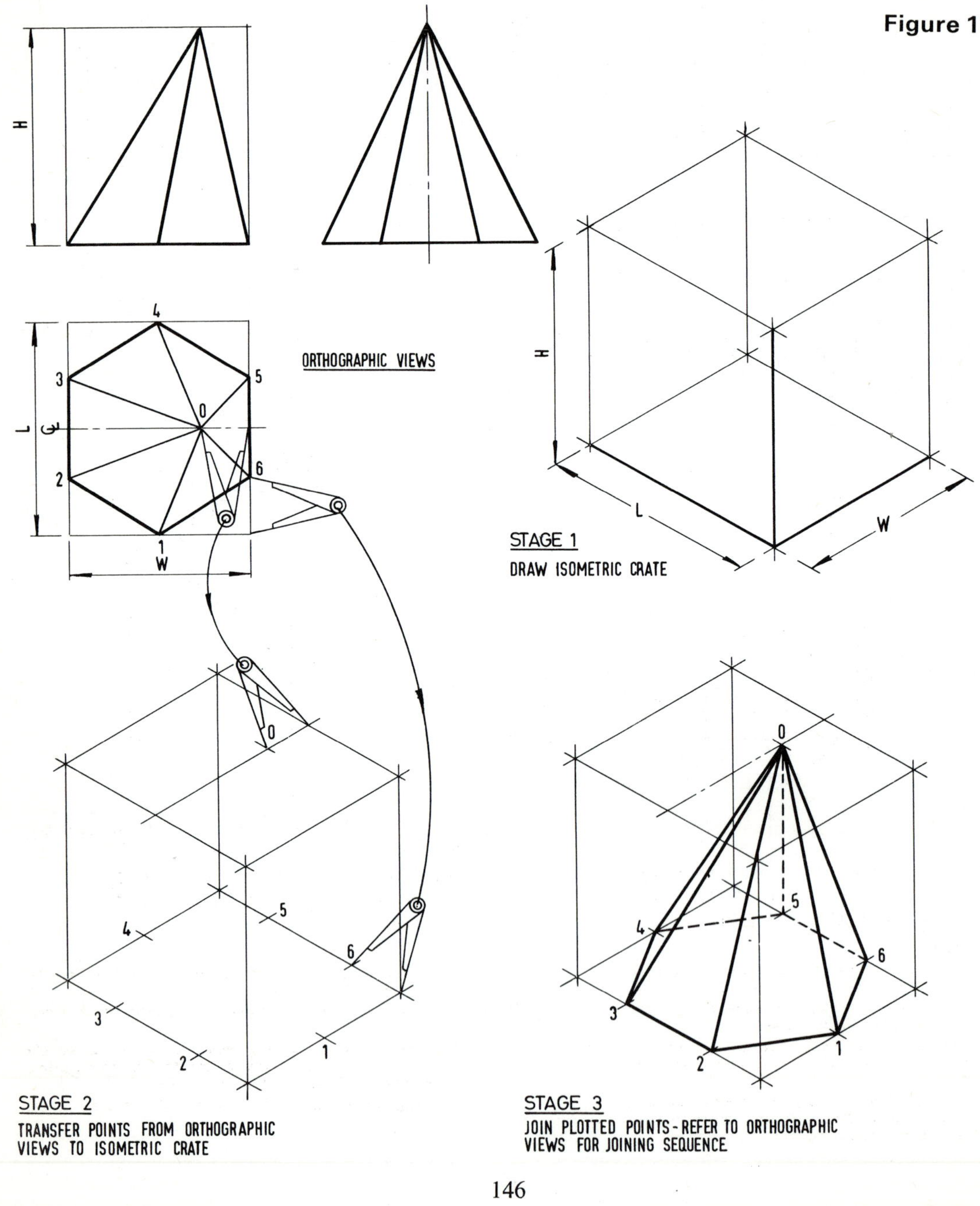

Drawing circles in isometric projection (*Figure 10.5*)

The object to be drawn is usually positioned so that circles representing holes and round bodies lie in the principal planes, or planes parallel to the principal planes. Circles on isometric planes appear as ellipses and can be drawn in either of two ways:

1. Plot points around the curve by the method of ordinates, transferring dimensions from the orthographic view, and join the points, using a french curve.
2. Construct a simple 2-arc ellipse and draw the ellipse profile with compasses.

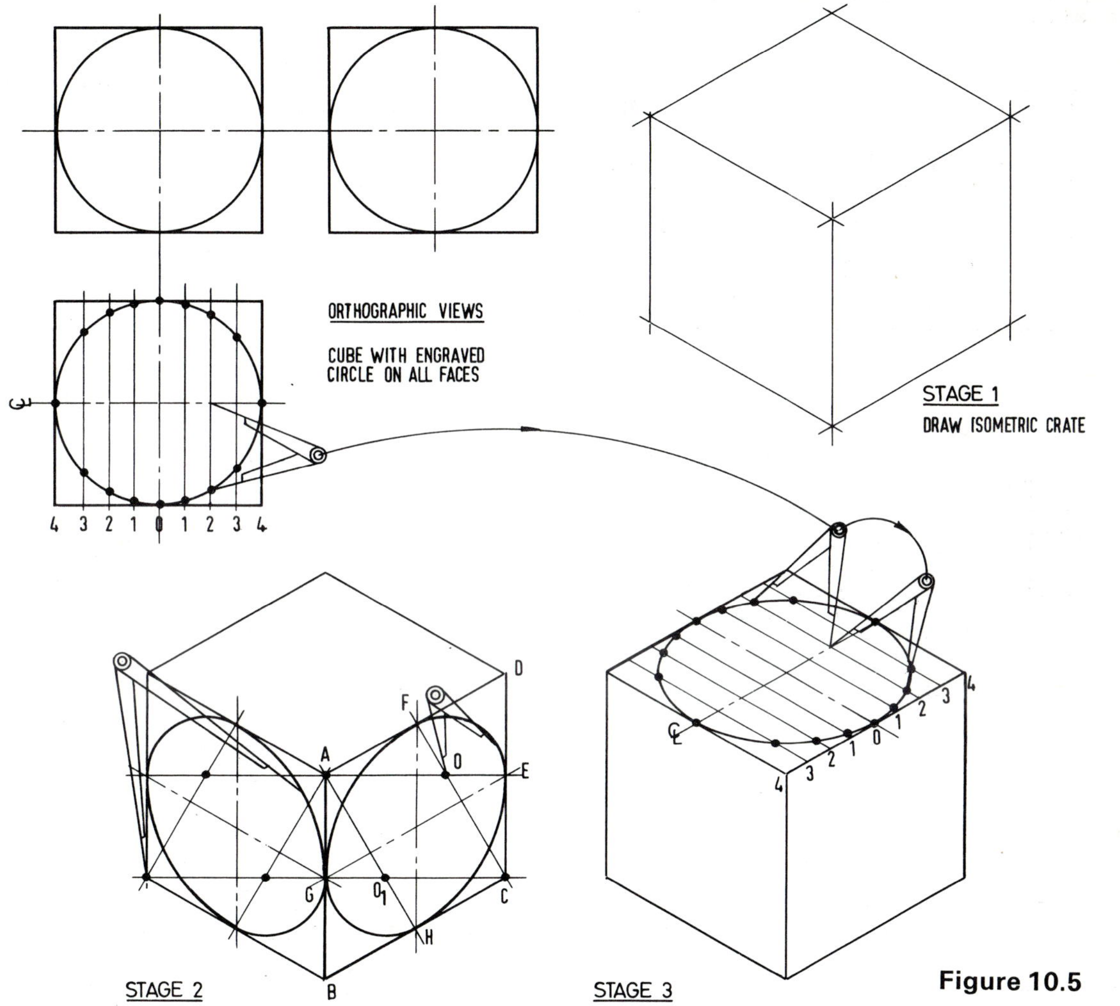

Figure 10.5

DRAWING CIRCLES IN ISOMETRIC PROJECTION BY CONSTRUCTING A SIMPLE TWO-ARC ELLIPSE

Method

1. In the isometric plane ABCD select the two end-points on the shortest diagonal, i.e. A and C.
2. Join A and C to the mid-points of the opposite sides, intersecting at points O and O_1.
3. With centre O and radius OE, draw an arc between E and F. With centre O draw a mirror-image arc between G and H.
4. With centre A and radius HE, draw an arc between E and H. With centre C draw a mirror-image arc between F and G to complete ellipse.

DRAWING CIRCLES IN ISOMETRIC PROJECTION BY PLOTTING A SERIES OF POINTS

Method

1. Draw equally spaced ordinates numbered 0 to 4 on the orthographic view as shown.
2. Draw corresponding ordinates on the isometric plane, and number 0 to 4.
3. Transfer ordinate lengths between centre line and circle from the orthographic view to the isometric drawing, thus plotting a series of points round a curve.
4. Using a french curve, join the points with a smooth curve.

Drawing curved profiles (*Figure 10.6*)

This is like drawing a graph. Establish several points on the curve within the isometric crate, then draw a smooth curve through them. Points on the curve are selected on the orthographic view and, by using ordinates, are transferred to the isometric view.

Remember that too few points will give an inaccurate profile but that, on the other hand, too many points will clutter the drawing with ordinate lines and plotting them will become tedious.

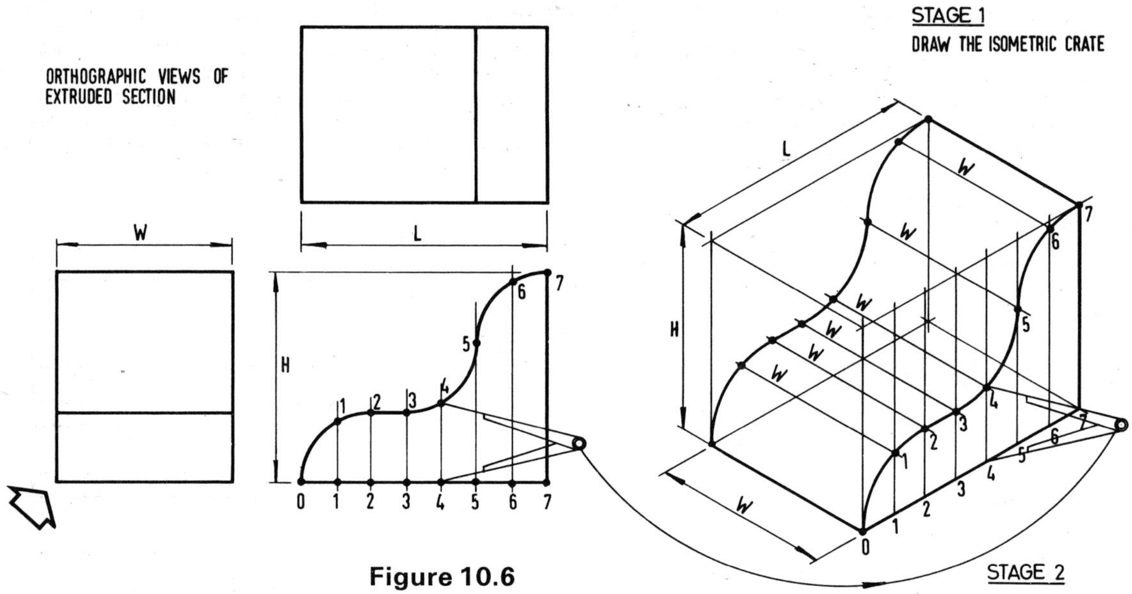

Figure 10.6

DRAWING CURVED PROFILES

1. Draw equally spaced ordinates on the curved profile in the orthographic view.
2. Draw corresponding ordinates on the isometric plane, and transfer heights 1—1, 2—2, 3—3 etc. from the orthographic view to the isometric drawing with dividers, to produce a series of points on a curve.
3. Using a french curve, join the points with a smooth curve.
4. To draw a parallel curve at the far end of the object, draw isometric lines through points on the curve and mark off dimension W on each line. Through these points draw a parallel curve.
5. With isometric lines, complete the shape of the object.

OBLIQUE PROJECTION

This is another pictorial method of drawing, similar to isometric drawing, and also produced by drawing instruments and direct measurement from the orthographic views. No account is taken of perspective and, wherever possible, the long axis of the object is drawn in the plane of the paper.

In the drawing, one face of the object, or a suitable cross-section which is established as a datum face, is shown in its true shape and size, and a sloping side view is attached. In many cases this is no more than adding a sloping side view to an orthographic elevation, provided the whole surface is flat and in the plane of the paper (Figure 10.7).

The side views may be drawn at any angle but, draughtsmen's squares being made with 30°, 60° and 45° angles, these are the angles most commonly used; the squares are convenient for drawing the many parallel lines that are necessary.

Figures 10.7, 10.8 and 10.9 show examples of oblique projection. It is usual, and simpler, where an object has a circular or contoured profile, to present this view as true shape and size in the plane of the paper.

The length of the lines on the sloping side is reduced to one-half, or sometimes two-thirds, actual length to give the object better visual proportions. If the full length of line is used for the sloping side, the drawing gives the impression that the object is much longer than it really is.

Figure 10.7

Oblique projection with the profiled view drawn true to scale in plane of the paper, and with long axis receding

STAGE 1

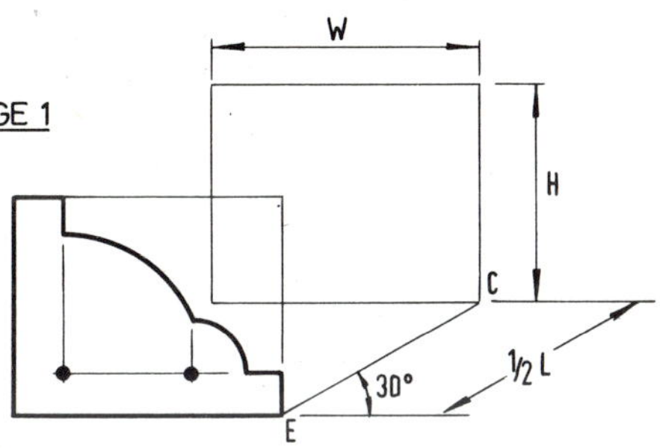

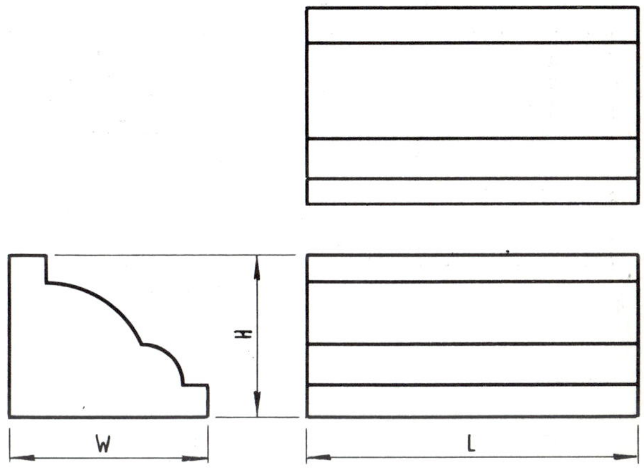

ORTHOGRAPHIC PROJECTION

1. Draw orthographic view of contoured profile.
2. Set off length of oblique view by drawing line EC at 30° (45° or 60°) from base corner. This length is usually drawn as half the actual length.

STAGE 2

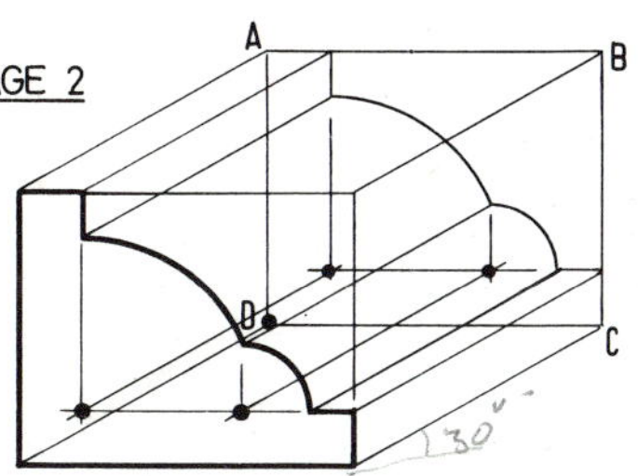

3. In rectangle ABCD construct contoured profile of end elevation. This can be done by projection and direct measurement within the rectangle.

STAGE 3

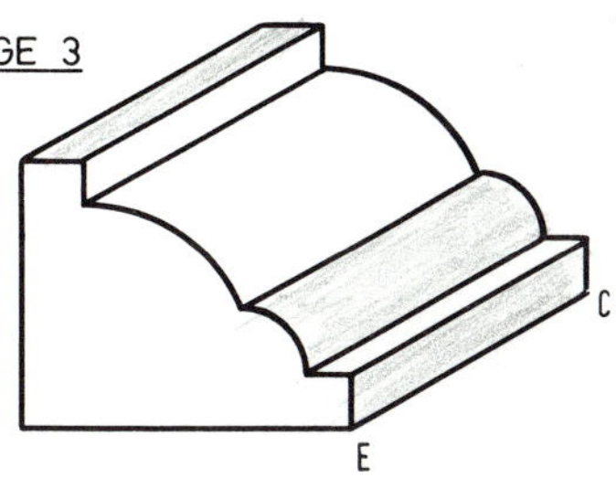

4. Line in all edges and corners across face of object, parallel to line EC.
5. Line in contoured end profiles.
6. Hidden detail may be shown if required.

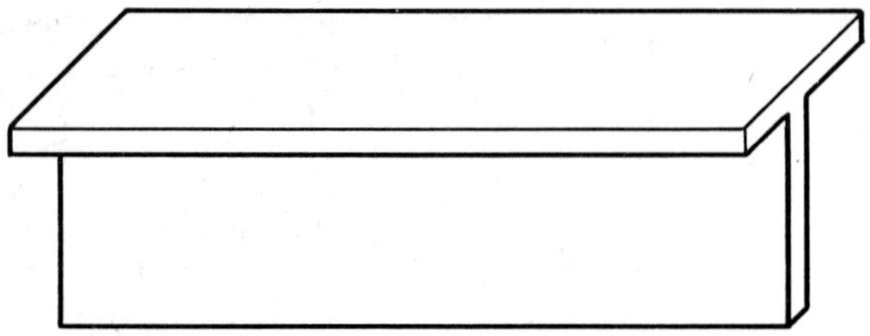

Where possible draw the long axis horizontal to give the object better proportions.

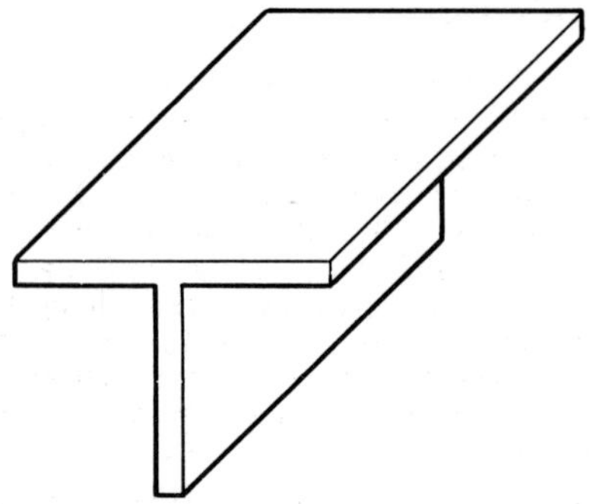

Oblique projection of same T-section with the long axis receding.

Figure 10.8

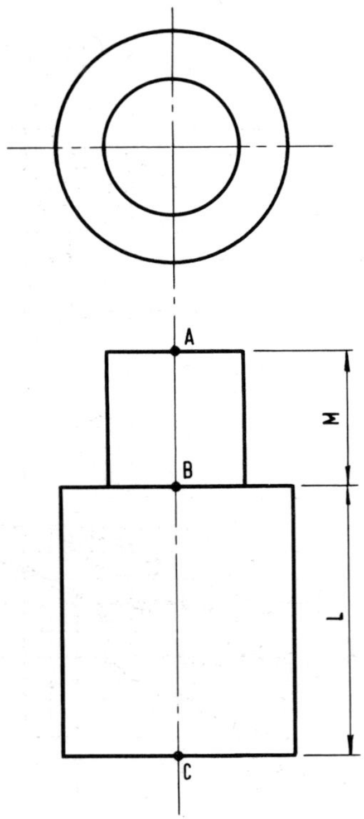

ORTHOGRAPHIC VIEWS

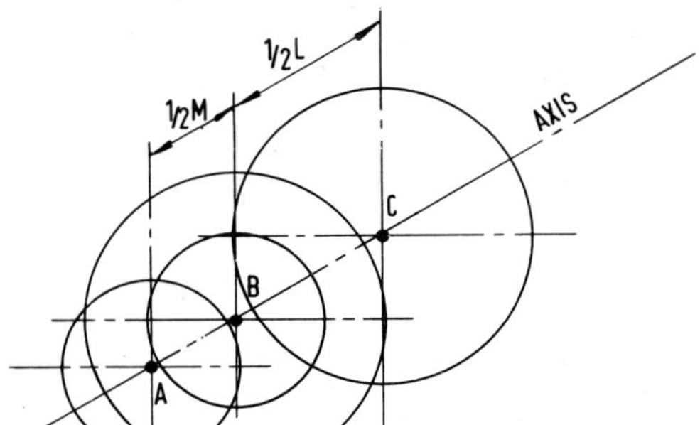

STAGE 1

1. Draw axis of cylinders inclined at 30° (45° or 60°).
2. Establish points A, B and C along the axis, using half-true lengths.
3. At points A and B draw full-size circles to represent ends of smaller cylinder.
4. At points B and C draw full-size circles to represent ends of larger cylinder.

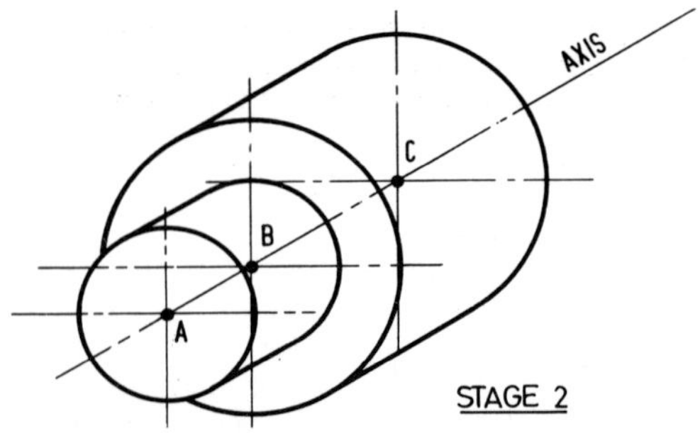

STAGE 2

5. Draw tangent lines to circles parallel to the sloping axis.
6. Line in all seen edges and corners.

Note: Oblique projection should always be considered for cylindrical objects because it is far simpler to draw than to construct ellipses.

Oblique projection of object with circular cross-section

Figure 10.9

FREEHAND SKETCHING

As mentioned before, sketching has very limited use in engineering drawing offices. It is sometimes used to convey what a person has in mind because words would be inadequate to describe the shape or function. In general, its use is limited to conveying simple ideas at the design stage, and this can be done either pictorially or orthographically. Most engineers and draughtsmen find it more convenient to sketch orthographically, thereby defining the complete shape by the use of related views. Pictorial sketching always leaves the rear side unshown (Figure 10.10).

Pictorial freehand sketching is best done in the manner of isometric or oblique drawing. First a 'crate' is sketched to represent the smallest rectangular block (length, width, height) from which the object could be cut, then the features of the object are built up within this framework. This results in a reasonably well-proportioned sketch.

Complicated sketches waste time. If the time is used to produce orthographic views which prove the feasibility of an idea or object, many of the snags and pitfalls that do not become evident when sketching (particularly on backs of envelopes) can be avoided, and time is saved.

However, sketches are sometimes necessary when collecting information on site and no drawing equipment is available. In these circumstances, sketches made on a sketch pad with a pencil, ruler and square—in orthographic projection with dimensions added—provide a necessary link between site and drawing office. Experience has shown that pictorial sketching is inadequate for conveying this information; it is far better to make orthographic related plans and elevations as near to scale as possible, particularly when the information is complex and there are numerous dimensions. This is of special importance when the site is a long way from the drawing office; it is futile to make only a rough sketch for guidance and to try to memorise the detail. A little extra time spent recording all details and measurements on site pays dividends when back in the drawing office.

It is often useful when collecting site information to supplement dimensioned sketches with a series of strategically taken photographs for reference. Remember that precise information in pictorial form is far better, and more convincing, than report writing, chit-chat and arm waving.

Figure 10.10

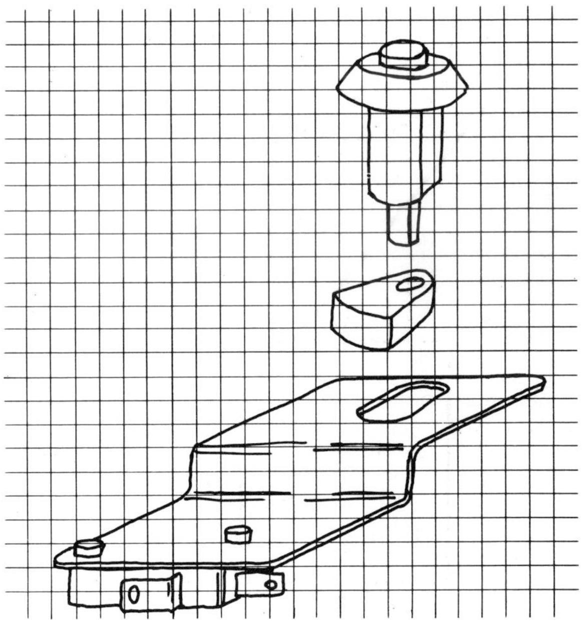

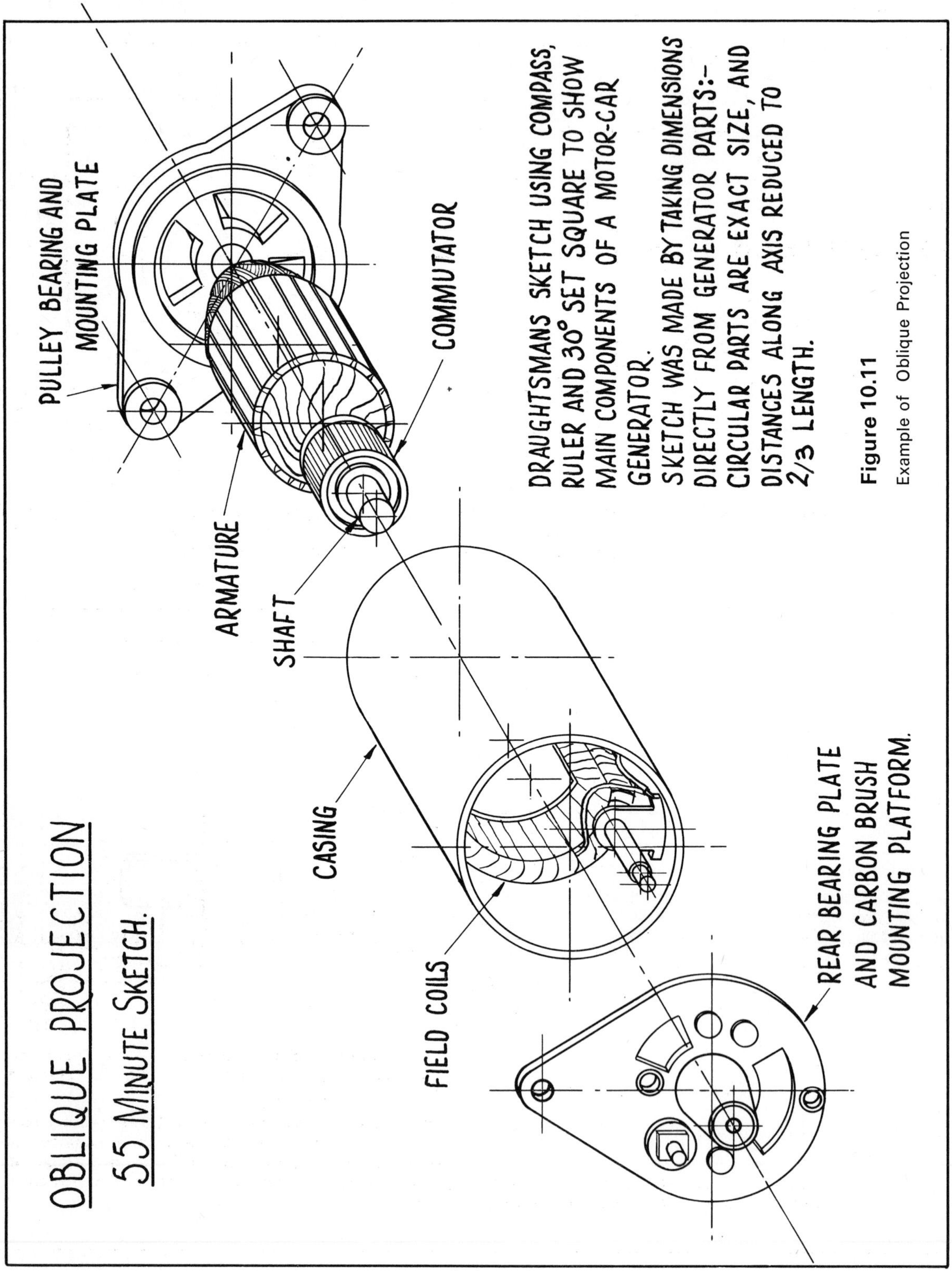

Figure 10.11
Example of Oblique Projection

EXERCISES

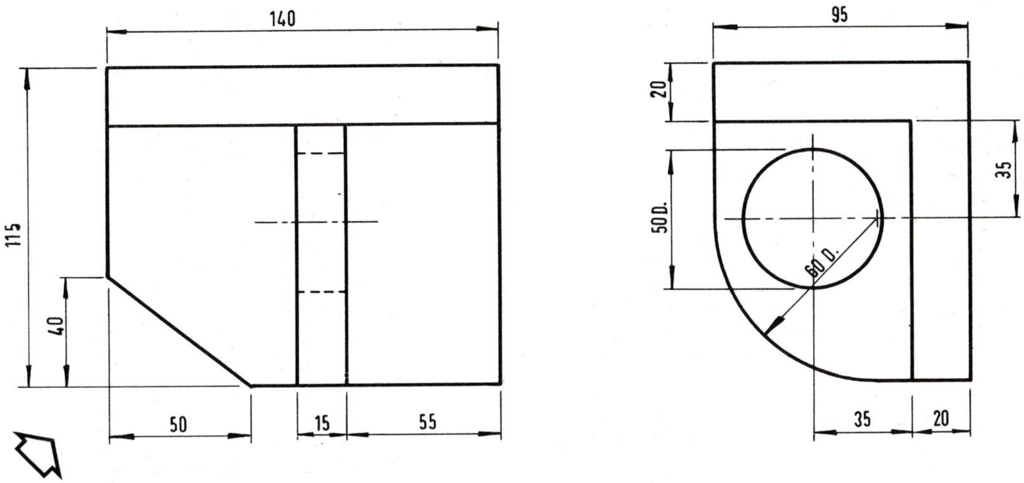

Exercise 10.1

Draw in isometric projection the bracket shown in the two accompanying orthographic views, viewed in the direction of the arrow. Do not use the isometric scale.

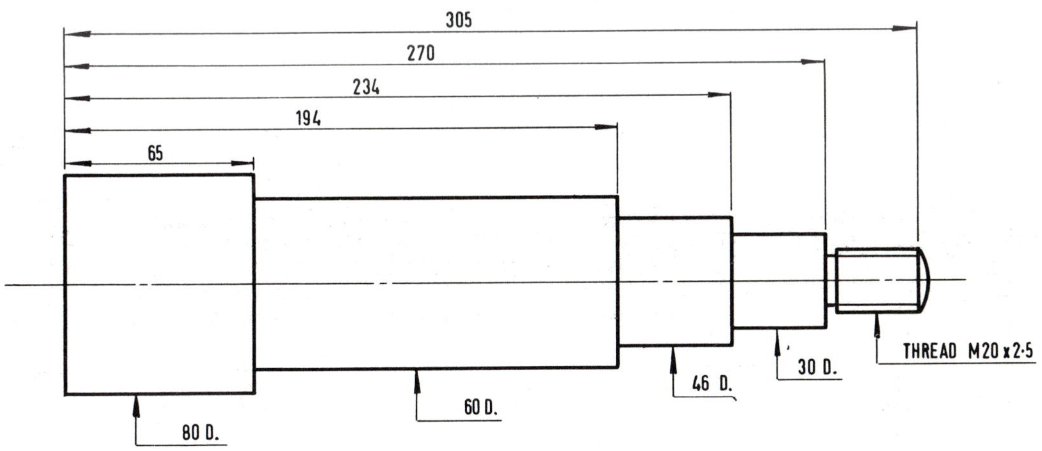

Exercise 10.2

1. Draw full scale and dimension the stub shaft as shown.
2. Draw the stub shaft in isometric projection with the shaft axis vertical. Do not use the isometric scale.
3. Draw the stub shaft in oblique projection with the axis sloping to the right at 30° and the threaded portion to the front.

Time yourself in making each of the three drawings and decide (a) which is the most satisfactory drawing for manufacturing purposes, and (b) which method of pictorial drawing is most satisfactory in this particular case.

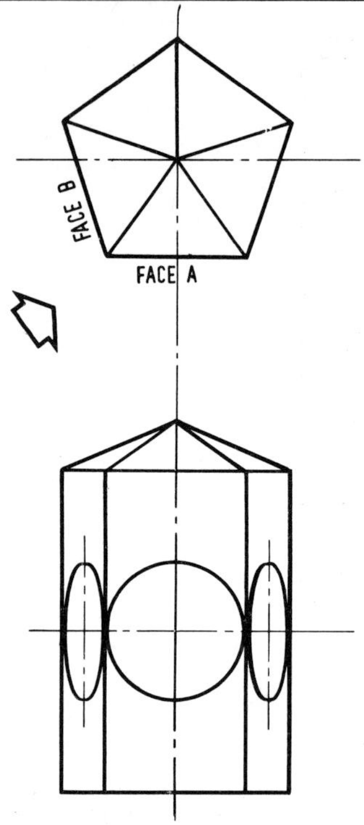

Exercise 10.3

The drawing shows two orthographic views of a regular pentagonal prism, with one end shaped to a point at the geometric centre. The side length of the pentagon is 60 mm. The overall height is 170 mm and the parallel section 150 mm. On each face of the parallel section, 80 mm from the base, a circle of 60 mm diameter is engraved.

Make an isometric drawing of the pentagonal figure viewed in the direction of the arrow, with face A positioned in the isometric plane.

Use hidden detail lines to show the full shape of the solid. Show engraved circles on faces A and B only. Do not use the isometric scale.

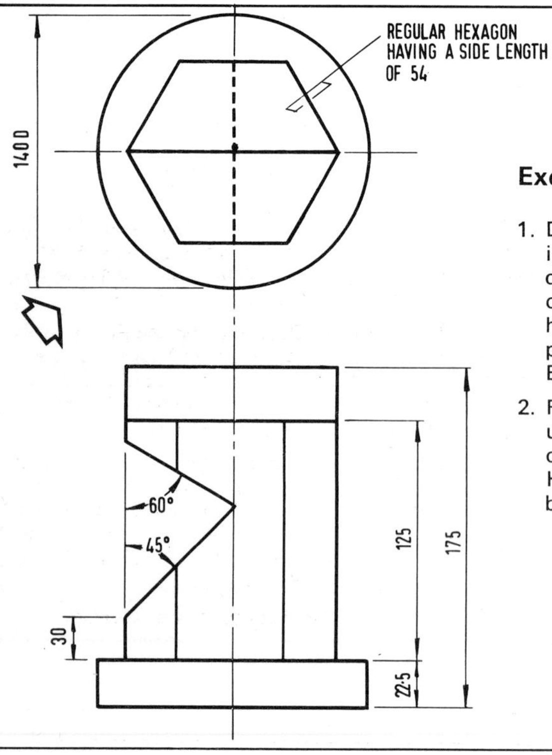

Exercise 10.4

1. Draw the two orthographic views as shown, full size in third-angle projection, and dimension. Use the quality of line recommended in BS 308 for visible outlines, centre lines, dimension lines etc. Arrow heads should be slim and thin, with figures and printing upright and in block capitals, as shown in BS 308.
2. Produce an isometric scale to measure 180 mm, and use this to draw a pictorial isometric view of the object, viewed in the direction of the arrow. Hidden detail only of the cut-away V-section is to be shown.

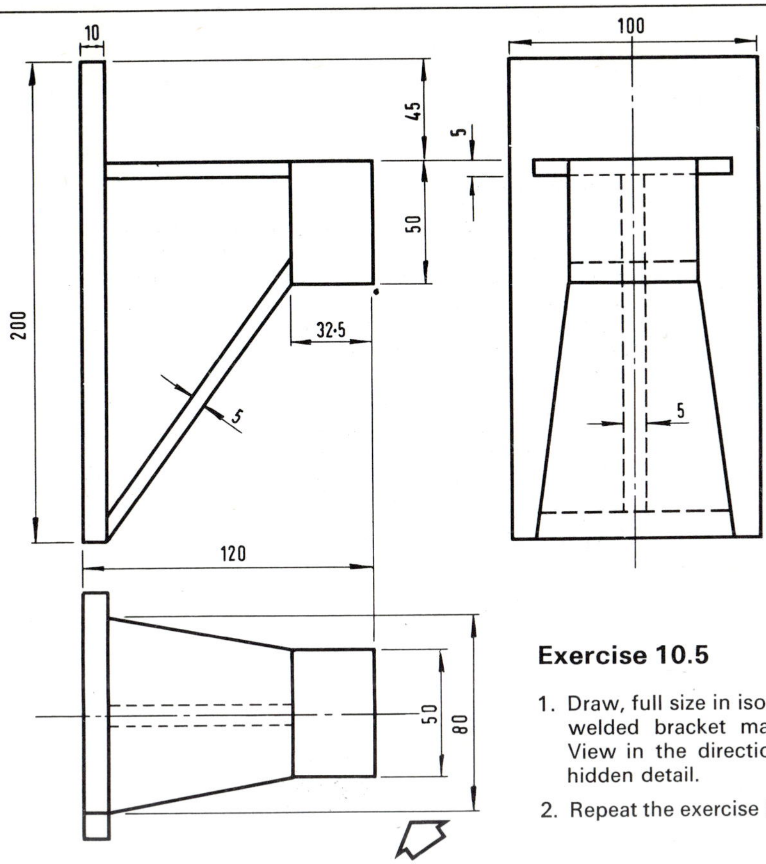

Exercise 10.5

1. Draw, full size in isometric projection, the fabricated, welded bracket made from steel plate and bar. View in the direction of the arrow. Do not show hidden detail.
2. Repeat the exercise in oblique projection.

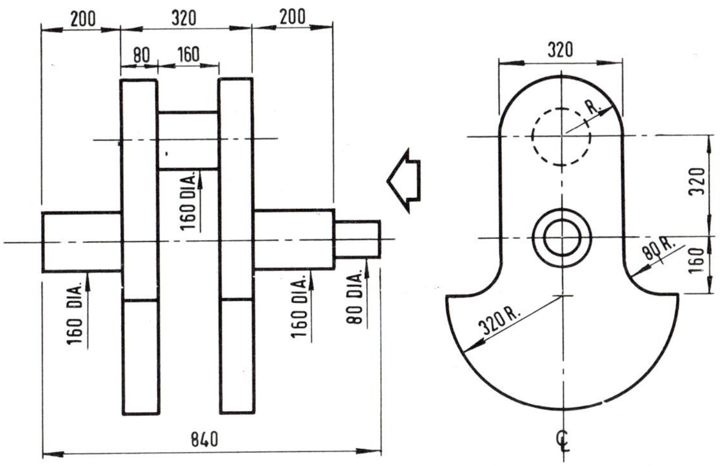

Exercise 10.6

1. Copy the two views of the crankshaft forging as shown, complete with dimensions. Use a scale of 1 : 4.
2. Draw the crankshaft forging in isometric projection, viewed in the direction of the arrow. Use the 2-arc construction for all curved surfaces.
3. Draw the crankshaft forging in oblique projection viewed in the direction of the arrow, with the main bearing axis sloping to the left at 30° to the horizontal.

Do not use isometric scale.

Jigs and fixtures

JIGS

A jig holds the component firmly and at the same time guides the cutting tool. It is applicable to drilling machine operations only. It may be small and easily manipulated by hand, or heavy and clamped to a machine table.

FIXTURES

A fixture holds the component firmly in position during the machining or assembly operation, but does not guide the tools. Fixtures can be small or extremely large. They may be clamped to machines, or free-standing for assembly purposes such as the accurate assembly of aircraft or rocket structures.

THE NEED FOR JIGS AND FIXTURES

Before 1760, which is the date generally accepted as the beginning of the Industrial Revolution, life in England was rural and localised, and village life was almost self-supporting. The village blacksmith met the demands for goods in metals, each article being individually make by the skill of the craftsman, with hand tools, at a time-consuming pace.

The development of transport, first by canals then by turnpikes, started the breakdown of rural life in England. This was rapidly speeded up by the development of the locomotive and railways, which permitted the fast and cheap transport of merchandise and people. Under the pressure of economic conditions, unskilled workers from the country were attracted to the rapidly expanding new industrial towns, where steam power had revolutionised the textile industry and the inventions of the age had created a vast manufacturing industry; Britain became one of the chief suppliers of manufactured goods of the world.

The later discoveries of other sources of power, such as electricity and oil fuels, and the development of the internal-combustion engine, cheapened and increased the speed of transport and communication still further. The skilled master craftsman of the medieval craft guild had been swamped out of existence by the rapid growth of machines, and the demand was for a new labour force of machine minders. This trend continues today. Manufacturing processes are divided into relatively simple tasks, among numerous operators whose limited skills must be guided to enable vast quantities of high-quality products to be manufactured, often with extreme dimensional accuracy to meet interchangeability requirements. This is where jigs and fixtures play an important part. They are holding devices that allow components to be assembled or accurately machined in quantity.

Even in small-quantity production, if interchangeability is important, skilled machine operators often use jigs and fixtures where workpieces are difficult to position and hold. Assembly and welding fixtures are provided where it is essential for parts to be accurately located, leaving the operator with both his hands free to use his tools.

JIG AND TOOL DESIGN

The cutting and removal of metal is usually carried out with small standard tools, the provision of which has become an industry. Much thought, money and research has been devoted to producing standard ranges of highly efficient tools for machining a multitude of shapes and materials. When designing jigs and features, it is essential to have knowledge of the machines and tools which will be used to do the actual cutting.

DRILLS

Twist drills (BS 328 : Part 1 : 1959 and BS 328a : 1963).
Tapping drills (BS 1157 : 1965 and Supplement No. 1 : 1967).

Drills are manufactured in various shapes to fit different chucking arrangements and to cut holes in many different materials speedily and efficiently. Drills should be machine-ground for the efficient production of holes of accurate size; hand grinding often produces drills with unequal lips and this results in the drilling of oversize holes.

To improve efficiency, the standard included point angle of 118° and the lip clearance angle are sometimes modified when drilling particular materials.

For many years British engineering has used standard drills manufactured in fractional inch sizes, increasing in 1/64th of an inch, together with another series of numbered and lettered drill sizes to supply the intermediate size holes necessary for clearance and tapping for small holes. With metrication the 'number' and 'letter' sizes have been replaced by millimetre sizes, with certain fractional inch sizes to complete the series. Universal metrication will finally make all inch sizes obsolete.

Figure 11.1 shows some types of drill which are available.

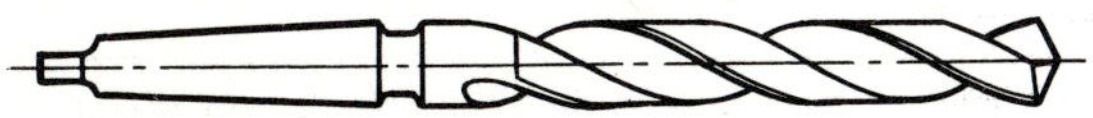

Morse taper-shank twist drill

Has two helical flutes with a standard morse taper shank for holding and driving.

Morse taper-shank core drill

Has three or four helical flutes and no drill point. Cuts on edges only and is used for opening out or enlarging holes.

Parallel shank twist drills

Manufactured in stub, standard and long series. Has two flutes, and the parallel shank has the same diameter as the cutting end.

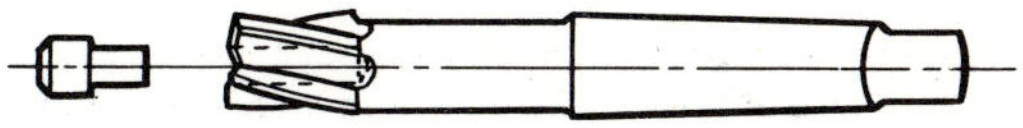

Counterbores

Have cutting teeth on both ends and sides to produce a flat machined seating face relative to a central hole for heads of bolts or set screws. Shanks may be either tapered or parallel. Pilots for small sizes are made integral with counterbore; for larger sizes they are usually detachable.

Figure 11.1 Drills

REAMERS (BS 122: Part 2: 1964)

To produce holes of accurate size and fine finish, reamers are used to finish off holes which have been drilled slightly undersize.

Figure 11.2 shows some of the common types of reamer used in industry.

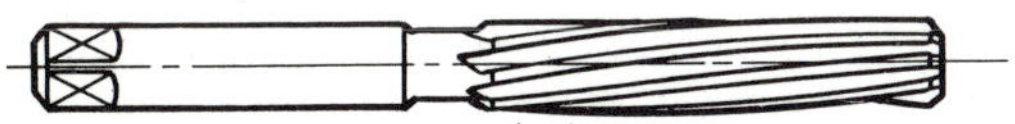

Hand reamer

Has virtually parallel cutting edges. The end is bevelled at 45° and the cutting edges have a 1° taper lead. The shank has a square end.

Parallel machine reamer

Has virtually parallel cutting edges, with a 45° bevel lead. Machined with a Morse taper shank for driving.

Machine chucking reamer

Parallel cutting edges, with 45° bevel lead and long body clearance. Fitted with taper or parallel shank in a lathe tailstock and turret.

Machine jig reamer

Short, virtually parallel cutting edges, with 45° bevel lead. Manufactured with guide section between cutting edges and taper shank for driving.

Taper pin reamer

Reamer having tapered cutting edges of 1 in 96 to produce holes for taper pins. May be fitted with either taper shank for machine use, or square end for hand use.

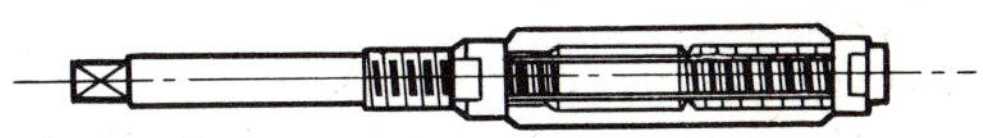

Expanding reamer

This is an adjustable hand reamer and is fitted with either 4 or 6 high-speed steel blades. The diameter is increased by tightening nuts down on blades.

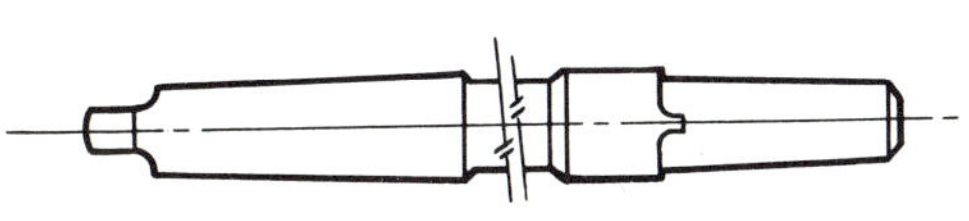

Arbor for shell reamer

May have either machine taper or parallel shank.

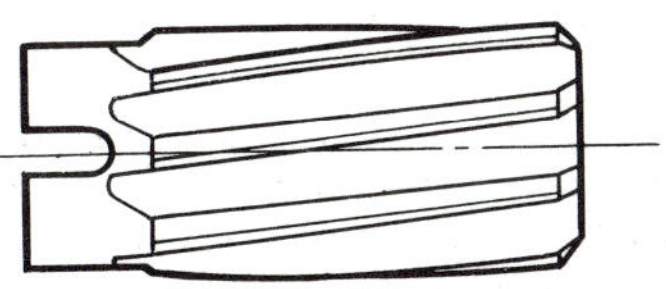

Shell reamer with taper hole

Short reamer with parallel cutting edges and 45° bevel lead. Has tapered axial hole for use on arbor.

Figure 11.2

Reamers

MILLING CUTTERS (BS 122: Part 1 : 1953)

Milling cutters are made in many shapes and sizes for a multitude of purposes. Standard ranges are listed in BS 122 but this Standard does not cover all the various types which are available.

Figure 11.3 shows various types of standard milling cutters, and Figure 11.4 shows various arrangements for using milling cutters. When two cutters are mounted on the machine arbor to cut two faces simultaneously, the arrangement is known as *straddle milling*; when three or more cutters are mounted on the arbor, cutting several faces simultaneously, the arrangement is known as *gang milling.*

When a number of components are mounted in a fixture along the length of a machine table so that they are all machined in one pass, using either a single cutter or a number of cutters, the arrangement is referred to as *line* or *string milling.*

Figure 11.4

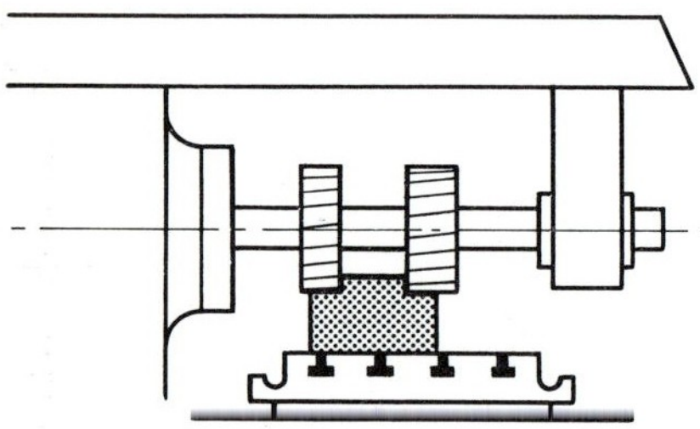

Straddle milling

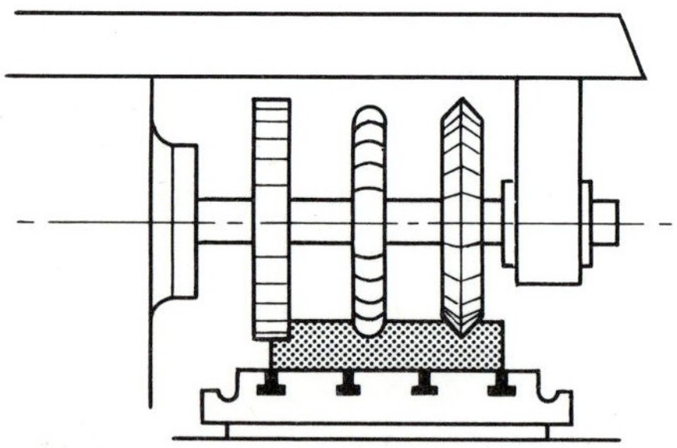

Gang milling

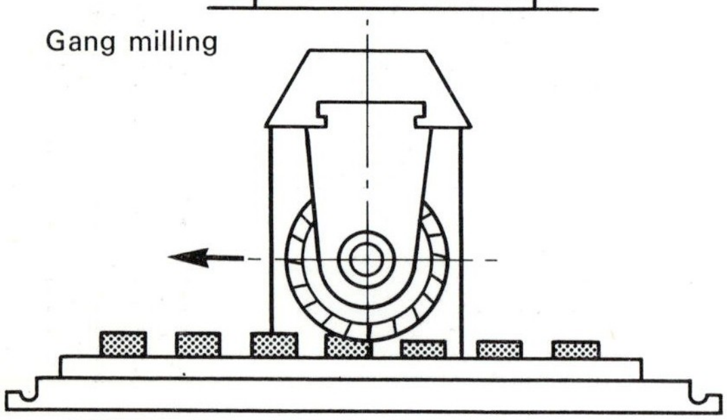

Line or string milling

Figure 11.3*a*

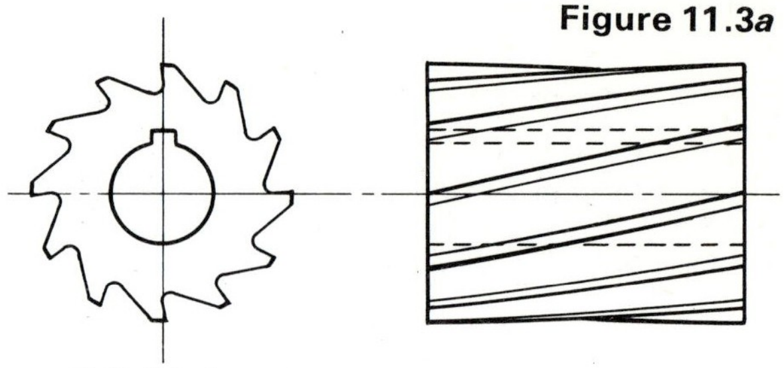

Cylindrical cutter

Cutter has teeth on cylindrical surface only, and is used to produce large flat surfaces. There is a large range of sizes available in both light-duty and high-power series. The teeth are sometimes of helical form.

Figure 11.3*b*

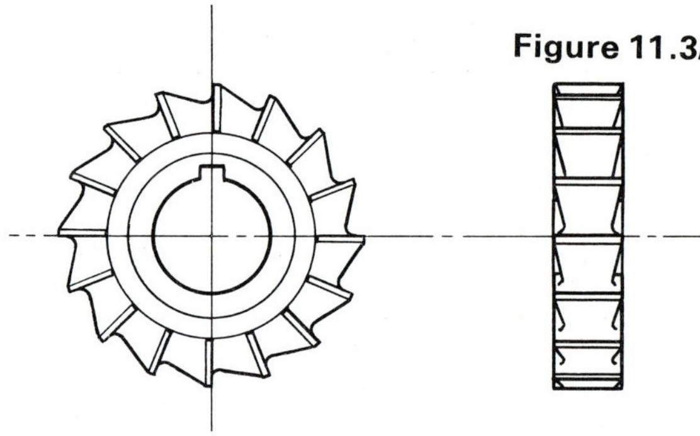

Side and face cutter

The cutters have teeth on periphery and on both sides. They can be used singly, but often two or more are mounted on a single arbor for straddle or gang milling.

Figure 11.3*c*

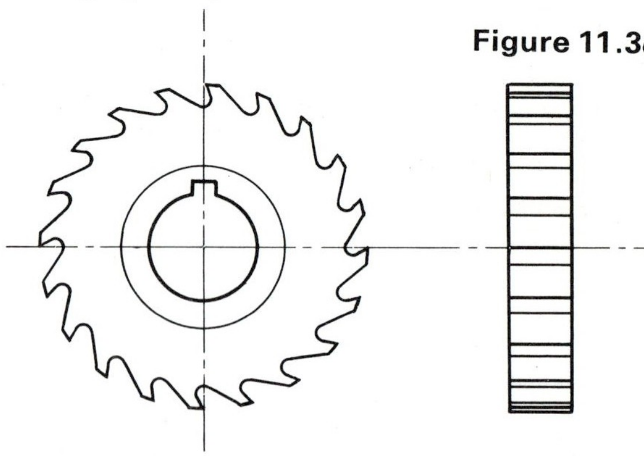

Slotting cutter or slitting saw

The cutter has teeth on the periphery only, and the teeth are ground back at a slight angle for clearance. Slotting cutters and saws are made in a range of diameters in varying widths. The term slitting saw is used for cutters of the thinner widths.

Figure 11.3*d*

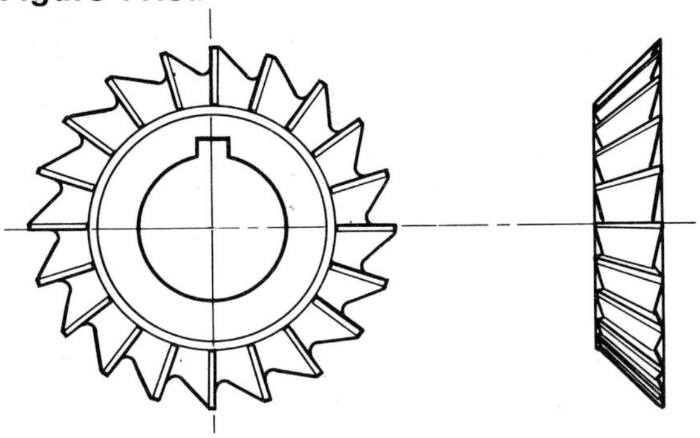

Single-angle cutters

This cutter has teeth only on the conical side. There is a range of diameters available, with varying teeth angles. Right-hand and left-hand cutters are available, and are used for angle milling.

Figure 11.3*e*

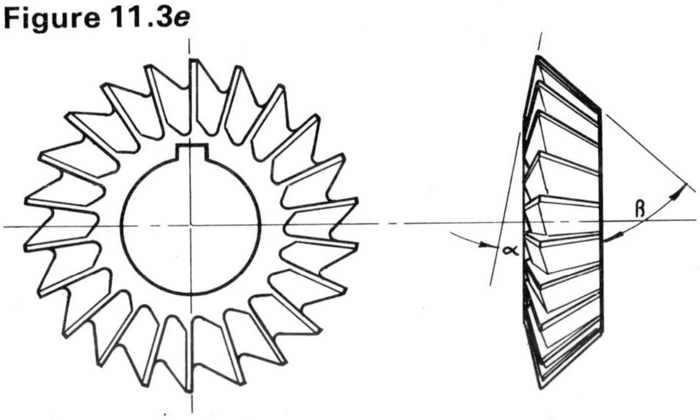

Double-angle cutter

This has teeth on two conical faces of unequal angle. In the standard series one conical surface has a fixed angle of 12° slope, and the included angle '$\alpha + \beta$' covers the range of 60° to 85° in increments of 5°. The cutter is used for machining angled faces and cutting V-shaped flutes.

Figure 11.3*f*

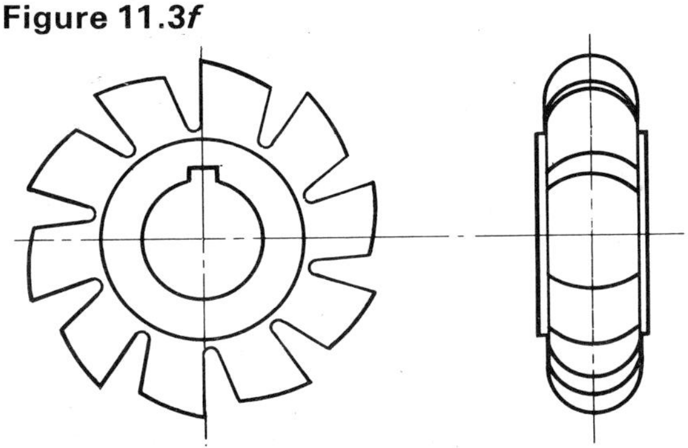

Convex cutter

This cutter has the form of a convex half circle and is available in a standard series of sizes similar to the concave cutter.

Figure 11.3*g*

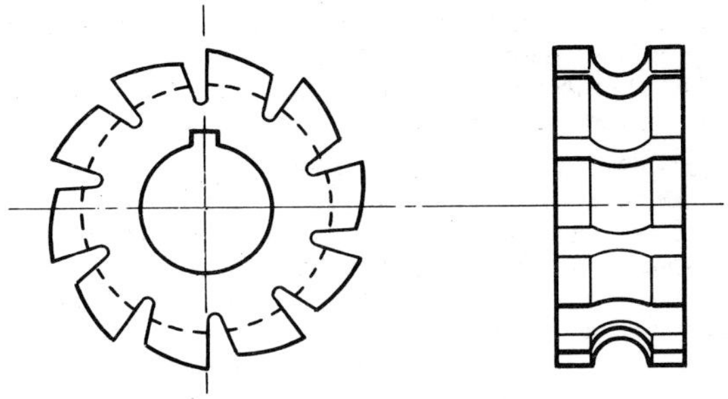

Concave cutter

This cutter is of a form which includes a concave semicircle on the centre line. It is available in a standard series.

Figure 11.3*h*

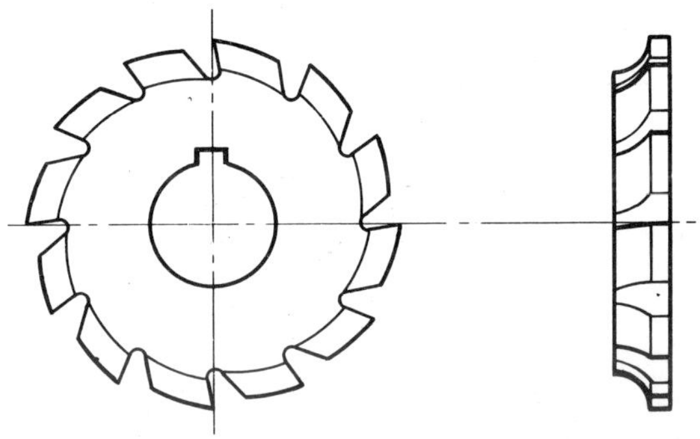

Single-corner rounding cutter

This cutter has concave teeth in the form of a quarter of a circle. Both left-hand and right-hand cutters are available in a range of sizes.

Figure 11.3*j*

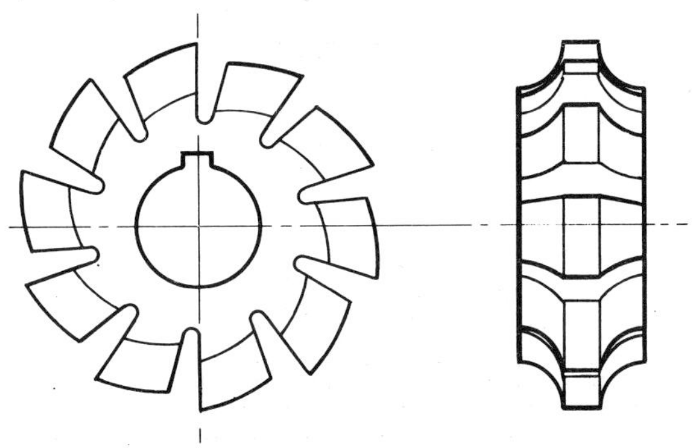

Double-corner rounding cutter

This is a double-sided version of the cutter shown above.

Figure 11.3*k*

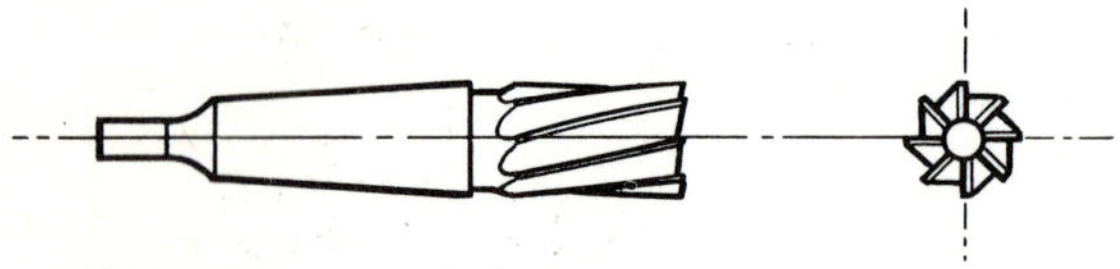

End mill with tapered shank

This cutter has teeth on both end and periphery. The standard series covers a range of diameters used for machining profiles, narrow slots and keyways.

Figure 11.3*l*

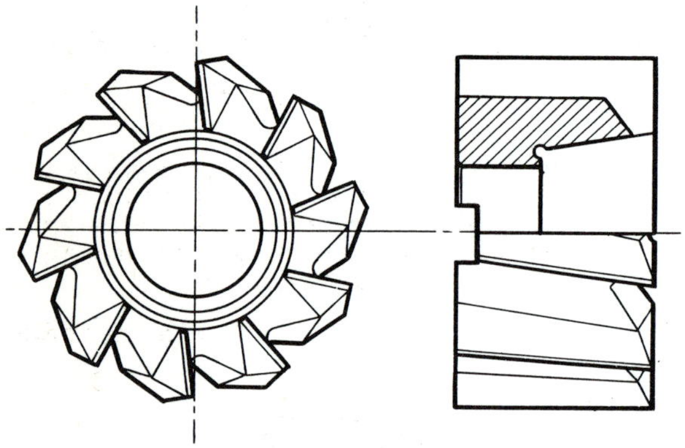

Shell-end mill

This cutter has teeth on the periphery and one end and is the larger version of the cutter shown at k. A standard range of sizes is available in both left-hand and right-hand versions. For holding and driving it is fitted on a machine arbor.

Figure 11.3*m*

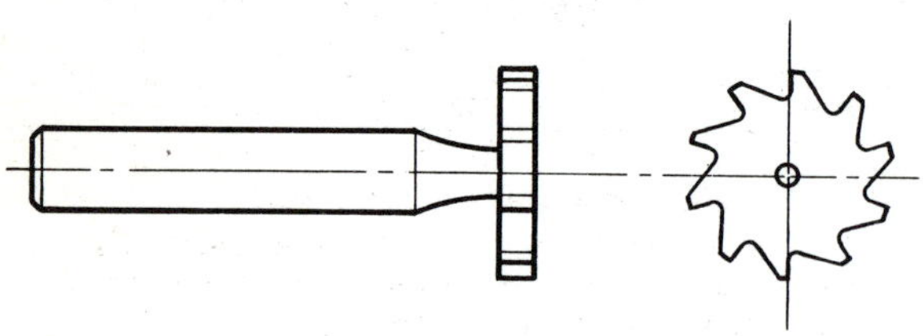

Woodruff keyway milling cutter

This cutter has teeth on the periphery only and the sides are slightly concave for clearance. The cutter is manufactured to definite diameter sizes and widths, and is used for cutting keyways in shafts to accept Woodruff keys.

T-slot milling cutters are of similar shape, but have side and face cutting teeth; they are used for machining T-slots in machine tables.

JIG AND FIXTURE COMPONENTS

Drill bushes (BS 1098 : 1967)

Drill bushes have been standardised and are usually purchased from suppliers who specialise in the production of these parts. They are classified as:

1. Fixed bushes, plain and headed.
2. Liner bushes, plain and headed.
3. Renewable bushes.
4. Slip bushes.

Figure 11.5 shows a selection of these types.

Drill bushes are made of hardened steel, and are pressed into accurately reamed holes in drill plates and jig bodies to give the required pattern of holes in the component when drilling.

The bushes have a generous 'lead in' at the entry point of the drill and, to prevent swarf jamming between the component and the drill plate, the jig is designed to bring the bottom of the drill bush either in contact with the component to allow swarf to escape through the drill bush or far enough away to allow the swarf to escape between component and drill plate. This distance should be not less than $0{\cdot}3D$ where D is the diameter of the drill, but on the other hand the gap should be kept to a minimum to prevent deflection of the drill; this is necessary to eliminate breakage and maintain positional accuracy.

The headless type of drill bush (Figure 11.5*a*) is used when the drilled hole depth is not critical, as in through holes.

The hard top surface of the headed drill bush (Figure 11.5*b*) is often used as a stop when the drilled hole depth in the component is critical, e.g. for blind tapped holes, or where drillways accurately connect inside a component. The depth is controlled by an accurately positioned stop nut attached to the drill chuck.

Where several operations, such as drilling, reaming and counterboring, have to be carried out on the one hole, a slip-bush arrangement is used, as shown in Figure 11.5*c*. For each operation, a slip bush of different size is used, being quickly removed and replaced by the next by partial rotation and screw-locking. For larger diameters, when spot-facing or counterboring is required, the hardened steel liner is used to guide the cutting tool.

In mass production the drill bush is subjected to wear which will ultimately produce positional inaccuracy of the drilled holes. To obviate this, a renewable bush arrangement (Figure 5.11*d*) is fitted to allow the bush to be simply and quickly replaced.

In some drilling jigs the workpiece is located some distance from the drill plate or has a surface not at right angles to the drill axis. Figure 5.11*e* shows one such arrangement requiring a special drill bush.

Figure 11.5*a*

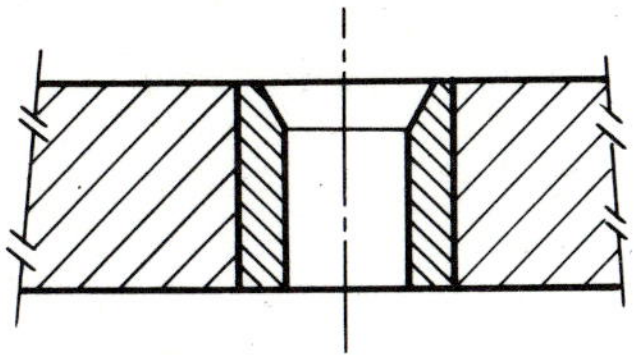

Headless drill bush

Used where depth control of drilling is unimportant.

Figure 11.5*b*

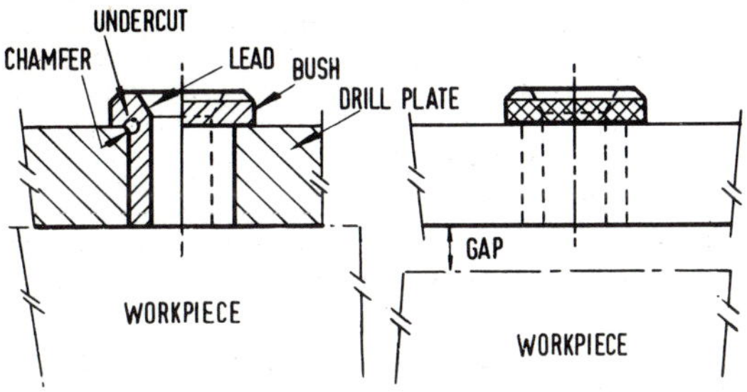

Headed drill bush

Used where hole depth must be controlled. Efficient seating of bush is facilitated by chamfer on drill plate and undercut on bush. Lead is to assist drill entry and swarf removal.

Figure 11.5*e*

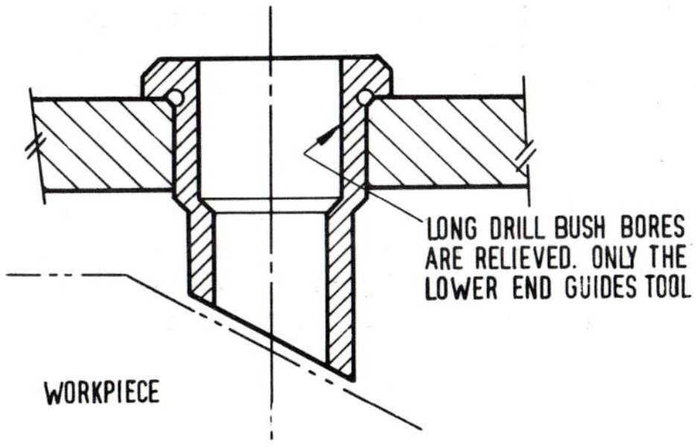

Figure 11.5*c*

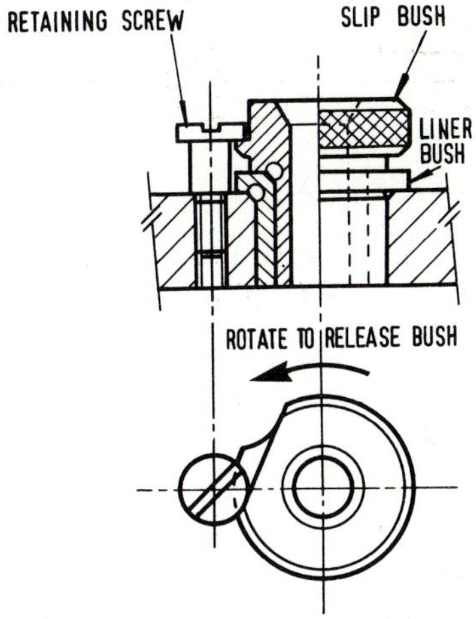

Slip bush arrangement

Used for drilling and reaming. Drilling bush is removed and replaced with reaming bush for second operations. For spot-facing, drilling bush is removed and liner bush is used.

Figure 11.5*d*

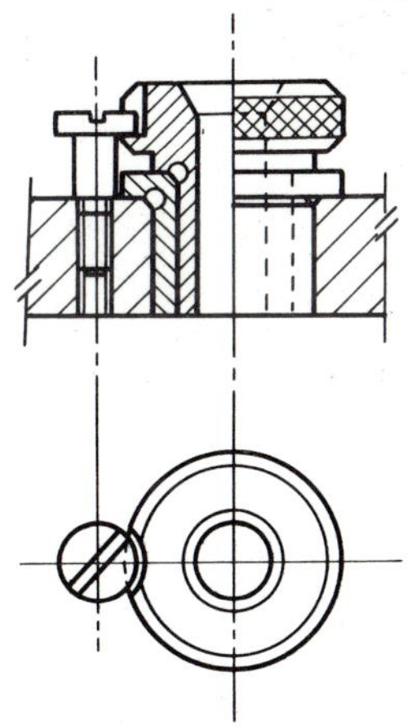

Renewable bush arrangement

On long production runs drill bushes wear. This arrangement allows quick replacement of bushes by removal of retaining screw.

Special drill bushes

The basic types of drill bushes shown above are often modified to accommodate the drilling of awkward workpieces. Opposite is shown an extended type of drill bush ground at an angle to prevent drill run on a sloping face.

LOCATION AND CLAMPING

When a component is placed in a jig or fixture it has to be held firmly and precisely in position relative to the cutting tools, to enable it to be machined to the required accuracy. To achieve this a system of location must be provided to position the component, and also a suitable clamping device to hold the component securely after it has been located in the jig or fixture.
Before considering specific examples of location, it is essential to grasp the fundamental freedoms of movement of an unrestrained object. It can move in six basic ways, three of which are rotation and three translation. This can be understood by considering the cube shown in Figure 11.6:

Figure 11.6(i)

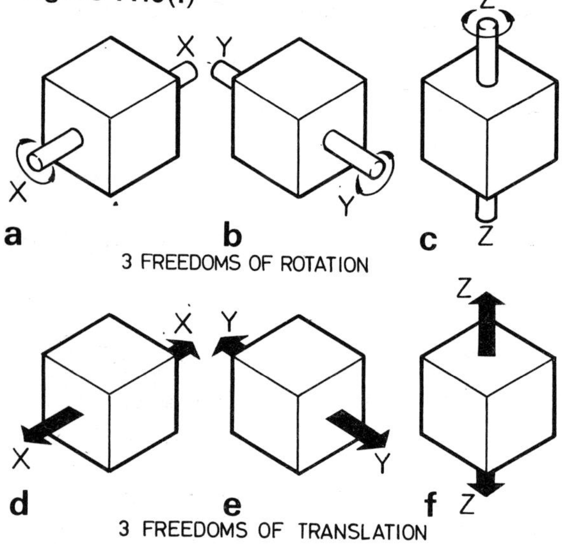

SIX FREEDOMS OF MOVEMENT

(a) shows a cube rotating on axis XX
(b) shows a cube rotating on axis YY
(c) shows a cube rot ıting on axis ZZ
(d) shows a cube moving linearly along XX
(e) shows a cube moving linearly along YY
(f) shows a cube moving linearly along ZZ

If these six freedoms can be completely restricted, then the object is immovable.

Figure 11.6 shows a cube drilled with a hole, fitted to a post, and retained by a nut and washer. This simple method eliminates five of the six freedoms of movement and the remaining freedom of rotation is partially restrained by the friction of the clamping nut and washer. This method is frequently used on components where a suitable hole is available. In a location and clamping system it is essential to restrict as many of the six freedoms of movement as necessary for the component to be held firmly and the required machining to be carried out.

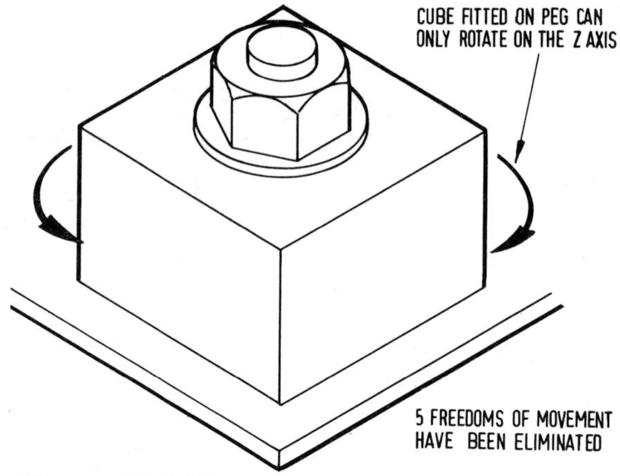

Figure 11.6(ii)

When designing a location system, the existing shape and size and the requirements of machining dictate the type of location to be used. Adjustable locations have often to be considered when dealing with unmachined castings, where a considerable variation in size is inevitable. Care must be taken to avoid redundant location, such as locating from two circular features on a component, whether they are holes or cylindrical bosses.

A typical example of redundant locating is fitting two holes in a component over two cylindrical posts in order to prevent rotation. One peg may be considered to be restricting rotation and at the same time restricting movement of translation in the same plane, hence one function is redundant. However, this would not be a practical solution because machining limits covering the hole centres and diameters would prevent loading of most of the components in the jig or fixture. Figure 11.7 shows a solution to this problem, using one cylindrical post and a locating peg.

Figure 11.8 shows some of the location methods used in jig and fixture design.

Peg locators

Figure 11.8*a* shows how pegs can provide simple locations for workpieces having rectangular, circular and other profiled shapes.

Button locators

Figures 11.8*b* and 11.8*c* illustrate two simple locating buttons to position or support workpieces. One is a simple round pad with an interference fit into the jig or fixture base, while the other is hexagonal and screwed into position. When a workpiece is contoured and requires supporting off two or more faces in different planes, adjustable locating buttons are fitted to give the required adjustment at the additional faces. Figure 11.8*d* shows a simple adjustable locating button.

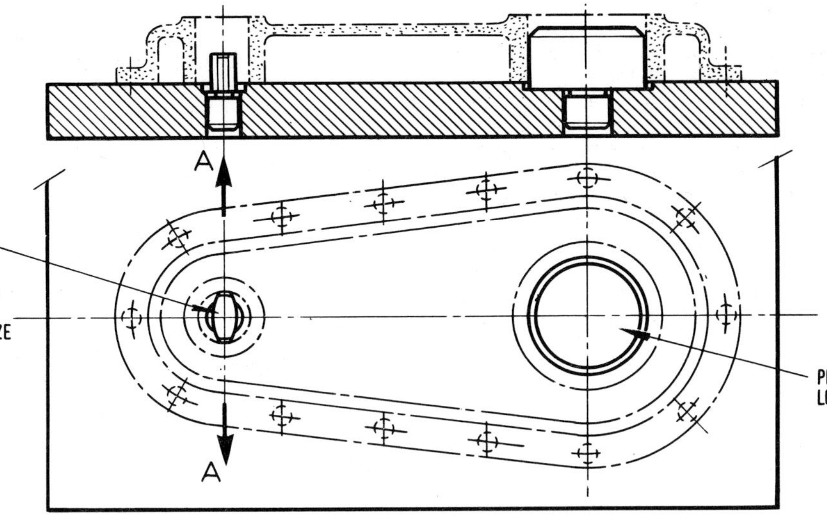

Figure 11.7

Location from two machined holes

Figure 11.8

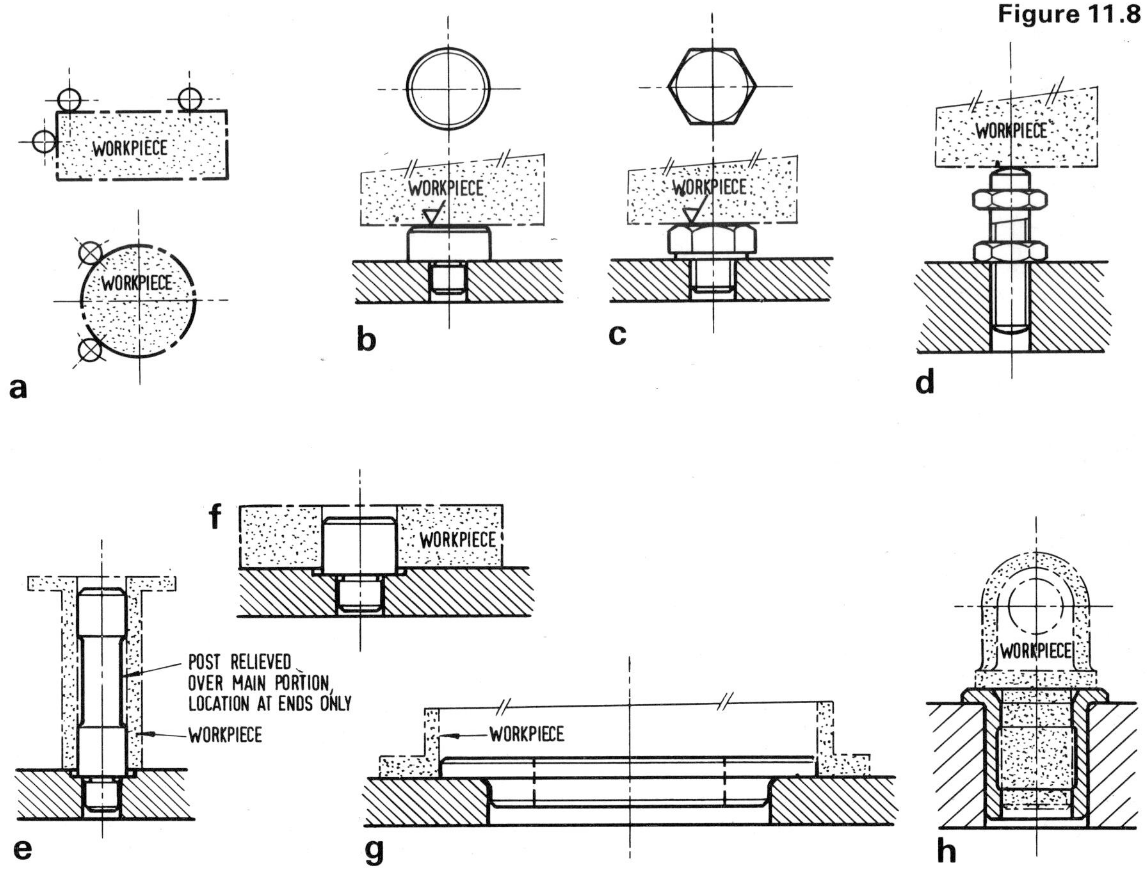

j

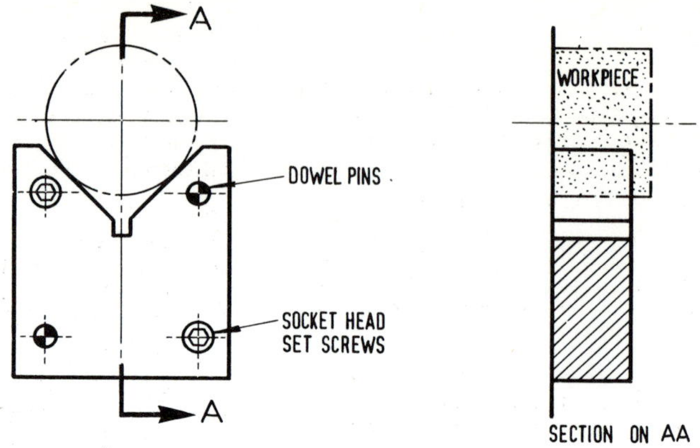

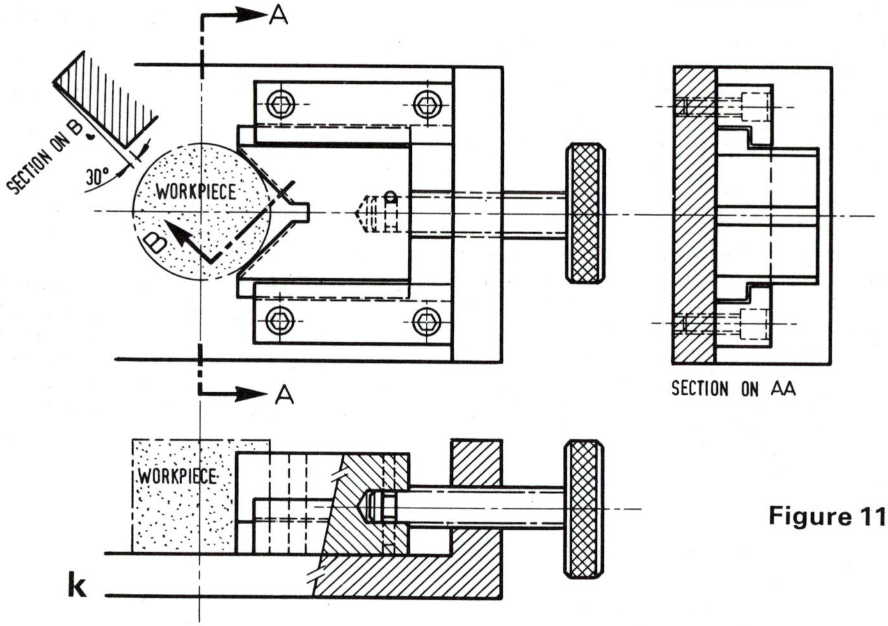

k

Figure 11.8 (*cont.*)

Post locators

One of the simplest and most effective methods of location is by a peg in a machined hole. This method restrains the workpiece in five of the six possible freedoms of movement (Figure 11.6). Figures 11.8*e*, 11.8*f* and 11.8*g* show variations of this method. Figure 11.8*e* is a tall post giving maximum support to workpieces which are long and narrow. The post is relieved over the centre section of its length, location taking place at top and bottom. This reduces friction and binding between workpieces and posts, thus assisting loading and removal. Figure 11.8*f* is a short post used where large-based workpieces require location only and not lateral support. Figure 11.8*g* shows a circular plate for locating workpieces with large circular bores. These large circular locators often have their centres bored out to eliminate unnecessary weight and also to give a swarf outlet. All locating posts are given a generous chamfer on the top edge to assist the operator in loading, and a swarf groove is machined at the plate to allow the workpiece to seat down squarely without interference from small swarf particles.

Figure 11.8*h* shows a location, which is the reverse application of the post, for workpieces having circular machined sections.

Vee locators

Figures 11.8*j* and 11.8*k* show examples of fixed and adjustable Vee locations used where workpieces are cylindrical or where it is required to locate off spherical surfaces. The Vee locators can also be used as clamping devices if the underside of the Vee is machined back at a small angle of about 3°. The movement of the sliding Vee block can be achieved by screw, cam, pneumatic or hydraulic methods.

Location devices are numerous. Special configurations are often required to produce the location necessary, but many of them are only variations on the basic examples shown above.

CLAMPING DEVICES

From a production point of view, clamping the component into position is time-consuming, so quick, simple and efficient clamping methods should be provided for during design.

Clamps should be arranged so as to apply their clamping force on solid sections of the component, in order to prevent distortion which would lead to inaccurate machining. The clamps themselves should be of such proportions that they will not deform under the clamping pressures.

The purpose of the clamp is to hold the component rigidly in position, and the clamp should in no circumstances be so positioned as to resist the direct forces of the cutting tool.

Clamping devices are numerous and each clamping system must be designed to suit the requirements of the particular application. Manufacturers of tooling devices supply ranges of standard clamping devices; for mass production pneumatic and hydraulic clamping devices are used, to reduce the time consumed in mechanical clamping.

Figure 11.9 shows examples of some simple clamping devices. Figures 11.9*a* and 11.9*b* illustrate simple clamps, one made of bent plate and the other being produced by welding two pieces of plate together. The spring gives immediate clearance of the clamp from the workpiece during removal. Figures 11.9*c* and 11.9*d* show simple clamps using a separate reaction peg; the clamp in Figure 11.9*c* allows a sliding action to clear the workpiece while that in Figure 11.9*d* clears the workpiece by rotation.

Figure 11.9*e* shows a double clamp which can be used to clamp two flat or two round workpieces. Figure 11.9*f* shows a three-point clamp with domed pressure pads to equalise the pressure on the three contact points. Spherical washers are fitted beneath the nut to allow full seating of the clamping nut.

Figures 11.9*g* and 11.9*h* show two arrangements of the quick-release C washer. To prevent loss, the washer in Figure 11.9*g* is fixed to the jig with a chain; the washer in Figure 11.9*h* is captive, and can be swung to one side for removal of the workpiece.

Figure 11.9*j* shows a clamp hook. This type of clamp is housed in a recess for support and is used in situations where space is limited. Figure 11.9*k* shows a swinging latch assembly. This assembly is well supported on the fixture by a mounting plate, and movement is restricted by a peg to obviate unnecessary movement by the operator during clamping operations. A toggle pad at the clamping point allows positive contact with the workpiece.

Figure 11.9*l* shows the arrangement of a simple quick-operation cam clamp.

Figure 11.9*m* shows a captive eye-bolt with a hand-operated clamping nut which swings clear to assist loading and unloading.

Figure 11.9*n* is a simple clamp made with a bolt to give sideways and downward components of force to hold a workpiece requiring a light cut. Figures 11.9*p* and 11.9*q* show two arrangements of wedge-type edge clamps. By tightening the nut, the serrated face is wedged against the workpiece with a downward force.

Figure 11.9*r* shows a simple screw clamp used in a channel jig.

Figure 11.9

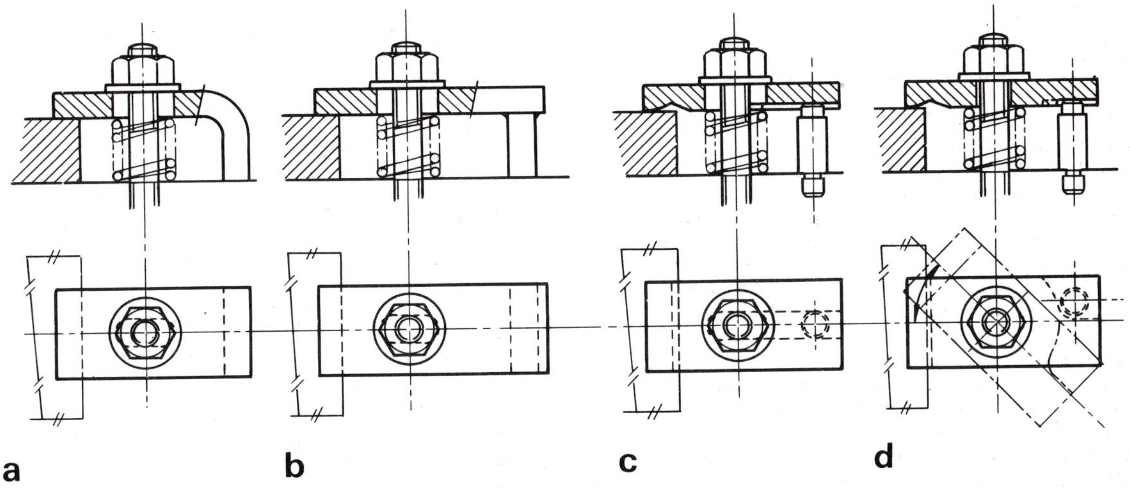

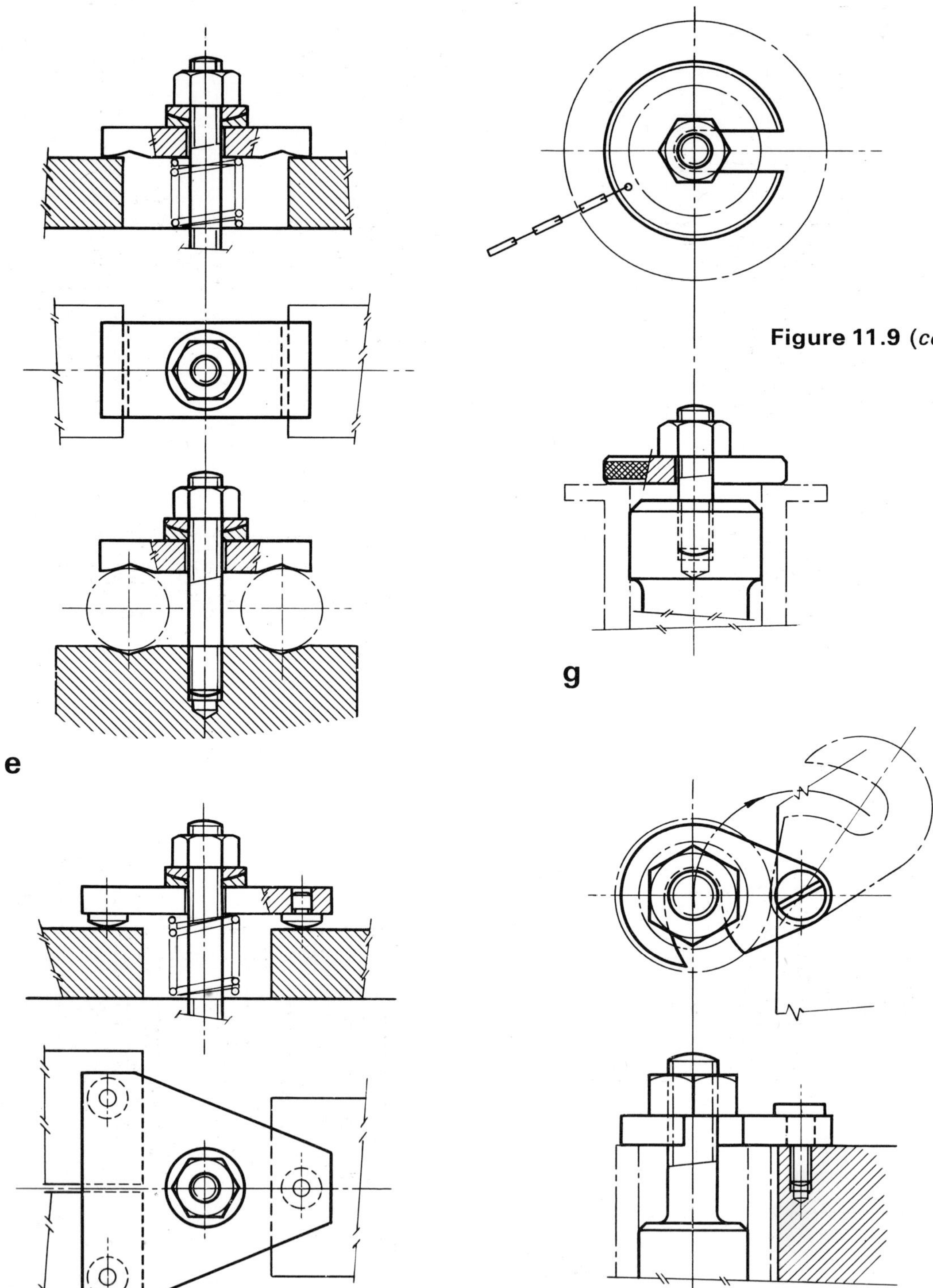

Figure 11.9 (*cont.*)

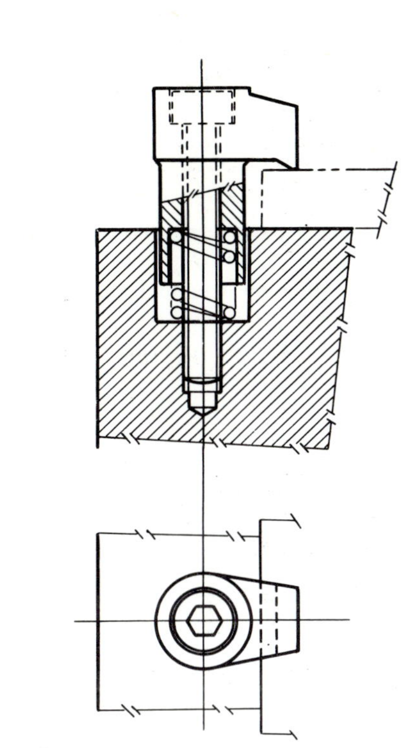
j

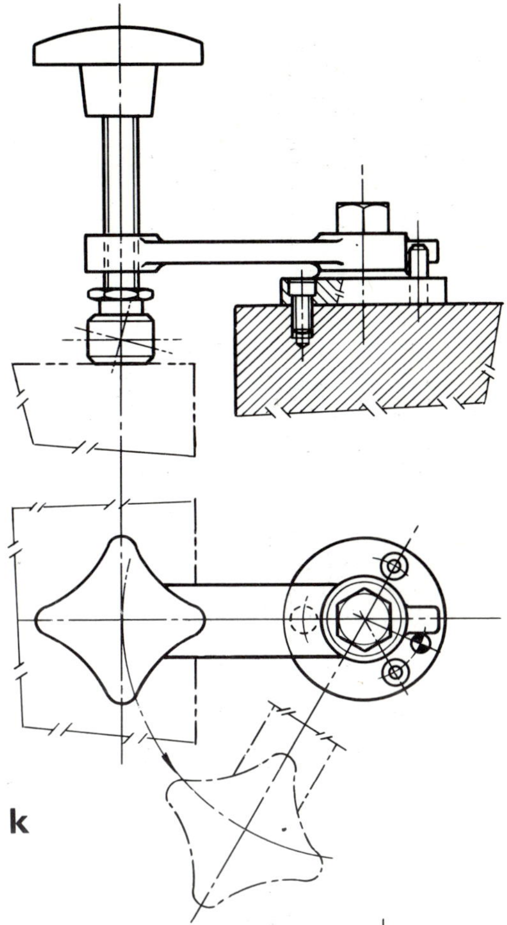
k

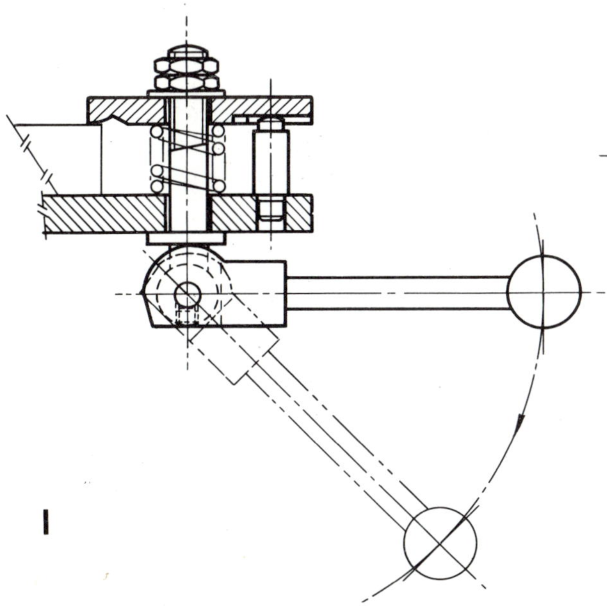
l

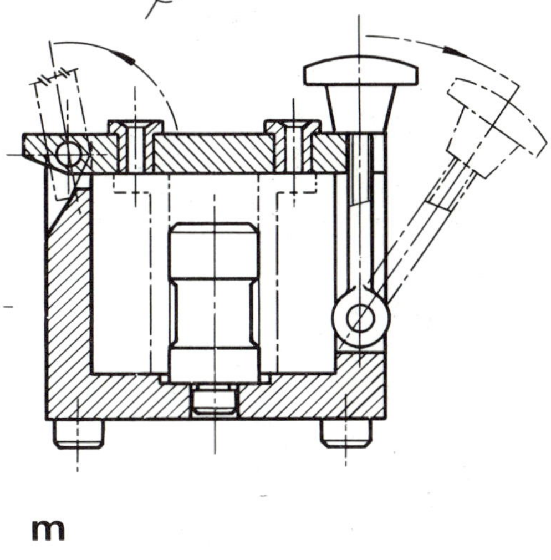
m

Figure 11.9 (*cont.*)

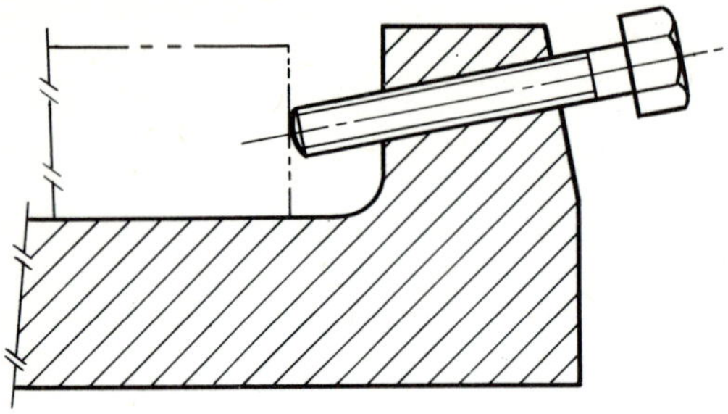

n

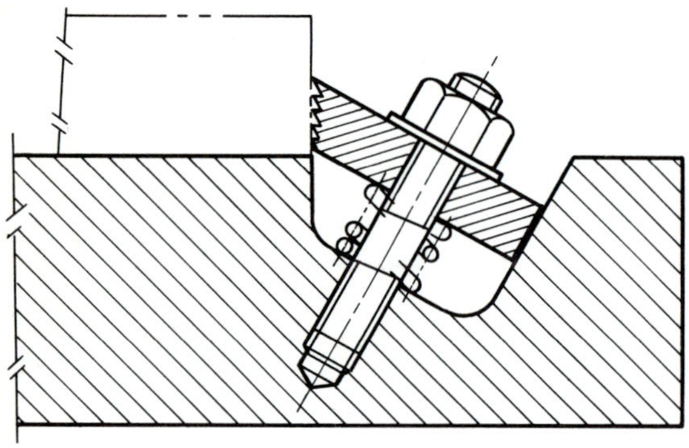

p

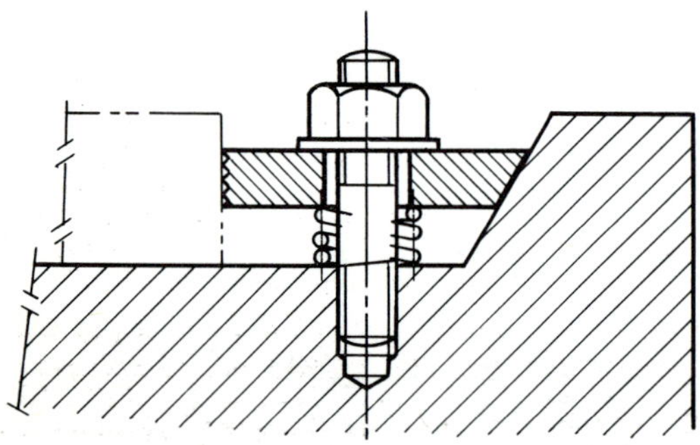

q

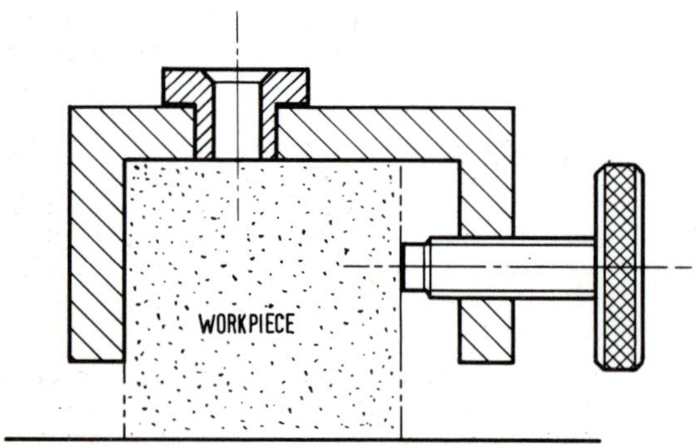

r

BASIC DESIGN PRINCIPLES

1. Decide on the operation to be carried out.
2. Decide whether the cost of a jig or fixture is justified:
   (a) for one off?
   (b) for limited run?
   (c) for mass production?
3. Ensure that the loading method is quick and simple.
4. Make the locating points adjustable if rough castings or forgings are being machined.
5. Ensure that the design is foolproof by introducing fouling pins and projections to prevent variations in loading positions.
6. Ensure that mating parts are located from a common datum.
7. Make all locations visible to the operator from the working position.
8. Clamping devices should have the following characteristics:
   (a) They should be simple, requiring only natural actions to apply.
   (b) They should clamp on to a substantial part of the workpiece to prevent deformation.
   (c) The clamps would be positioned to give maximum resistance to the cutting force.
   (d) The clamps should be integral with the body of the jig or fixture.
   (e) Tool pressure should not be taken on the clamps.
9. Allow clearance for varying size of workpiece, and also clearance for manipulation of the workpiece by the operator.
10. Ensure that there is ample swarf clearance.
11. Make the equipment rigid but compact in design, using standard components and proprietary parts as much as possible.
12. Provide means for attachment to machine tables and spindles.
13. Four feet must be provided on all jigs:
    (a) to reduce contact areas with the machine table and ensure even seating,
    (b) to clear swarf on the table.
14. Make the equipment as light as reasonably possible without loss of rigidity, with no sharp corners. Provide handles for lifting large fixtures.

MATERIALS USED IN JIG AND FIXTURE CONSTRUCTION

Jigs and fixtures are usually fabricated from standard structural steel sections and plates, cut as required and welded or bolted together, with specially machined parts for locating purposes welded or pressed into position. Sometimes it is found more convenient to cast in iron. The method used depends on cost, shape, size and time available.

Location faces are made from surface-hardened steel, and knobs and handles are often made from plastic materials.

Proprietary parts are used wherever possible. On request, manufacturers supply well-illustrated brochures of standard tooling aids which can be purchased comparatively cheaply because of specialised production.

ASSEMBLY DRAWING

The illustrated worked example (Figure 11.10) of a plate-type drilling jig shows the dimensioned detail parts which go to make the complete jig. The exercise is to make an orthographic engineering assembly drawing with the necessary views, labelling each individual detail with a balloon reference, and cross-referencing it to a parts list.

When making assembly drawings it is advisable to position balloon references in horizontal or vertical lines around the drawing. This allows for easy reference when reading the drawing, as well as adding a professional finish. Balloons placed at random over the entire drawing result in a time-wasting 'find the number' game, especially on large drawings with many references. Balloon references are usually placed on the sectional view, where the component parts appear most clearly.

Balloons are drawn about 17 mm (5/8″) in diameter, with a leader line drawn from the circumference to the outline of the component part and ending as an arrow head or dot.

All assembly drawings are cross-referenced to a parts list, which is a very important part of the drawing. Sometimes the parts list is made on a separate sheet because it is used by departments outside the actual manufacturing sections of an organisation, such as the estimating, costing and buying departments, and stores; drawing stores require it for issuing the necessary drawings to subcontractors and customers. The parts list should be drawn up with a strong visible outline grid, having columns for balloon reference number, description, number required, and material. Other columns may be added if required.

When drawing up a parts list it is usual to draw the horizontal spaces about 8 mm or 5/16″ apart, with the spaces for headings about 12 mm to 16 mm, or 1/2″ to 5/8″, wide. The width of the vertical columns depends on the best allocation of available space, the *Item* and *No. Off* columns requiring about 12 mm.

Printing on the parts list should be about 4 mm or 1/8″ high, in upright block capitals and printed between guide lines. The first letters in every line are positioned vertically below each other, and numbers must always be vertically positioned in columns of tens and units. Faint vertical guide lines can assist in achieving this, and help to produce a professional finish on a drawing. Students' work often lacks a professional finish because the importance of printing, dimensioning, title blocks and parts list is not realised until the drawings have to be used in actual manufacture.

It is on the assembly drawing that the complete outline of the finished object is shown, so it is most important to remember to use the line quality described in Chapter 2 and recommended in BS 308.

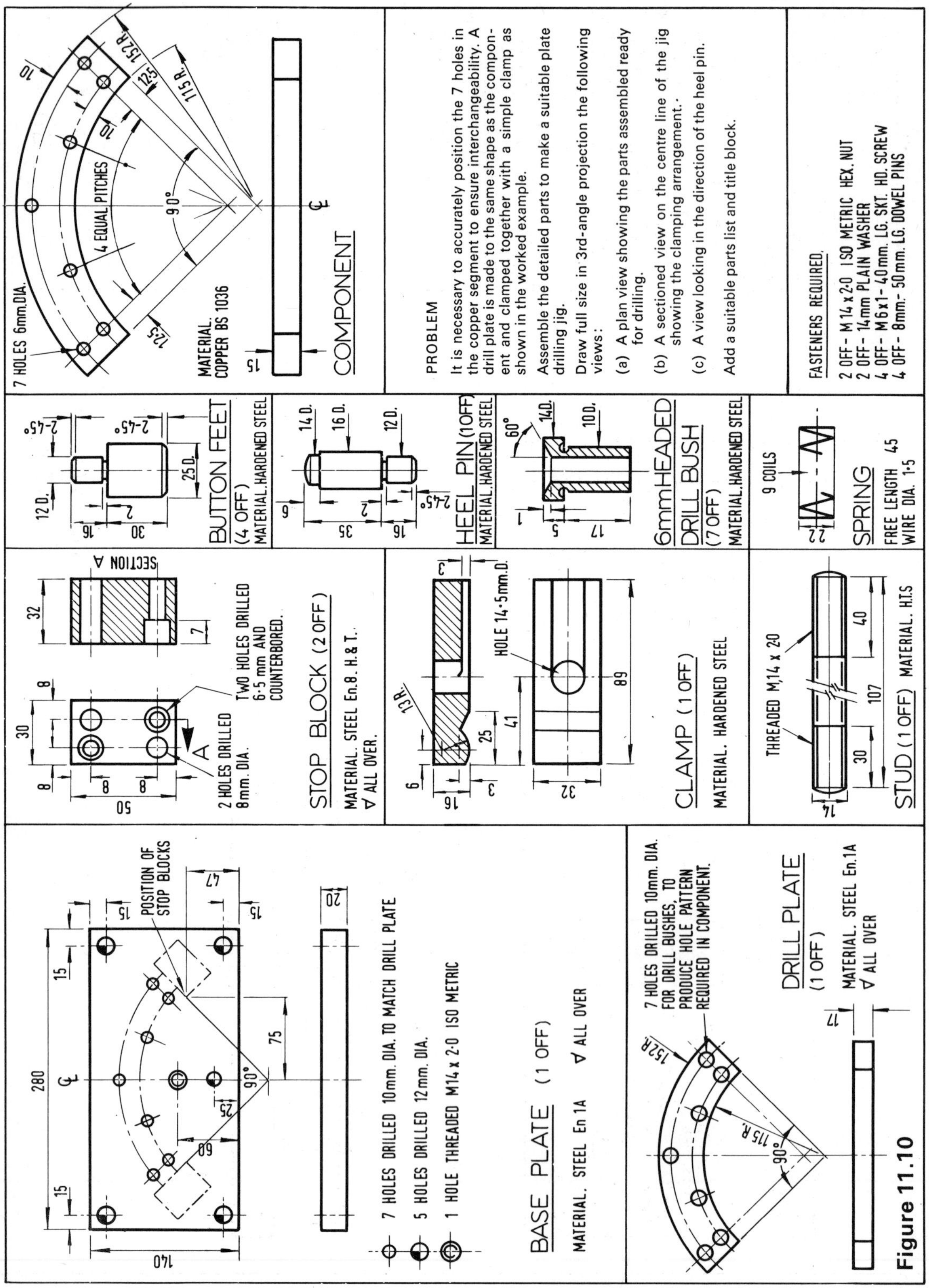
NOTE. ALL DIMENSIONS ARE IN MILLIMETRES. SCREW THREADS ARE ISO METRIC BS.3643
280
15
15
140
15
15
47
POSITION OF STOP BLOCKS
75
90°
25
60
20
7 HOLES DRILLED 10mm. DIA. TO MATCH DRILL PLATE
5 HOLES DRILLED 12mm. DIA.
1 HOLE THREADED M14 x 2·0 ISO METRIC
BASE PLATE (1 OFF)
MATERIAL. STEEL En 1A ∇ ALL OVER
152R.
115 R.
90°
7 HOLES DRILLED 10mm. DIA. FOR DRILL BUSHES, TO PRODUCE HOLE PATTERN REQUIRED IN COMPONENT.
DRILL PLATE
(1 OFF)
MATERIAL. STEEL En.1A
∇ ALL OVER
12
50
8
30
8
8
8
32
7
SECTION A
A
2 HOLES DRILLED 8mm. DIA.
TWO HOLES DRILLED 6·5 mm AND COUNTERBORED.
STOP BLOCK (2 OFF)
MATERIAL. STEEL En.8. H.& T.
∇ ALL OVER.
13R.
6
16
3
25
41
3
HOLE 14·5mm.D.
89
32
CLAMP (1 OFF)
MATERIAL. HARDENED STEEL
THREADED M14 x 20
14
30
107
40
STUD (1 OFF) MATERIAL. H.T.S
12 D.
2-45°
2-45°
25 D.
16
30
2
BUTTON FEET
(4 OFF)
MATERIAL. HARDENED STEEL
14 D.
16 D.
12 D.
9
35
2
16
2-45°
HEEL PIN (1 OFF)
MATERIAL. HARDENED STEEL
60°
14 D.
10 D.
1
5
17
6mm HEADED DRILL BUSH
(7 OFF)
MATERIAL. HARDENED STEEL
9 COILS
2·2
SPRING
FREE LENGTH 45
WIRE DIA. 1·5
7 HOLES 6mm.DIA.
10
152 R.
12·5
10
115 R.
4 EQUAL PITCHES
90°
12·5
MATERIAL. COPPER BS 1036
15
COMPONENT
PROBLEM
It is necessary to accurately position the 7 holes in the copper segment to ensure interchangeability. A drill plate is made to the same shape as the component and clamped together with a simple clamp as shown in the worked example.
Assemble the detailed parts to make a suitable plate drilling jig.
Draw full size in 3rd-angle projection the following views:
(a) A plan view showing the parts assembled ready for drilling.
(b) A sectioned view on the centre line of the jig showing the clamping arrangement.
(c) A view looking in the direction of the heel pin.
Add a suitable parts list and title block.
FASTENERS REQUIRED.
2 OFF - M14 x 2·0 ISO METRIC HEX. NUT
2 OFF - 14mm PLAIN WASHER
4 OFF - M6 x 1 - 40mm. LG. SKT. HD. SCREW
4 OFF - 8mm.- 50mm. LG. DOWEL PINS

Figure 11.10

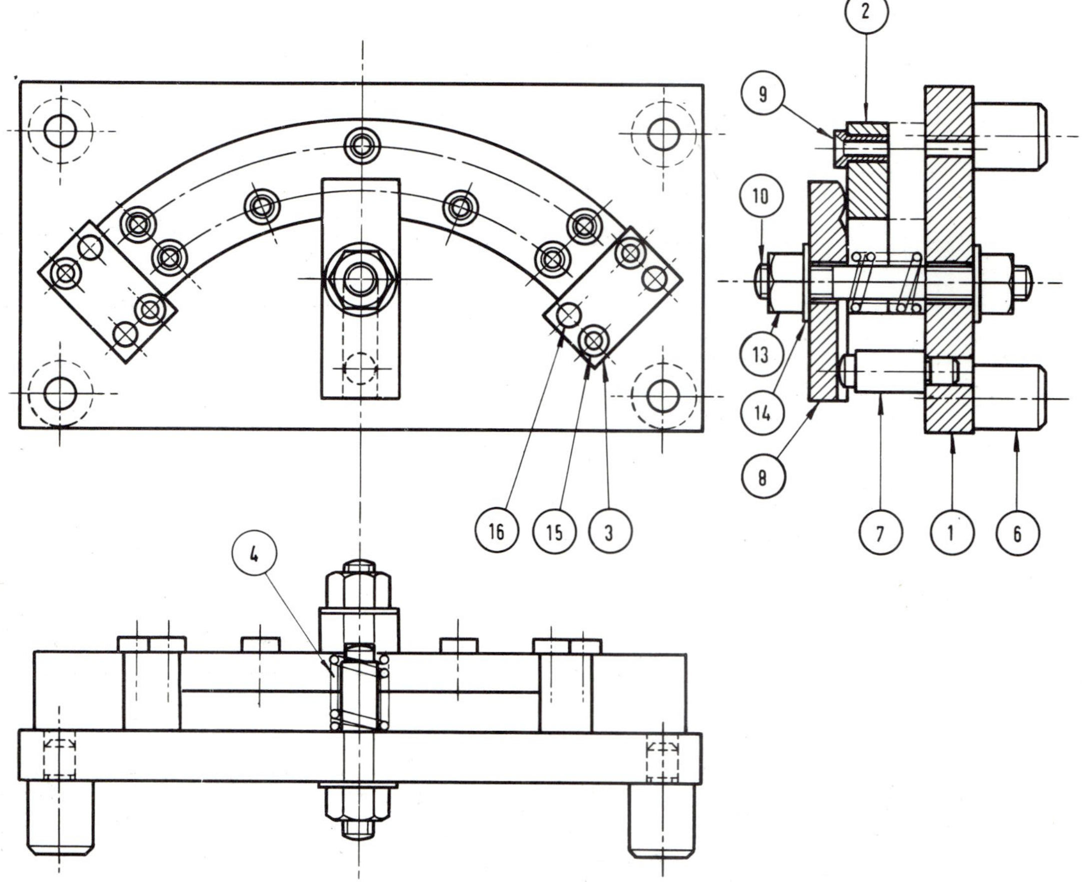

PARTS LIST			
ITEM	DESCRIPTION	NO. OFF	MATERIAL
1	BASE PLATE	1	STL En 1A
2	DRILL PLATE	1	STL En 1A
3	STOP BLOCK	2	En 8 H&T
4	COMPRESSION SPRING	1	
5			
6	BUTTON FEET	4	BOUGHT OUT
7	HEEL PIN	1	
8	CLAMP	1	HARDENED STEEL
9	20mm HEADED DRILL BUSH	7	
10	STUD	1	H.T. STEEL
11			
12			
13	M 14 x 2,0 HX NUT	2	STL
14	WASHER	2	STL
15	M6 x 1,0 SKT. HD. SCREW	4	UNBRAKO
16	8 mm ROLL PIN x 50 mm	4	UNBRAKO
17			
18			
19			
20			

ISS.	REVISIONS	date
1	———	APR 70

Gregory F Whiteley Ltd.,
34 Bedford Row, London, W.C.1.
telephone 01-405 2371
gfw

CLIENT

TITLE
PLATE DRILLING JIG

scale	1-1	DRAWING No.
drawn	A.S.C.	
checked	VEB.	
approved	GFW.	
date	JAN 70	

Figure 11.10*a*
Worked Example

NOTE. ALL DIMENSIONS ARE IN MILLIMETRES. SCREW THREADS ARE ISO-METRIC (BS. 3643)

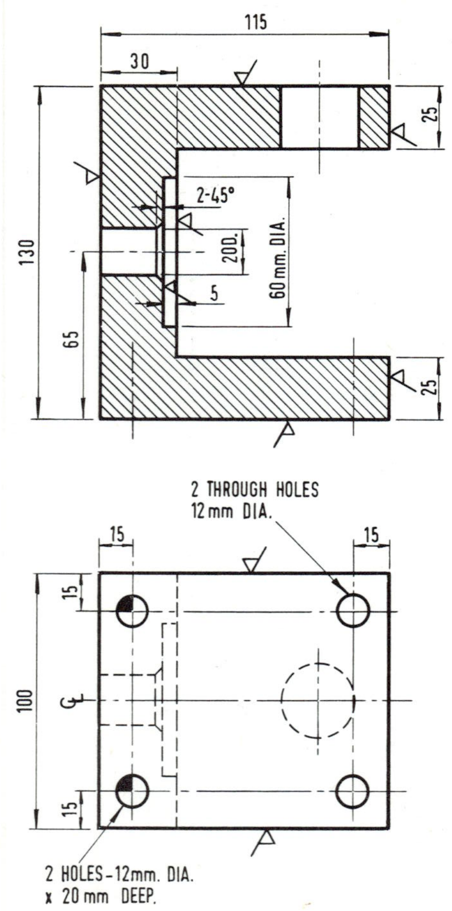

CHANNEL (1 OFF)

MATERIAL. CAST IRON

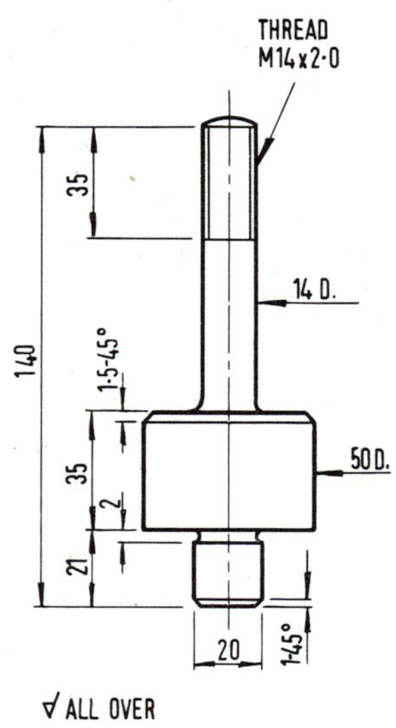

LOCATION PIN (1 OFF)

MATERIAL. STEEL En.8. H.& T.

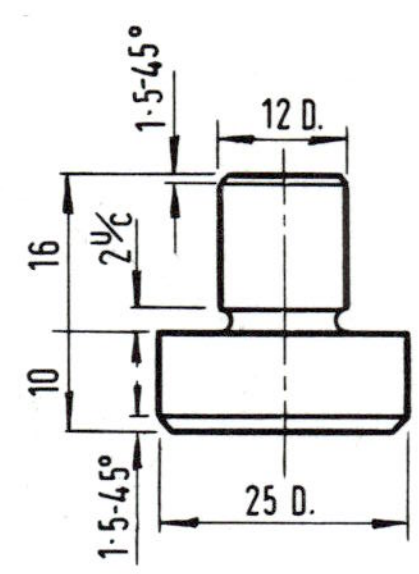

BUTTON FOOT

(4 OFF)

MATERIAL. HARDENED STEEL

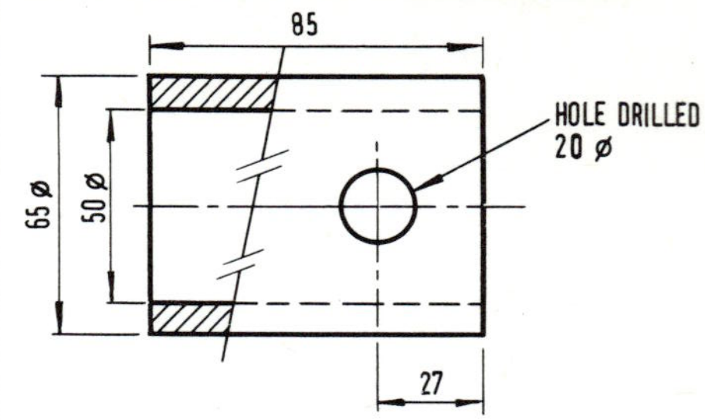

CHANNEL TYPE DRILLING JIG

COMPONENT

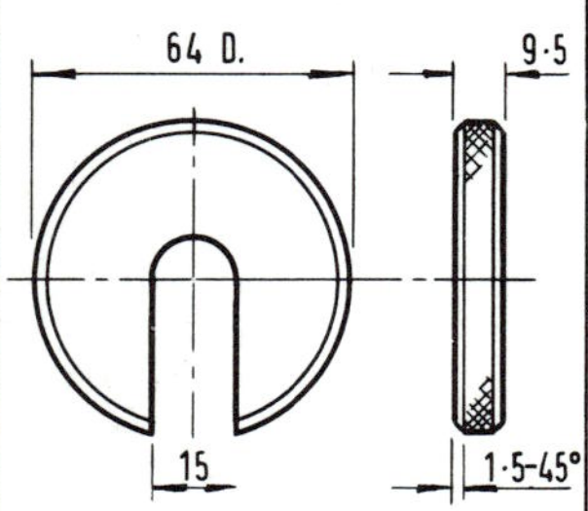

'C' WASHER (1 OFF)

MATERIAL. HARDENED STEEL

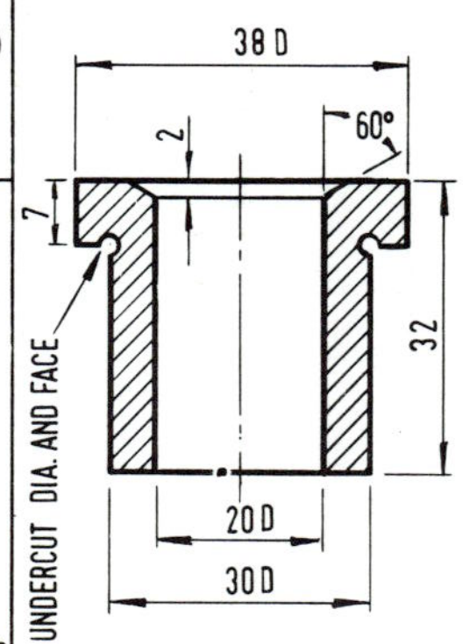

HEADED DRILL BUSH (1 OFF)

MATERIAL. HARDENED STEEL

Exercise 11.1

PROBLEM

The tubular component shown above is required in quantity. A drilling jig is required to drill the 20mm diameter hole.

Assemble the detail parts shown to form a channel-type drilling jig as shown in the pictorial view.

Draw full size in 3rd-angle projection the following views.

(a) A sectional view in the direction of arrow 'A'. Show the component in chain-dot outline.

(b) A view in the direction of the 'C' washer.

(c) A plan view looking down on the drill bush.

Also, add a suitable parts list, and a title block, showing title, scale, date, projection and name.
Draw the 'C' washer held in position by an M.14 × 2,0 ISO-metric hexagon nut and plain washer.

Exercise 11.1

Pictorial assembly

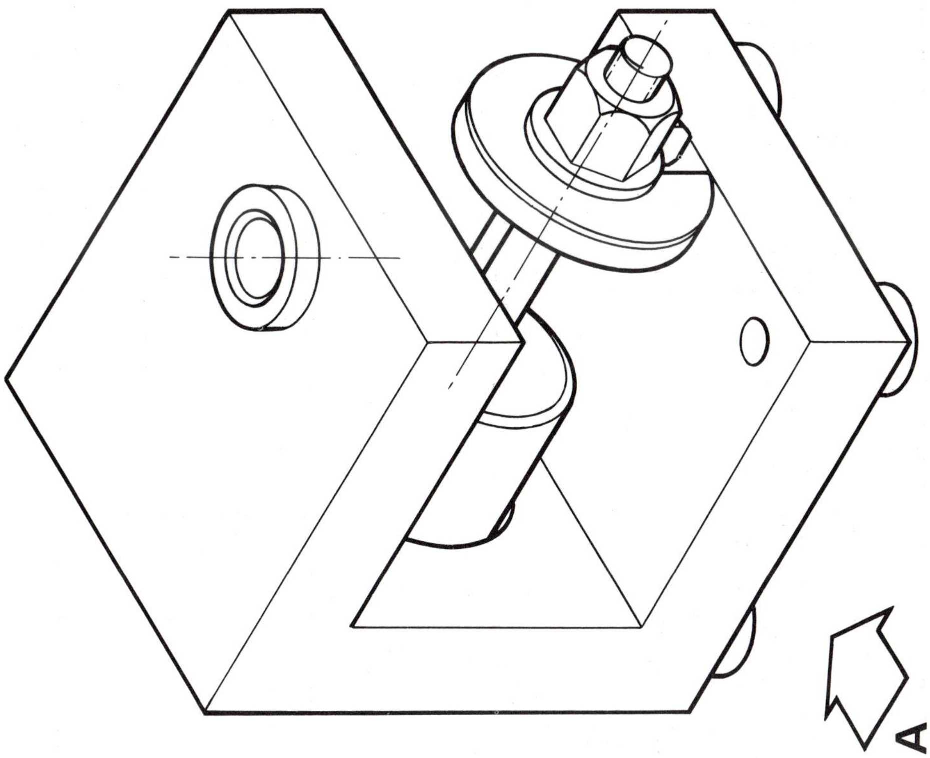

Exercise 11.2

PROBLEM

The component drawing shows a fork end fitting for a tubular member of an airframe structure. It is manufactured from an alloy bar and the first machining operation by turning produces a spherical ended part, as shown by the chain-dot outline.

A fixture is now necessary to enable the fork end to be gang-milled from the spherical end. Assemble the detailed parts shown to form a *string milling fixture* in which 5 components at a time may be loaded for machining. Position the setting block as instructed on the detail, and fit tenons to the underside of the base plate, suitable to fit the milling table in your local workshop.

Draw half-size in 3rd-angle projection the following views:

(a) A view in the direction of the top-side of the base plate showing all the parts assembled.

(b) A sectional view taken on the long centre-line of the base plate.

(c) 3 scrap sections suitably placed and annotated to show the following:

1. Screw fixing of side plate to base.
2. Screw fixing of end plate to side plate.
3. Screw and dowel fixing of setting block to base plate.

Balloon reference all parts and add a suitable parts list and title block.

Show component outlines in red in both views 'a' and 'b'. Estimate any missing dimensions.

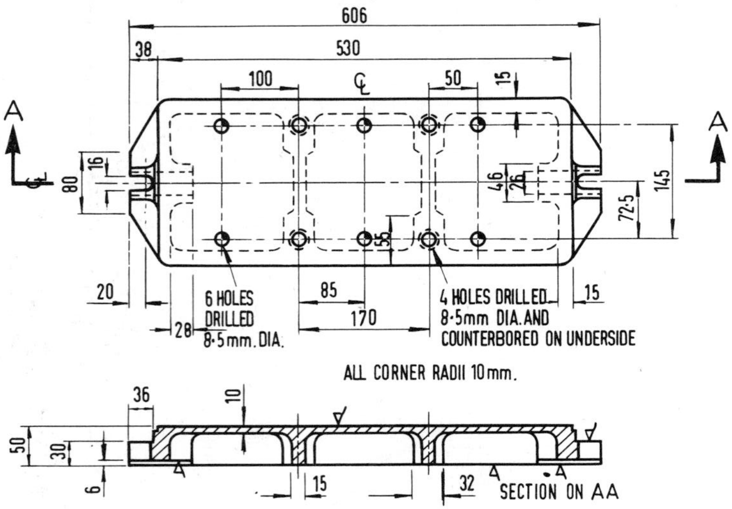

BASE PLATE (1 OFF) MATERIAL. CAST IRON

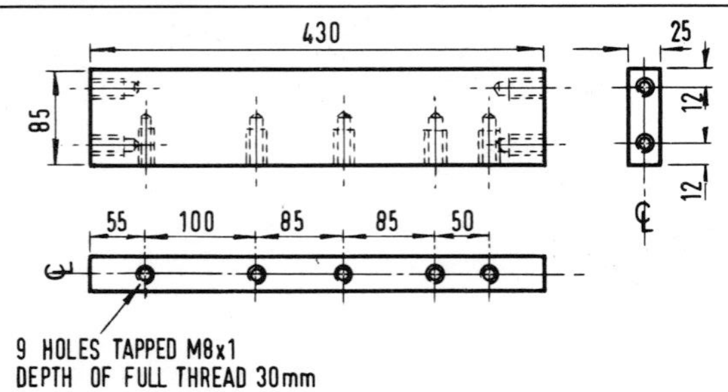

SIDE PLATE (2 OFF) MATERIAL. STEEL En.1

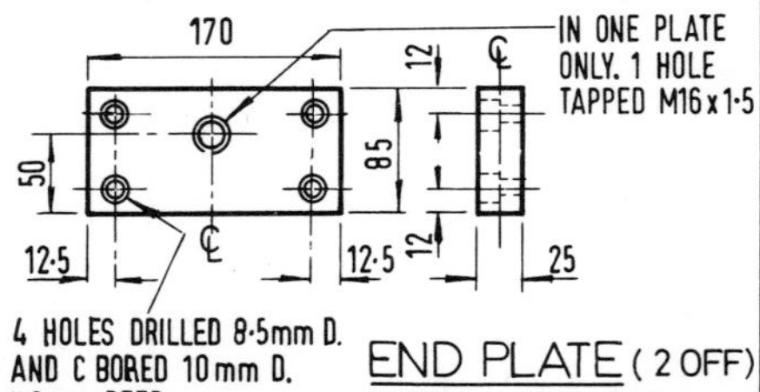

END PLATE (2 OFF)
MATERIAL. STEEL En.1

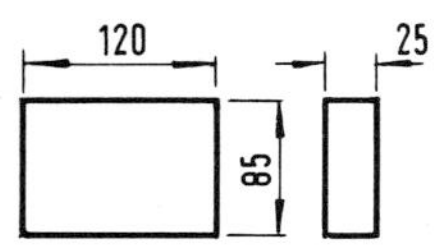

CLAMPING PLATE
(1 OFF)
MATERIAL. STEEL En.32 CASE HARDENED

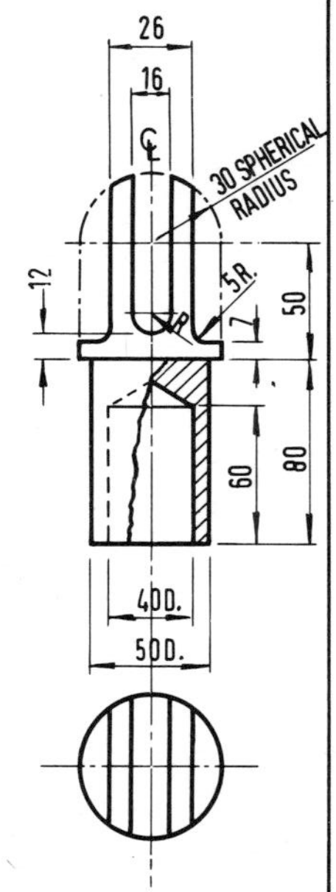

COMPONENT
FORK END

MATERIAL. ALL ALLOY H.E.30 WP.

FASTNERS REQUIRED

4 OFF- M8x1 SOCKET HEAD CAP SCREWS 65mm LG.
14 OFF- M8x1 " " " " 35mm LG.
2 OFF-M6x1 SOCKET HEAD CAP SCREWS 20mm LG.
1 OFF-M16x1·5 HEXAGON HEAD BOLT 80mm LG.
2 OFF-6mm. DIA. DOWEL PINS 25mm LG.

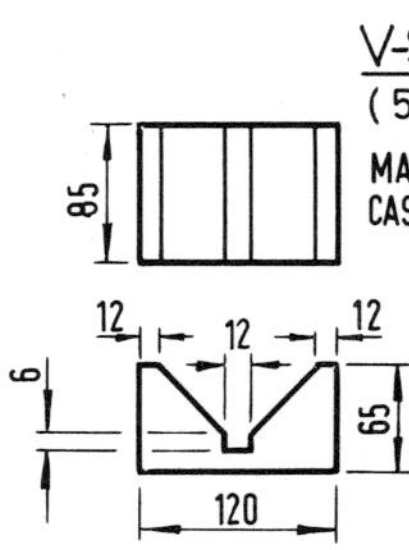

V-SECTION
(5 OFF)
MATERIAL. En.32 STEEL CASE HARDENED

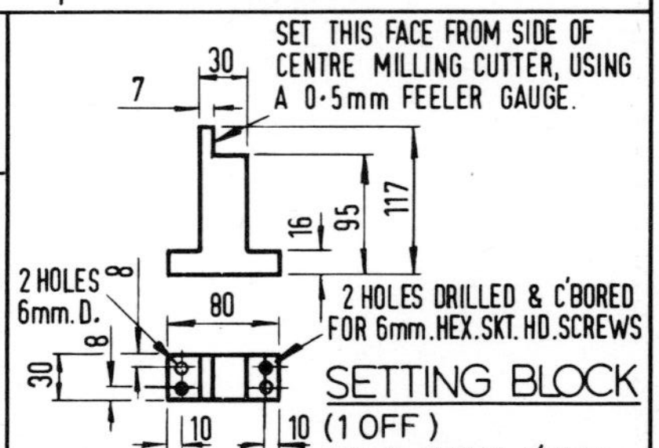

SETTING BLOCK
(1 OFF)
MAT. En.32 STEEL C'HDND

Exercise 11.2

Pictorial assembly

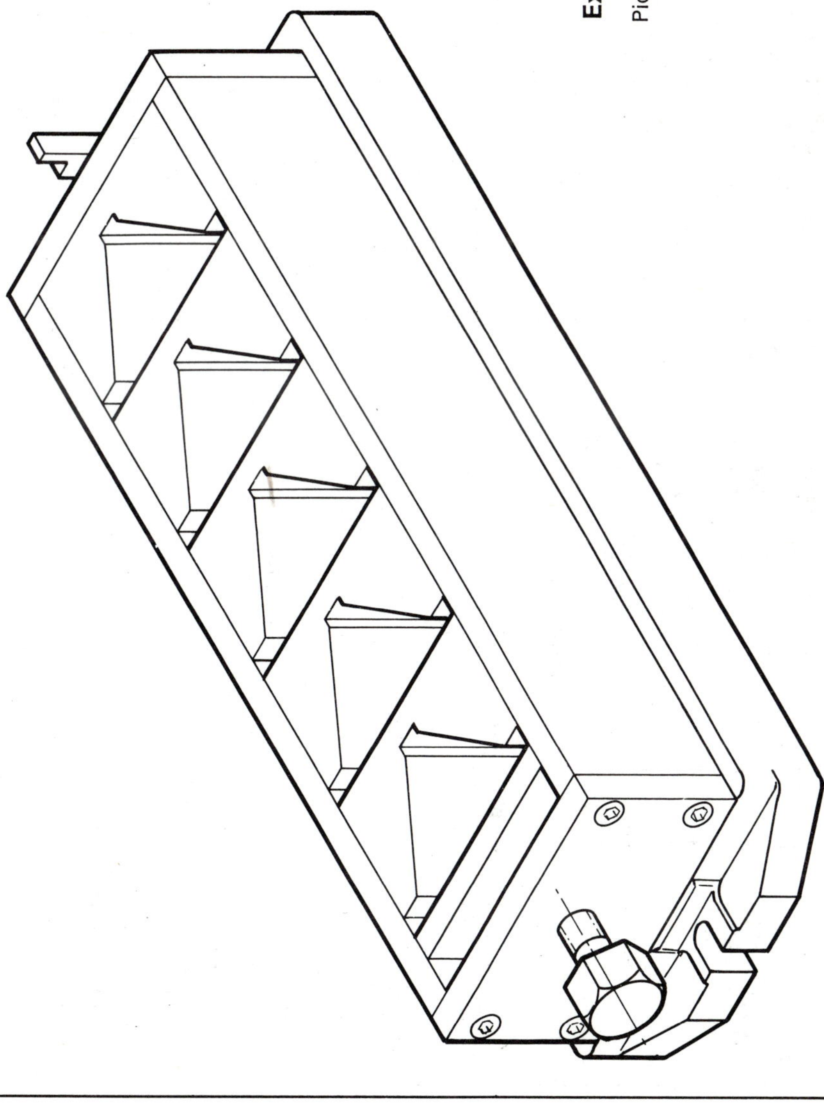

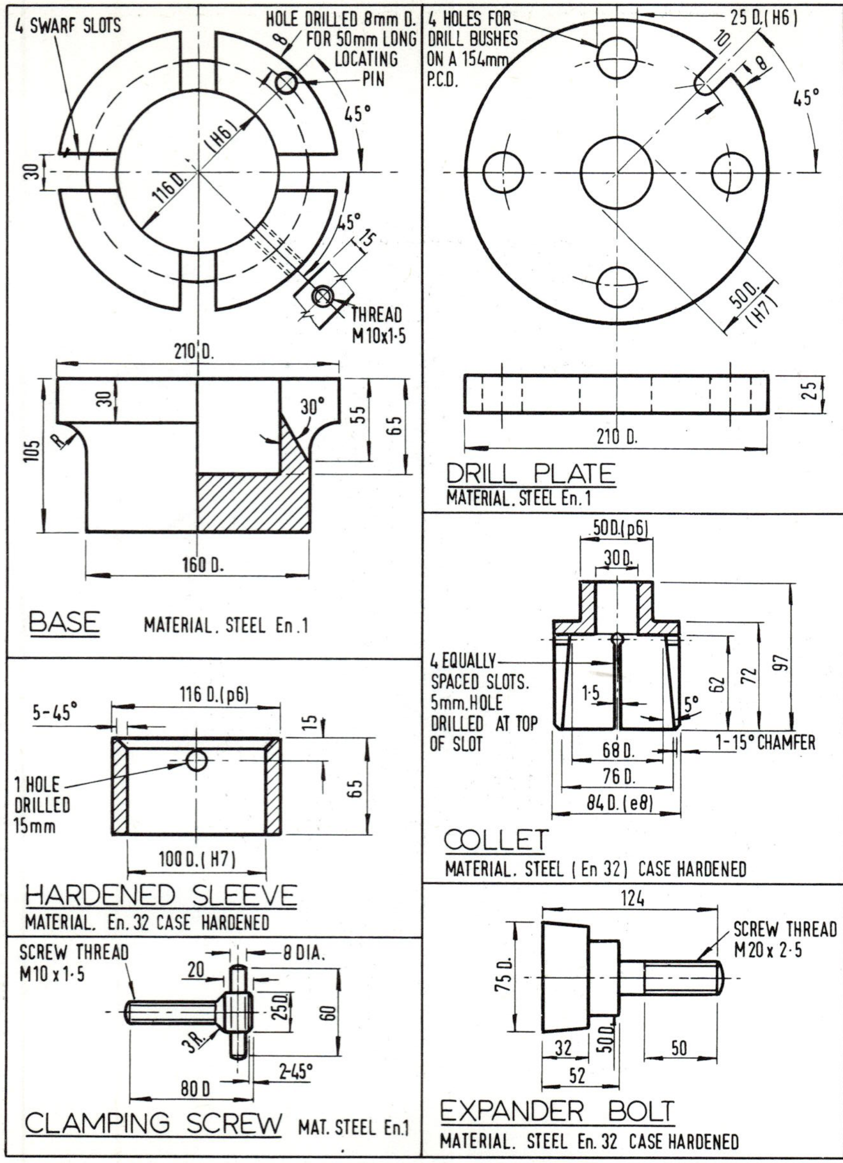

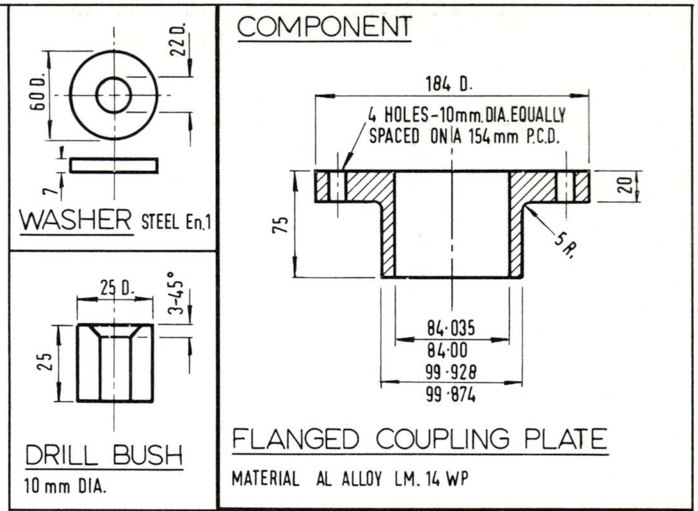

Exercise 11.3

PROBLEM

The flanged coupling plate is to be produced in limited quantity. It is cast in aluminium alloy then machined to size. A drilling jig is necessary to drill the four 12-mm diameter holes in the flange. Assemble the detailed parts shown to form a pot-type drilling jig using the two accurately machined faces for location.

Draw full size in 3rd-angle projection the following views:

(a) A view in the direction of the drill plate and clamping nut.

(b) An aligned sectional view through the assembly showing the fitting of the drill bushes and the clamping screw.

(c) A scrap sectional view to show the detail of the 10-mm diameter locating pin.

Balloon reference all parts and add a suitable parts list and title block.

Estimate any missing dimensions, and secure the expander bolt with a standard M20 × 2·5 hexagon nut.

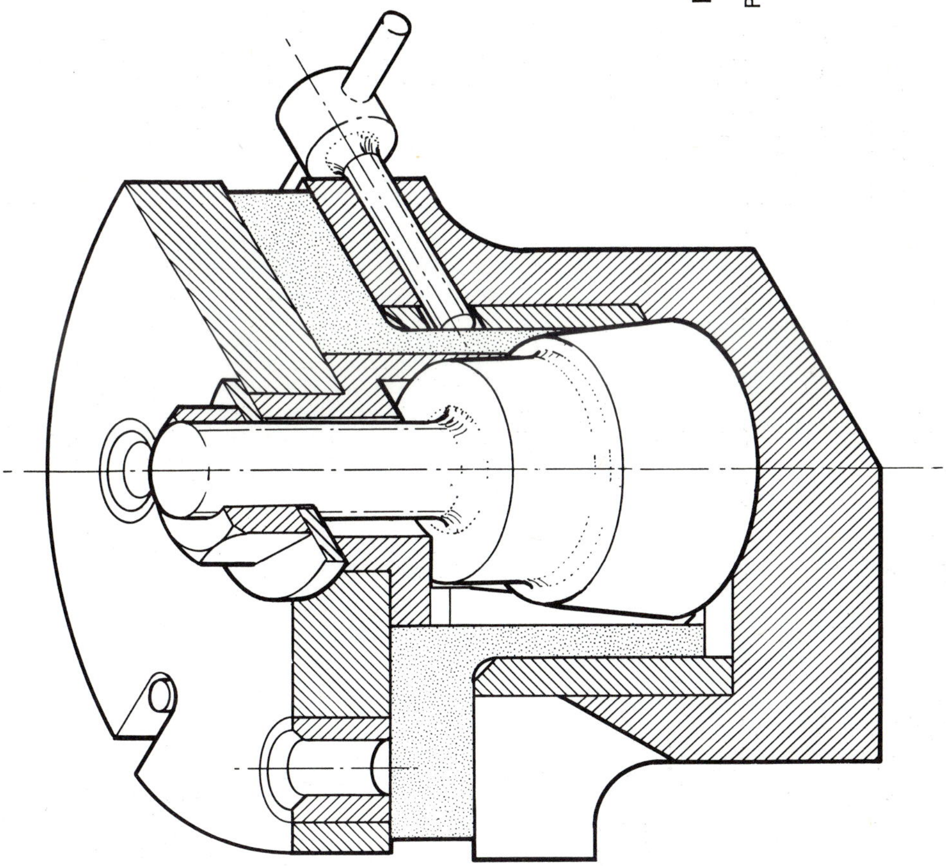

Exercise 11.3
Pictorial assembly

When a new product is being designed, there may be no record of sizes and materials used previously, or no one can suggest the most suitable material for a particular product. Calculations of size, strength and function, as well as decisions on materials, have then to be made by the designer.

A competent designer should have a wide knowledge of available materials, their mechanical and physical properties, and the methods of manufacture. Consultations with metallurgists, manufacturers of materials and production engineers will, if necessary, assist him in making his selection.

There are three main problems in putting a design on paper; in order of increasing difficulty and importance these are:

1. Making the drawing orthographically.
2. Dimensioning the drawing and interrelated parts with all the fits and clearances necessary for manufacture.
3. Selecting the right materials for each particular component.

Assuming that drawing and dimensioning are fully understood, how is a material selected? The designer of today has a multitude of different engineering materials at his disposal, the number increasing each year with the discovery of new chemical and alloying techniques using, for example, elements such as the rare earths, cerium, ytterbium, europium, lanthanum and hafnium. Many new materials have been developed for the extreme conditions encountered in space flights and electronics.

The selection of the material is, in the first instance, the responsibility of the designer who first conceives the idea for a new product. His years of experience and study, kept up to date by reading of new discoveries, and his close contact with materials and machines, enable him to choose a material which, even if not precisely correct, is of the correct type.

When selecting a material, cost is often an overriding factor imposed on the designer by company policy.

Selection of a material must be governed by the conditions to which it will be subjected in service. These may vary widely, and every aspect of operation should be scrutinised. Each component must also be examined for its importance in the overall design, and with a view to minimising its mass without endangering safety if it should fail mechanically. This safety check may involve calculations and physical tests for rigidity, resilience, flexibility etc., and for resistance to stress, shock, overload and similar conditions.

Other important features to be considered are friction of moving parts, the temperature at which components must operate (bearing in mind expansion and contraction with temperature fluctuations) and exposure of the parts to corrosive liquids or atmosphere.

There are also the requirements of economic production. A small quantity of manufactured components would not justify elaborate and expensive tooling. The use of stock raw materials should be considered, and the expensive labour of skilled craftsmen should be minimised. Thought should be given to whether fabricated parts might sometimes be used instead of castings. Machine tools for operations for turning, milling, shaping and planing must be worked out, and also the amount of finishing.

Large-quantity production completely changes the picture—the type of materials used, and the method of manufacture. The high cost of patterns and tooling required may be negligible when spread over the whole production run. Casting in iron or non-ferrous metals, diecasting in zinc-based or aluminium alloys, investment casting or even plastic moulding, extrusion or dipping, will often produce a satisfactory article with all the necessary physical and mechanical properties at much less cost.

Other factors which may have to be considered when selecting materials for a component are response to heat treatment, forgeability, weldability, suitability for hot or cold forming, whether casting is practicable and whether machining is possible. There are also the problems of finishing operations such as plating, painting, plastic coating or anodising and, in considering these problems, cost, customer appeal and feasibility must all be taken into account.

As regards mechanical properties, manufacturers usually quote, for metals: ultimate tensile strength (U.T.S.), yield stress or 0·1% proof stress, percentage elongation, Izod impact value for a notched specimen, and a hardness number. These values give the designer useful information for calculating sizes required for adequate strength. Although percentage elongation can give an indication of ductility, and impact and hardness figures can give an indication of toughness and hardness, no direct use of these figures is made in calculations, although they are employed in assessments to safeguard against premature failure. A difficulty that designers have to face in calculations of stress is that assumptions of operating conditions are often based on insufficient data, and these conditions are sometimes too difficult to analyse.

The designer has to choose between different methods of approach according to the purpose of a particular component. Is the main consideration to be strength or wear? Is the problem one of bearing or fatigue stress? In all these situations some allowance is necessary, and it is only by long experience that reasonable safety factors have been evolved to ensure against failure caused by overload.

Bearings with sliding or rolling contact cannot be designed simply on strength considerations. Toughness and hardness of material, speeds, lubrication, conditions of bearing, and clean or dirty working conditions must be considered; operating values can only be obtained by test and experience.

Materials working at high temperatures are subject to creep and loss of strength. At abnormally low temperatures the mechanical properties of materials are often radically changed, for example, at very low temperatures rubber tubing can be broken like a rotten stick and tin can be crushed into a grey powder.

An engineering designer in any field must have a wide knowledge and appreciation of the properties of available materials, and from his experience (because there is no universal reference for every problem) he will be able to select the material most suitable for his purpose. With his acquired skill, which is a combination of ability, training and experience, he will so proportion the article as to give the best economy and use of the material chosen.

The final test of any new design is how it functions in service.

PHYSICAL PROPERTIES OF A MATERIAL

Properties from the following list are quoted by manufacturers to assist designers in their use. These properties, as distinct from the mechanical properties, are inherent in the physical structure of the material:

Melting point
Boiling point
Density
Electrical conductivity
Thermal conductivity
Magnetic properties
Specific heat
Coefficient of thermal expansion
Latent heat of fusion
Latent heat of vaporisation
Water absorption rate
Atomic weight

MECHANICAL PROPERTIES OF MATERIALS

The fundamental mechanical properties of a metal are:

Ductility
Malleability
Toughness
Strength

Ductility refers to the ability of a metal to undergo deformation without breakdown under the influence of tensile loading, such as in the processes of wire or tube drawing.

Malleability is the ability to withstand rupture under the influence of compressive loading, as in the processes of pressing, stamping, forging, extruding, rolling etc.

Although malleability is similar to ductility, both properties cannot be assumed for any particular material. When a metal is hot its malleability is increased, and advantage is taken of this property in hot forging, stamping and rolling, whereas in the cold condition the metal would be weak in tension. Ductility of a metal, however, is generally decreased on heating because the tensile strength is reduced by the heating process.

Toughness is the ability of the metal to withstand shock, and to withstand without fracture the effect of bending and shear stresses. A hammer head, although hard on the outside, has a tough core; under repeated shock conditions it does not shatter as it would do if made from hard and brittle cast iron. Copper and aluminium are, by the above definition, extremely tough where cast iron is not. Toughness should not be confused with strength or hardness.

Since ductility, malleability and toughness cannot as yet be expressed directly in numerical terms, certain mechanical properties based on mechanical tests on specimens are quoted by manufacturers. This allows designers to make a comparison between materials.

TYPICAL MECHANICAL PROPERTIES QUOTED FOR A MATERIAL

Ultimate tensile strength (U.T.S.) N/m² (tonf/in², lbf/in²)
Yield stress " " "
0·1% proof stress " " "
Elongation %
Vickers Diamond Pyramid, Brinell or Rockwell hardness No.
Young's modulus of elasticity N/m² (tonf/in², lbf/in²)
Modulus of rigidity " " "

TENSILE TESTING

Specimens are cut from materials and subjected to tensile loading on a tensile test machine. The results of load applied against extension of specimen are plotted on a graph to give the typical load/extension diagram familiar to all students of mechanical engineering, who carry out such tests in their laboratory work for mechanical science.

Figure 12.1 shows a Hounsfield tensile-testing machine and an Avery tensile-testing machine; Figure 12.1*a* shows the shape of a typical metal specimen used for tensile testing.

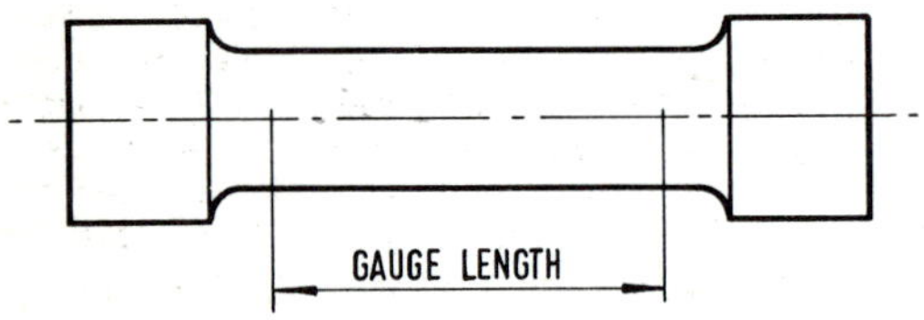

Typical metal specimen used for tensile testing

Figure 12.1*a*

Figure 12.2 shows a typical load/extension diagram for annealed carbon-steel, which is plotted out on the revolving drum of the machine as the test is carried out.

With most metals that have been heat-treated or cold-worked, load/extension diagrams do not show a definite elastic limit or yield point, particularly non-ferrous metals. As the yield point is important to the designer, it becomes necessary to specify a stress which corresponds to a definite amount of permanent strain.

Figure 12.3 shows a typical load/extension diagram for a non-ferrous alloy, and a method of obtaining the 0·1% proof stress which is used as a substitute for yield stress.

PERCENTAGE ELONGATION

When broken on the tensile-strength test, the two parts of the specimen are carefully fitted together and measured to obtain the total increase in length. The increase in length is worked out as a percentage and this figure gives an indication of the ductility of the material.

$$\text{percentage elongation} = \frac{\text{increase in gauge length} \times 100}{\text{original gauge length}}$$

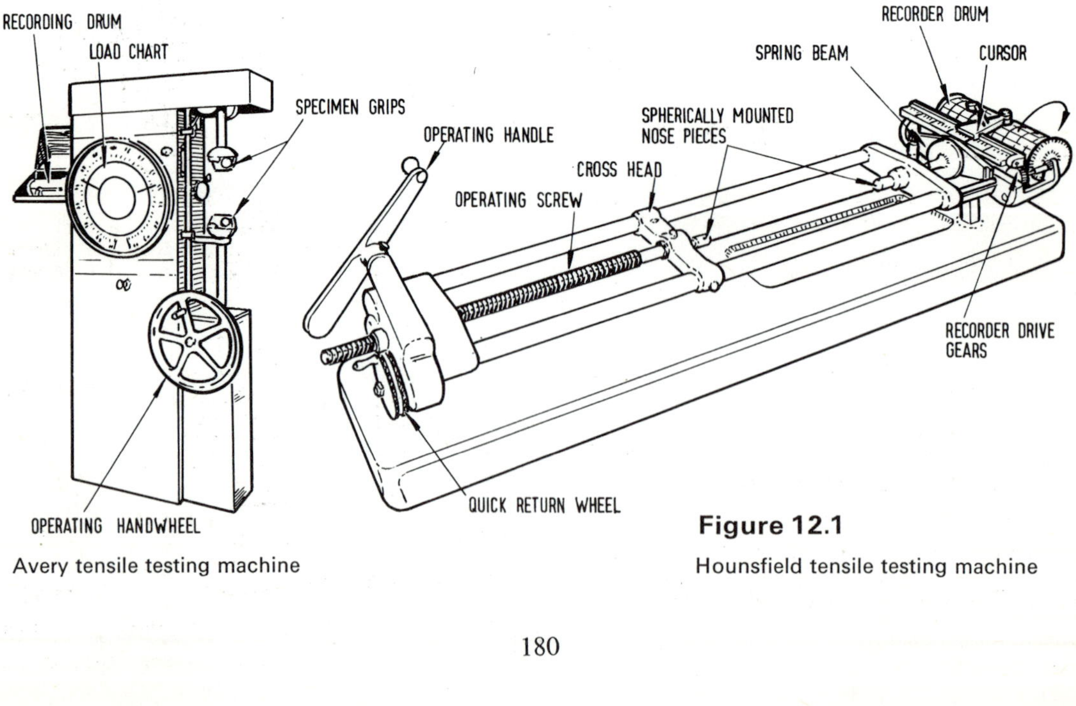

Avery tensile testing machine

Figure 12.1

Hounsfield tensile testing machine

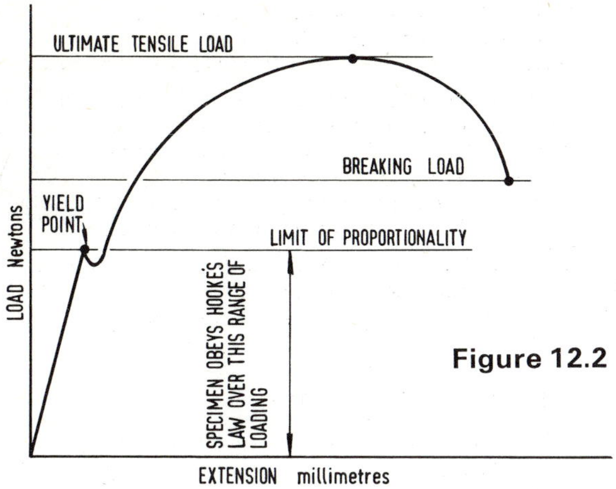

Figure 12.2

Typical load: extension curve for annealed carbon steel

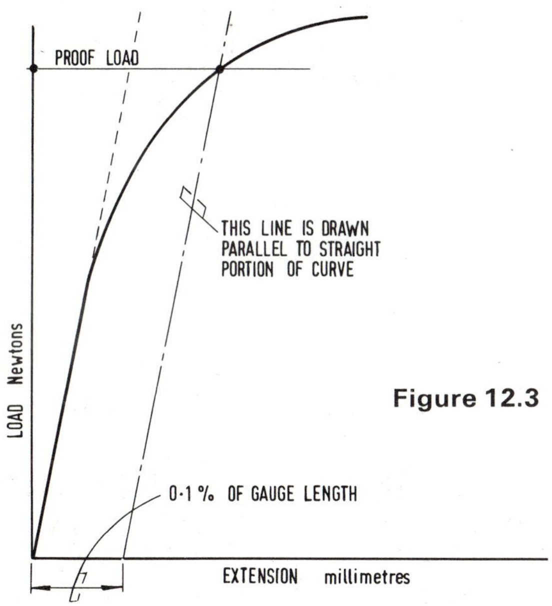

Figure 12.3

Typical load: extension curve for non ferrous material, and method of obtaining 0·1 per cent proof load

HARDNESS TESTING. Figure 12.4.

The three principal methods of measuring hardness of materials are:

1. The Brinell hardness test
2. The Vickers Pyramid hardness test
3. The Rockwell hardness test

The result of each test is quoted as a number which is related to the hardness numbers of other known materials. The number increases with the hardness.

In the *Brinell* test a hardened steel ball is forced into the surface of a test piece of the material. The area of impression left by the steel ball is calculated and the hardness number H is found from:

$$H = \frac{\text{load applied}}{\text{surface area of impression}}$$

In the *Vickers Pyramid* hardness test an impression is made in the material with a diamond square-based pyramid. The square impression is viewed optically and measured diagonally on a rotatable measuring screen incorporating a digital micrometer, the reading is then converted into the V.P.H. number by referring to tables.

The V.P.H. test is more reliable than the Brinell test for harder materials because the diamond used does not deform under heavy loading to the same extent as the steel ball used in the Brinell test.

The Vickers Diamond Pyramid and the Brinell hardness testing machines are mainly laboratory test equipment and are used to check the quality of materials used in manufacture.

Rockwell hardness testing can be satisfactorily used in rapid routine testing of finished material. There are three Rockwell scales, A, B and C, and the values of hardness are read directly from scales on the machine. For scale A and scale C readings, a diamond cone is used, with loadings of 60 kgf and 150 kgf respectively. For scale B readings, a steel ball of 1/16″ diameter with a 100 kg load is used for indentation. (Figure 12.5).

Figure 12.4

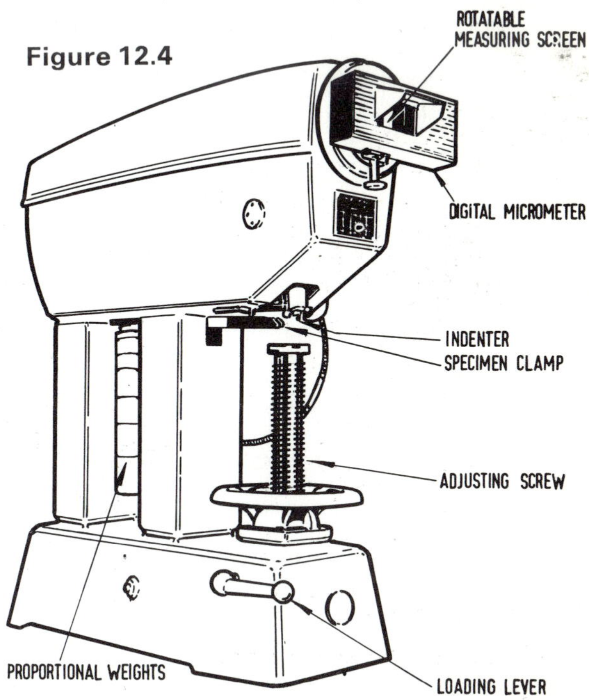

Avery testing machine (Brinell hardness)
(Vickers diamond-point hardness)

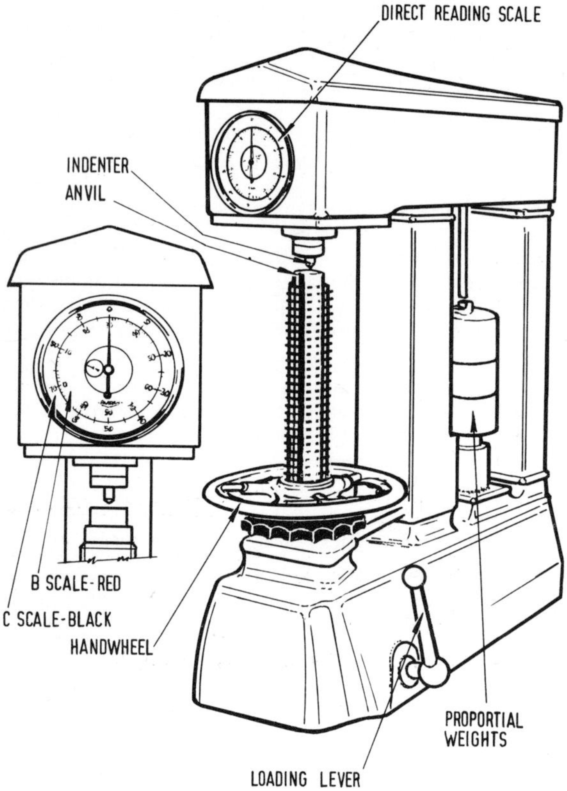

Figure 12.5

Avery direct-reading Rockwell-hardness testing machine

IMPACT TESTING. Figure 12.6.

The *Izod impact test* is carried out to measure the capacity of a material for resisting shock. This test will show up the brittleness of a material caused by wrong heat treatment or faulty alloying, which the tensile test might not reveal.

The specimens for test are made to a standard section and are notched with a standard-shaped notch; they are then clamped at the lowest point of the swing of a heavy pendulum-type hammer. The hammer is released from a fixed height and when it strikes the specimen some kinetic energy is absorbed in breaking the specimen. The remaining energy carries the hammer to a height beyond the specimen indicated by a pointer on a scale graduated to show the energy used in breaking the test piece.

Izod values are not always easy to interpret, but the test gives a good indication of shock resistance.

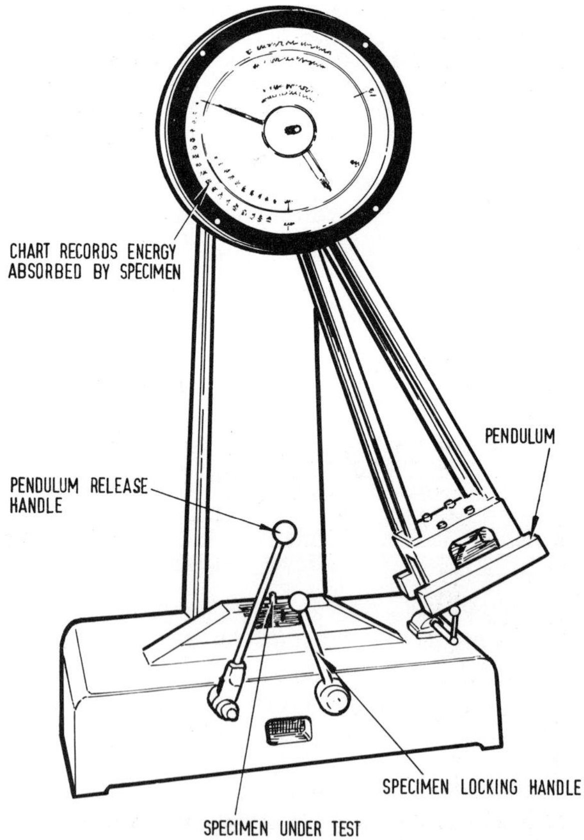

Figure 12.6

Avery Universal impact testing machine for Izod and Charpy impact tests

FATIGUE TESTING

Fatigue is a phenomenon which causes failure of a component subjected to repeated alternating stresses. It occurs, for example, in the revolving half-shafts supporting the road wheels of a car. At every 180° of revolution, tension stress changes to compression stress, peak values occurring in the surface layers of the shaft material. This condition also occurs in aircraft wings, due to air buffeting and to landing and take-off conditions. Aircraft fuselages, because of pressurisation, expand and contract according to the altitude of flight, again causing conditions of alternating stress. These conditions have to be provided for by fixing a maximum fatigue limit of stress or deciding on a specified life for affected components.

Figure 12.7 shows the principle of a fatigue-testing machine for carrying out tests on materials by revolving specimens in a chuck, the free end being loaded until breakdown occurs or until about 20 million reversals of stress have been endured.

Many specimens are tested at different loadings and the results plotted as a stress/reversal curve. There is usually a wide scatter or points, but a fair curve through the points produces a curve which flattens and becomes horizontal at a stress known as the endurance or fatigue limit. Figure 12.8.

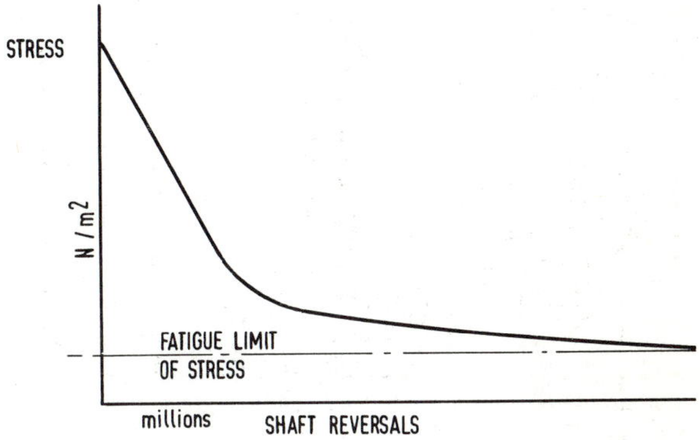

Figure 12.8

Stress: reversal curve for assessing fatigue limit

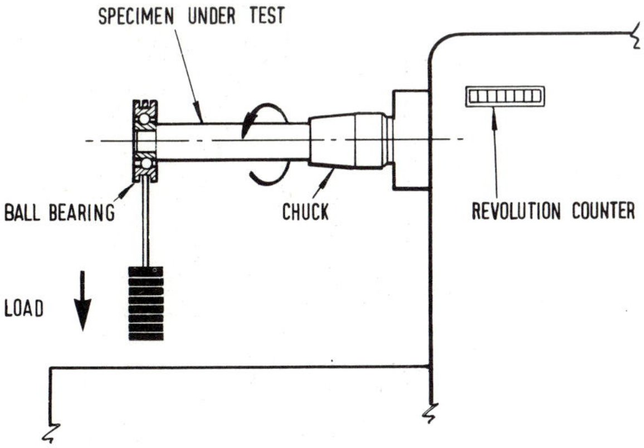

Figure 12.7

Diagram showing principle of common fatigue testing machine

NON-DESTRUCTIVE TESTING

X-ray. Where finished products or structures need to be tested for flaws in welds, as in shipbuilding, pressure vessels, atomic structures, or for cavities or cracks in large forgings and castings, portable X-ray apparatus is widely used.

Magnetic dust method. Crack detection in steel machinings is carried out by magnetising the component and immersing it in a bath of paraffin containing fine iron dust. The sides of the cracks, when lying across the magnetic field, become magnetic poles and collect the iron dust, thus tracing out the extent of the flaw or crack.

An introduction to plastics

Chapter 13

This chapter can give only a brief outline of the complex field of plastics technology and the fast-expanding field of engineering in which the many plastic materials available today are used.

HISTORICAL NOTE

The plastics industry can be said to have begun in 1846 when a German scientist, Dr Friedrich Schönbein, discovered that he could convert cellulose (a natural polymer) from wood and plants into a clear, tough, horn-like material by treating it with nitric acid. The resulting material was called nitrocellulose or, more correctly, cellulose nitrate, which in one form is the explosive known as gun-cotton.

The next discovery was made by Alexander Parkes of Birmingham, a metallurgist who was one of the inventive geniuses of the nineteenth century. Parkes discovered that cellulose nitrate could be dissolved in molten camphor which, on cooling, went through a putty-like stage when it could be moulded into shape. When cold it set as a flexible horn-like material. He described his discovery of the material, which he called *Parkesine*, in a lecture to the Royal Society of Arts in 1865.

Parkesine was not successfully exploited in England but an American, John Wesley Hyatt, won a prize of $10 000 offered by a billiard-ball manufacturer to the person discovering a substitute for ivory, which was expensive and in diminishing supply. Using Parkes' method, he ground cellulose nitrate and camphor together and fused the mixture into a mouldable plastic by heating. He took out a patent and named the material *Celluloid*. Hyatt quickly built up a thriving business manufacturing celluloid products, in spite of the fact that celluloid had the great disadvantage of being highly flammable.

About the same time, a similar material called *Xylonite* was produced in London, and the British Xylonite Company was formed in 1877 to exploit it. This laid the foundations of the plastics industry in Britain.

Other naturally occurring materials used in the early days of the plastics industry were shellac and bitumen. Shellac, when mixed with asbestos and shaped, provided insulators in the early days of electrical engineering, and also the material from which gramophone records were pressed. Bitumen was also used in another familiar plastic material to make screw stoppers for beer bottles.

Towards the end of the nineteenth century, another important plastic using the casein readily available in skimmed milk was discovered.

Casein plastics are produced in small sheets and extruded into tubes and rods from which buttons, dress buckles, knitting needles and similar articles are manufactured. These plastics are made in a range of delicate colours and, despite strong competition from synthetic plastics, are still used today in the decorative and fashion fields.

For some fifty years the plastics industry remained relatively unimportant because the availability of materials was limited and any progress was dependent on empirical chemistry. Then, in 1909, a Belgian chemist, Leo Bakeland, took out patents on a material which he called *Bakelite*. This resinous material was formed during a reaction between phenol and formaldehyde, both derivatives of coal. It was found that softening took place when the material was heated, and that it could be moulded; further heating in the mould hardened the material in the final permanent shape.

Bakeland had discovered the first synthetic thermosetting plastic, based on phenol-formaldehyde resin. In a short time the study of plastics became a specialised field of theoretical chemical research, and polymer chemistry has now become a most important branch of modern science.

Until World War II, when the plastics industry was given a tremendous stimulus, phenolic plastics were the backbone of the industry. Even today, in spite of the multitude of other plastic materials that are available, they retain considerable commercial importance as moulding materials in the manufacture of electrical fittings and the preparation of laminates, and as adhesives in joinery and foundry practice.

WHAT ARE PLASTICS

Plastics are a group of materials synthetically produced from naturally occurring basic raw materials such as coal, petroleum, water, air, salt, fluorspar, limestone, wood, vegetable matter and sand. At some stage during manufacture the materials are in a plastic condition and the shaping process is carried out with the aid of heat and pressure, usually in a mould.

In more scientific terms, plastics are *organic high polymers* (*organic*—having the element carbon as the basic structural unit; *high polymer*—consisting of large chain-like molecules which are formed during processing by hooking together the small-chain molecules of a monomeric chemical compound). This kind of chemical reaction is known as polymerisation. When the short-chain molecules of a monomer are linked together in chemical reaction, the resulting plastic is a *polymer*, such as polythene produced from an ethylene monomer.

When the linking of short-chain molecules results from chemical reaction between chemical dissimilar monomers, the process is known as copolymerisation and the resulting plastic is a copolymer. In general, the object is to make polymer molecules as large as possible to increase the favourable physical properties of the resulting plastic.

Plastics are valuable materials in their own right. They are not cheap substitutes or by-products from something else. They have been produced because they fill a need in our modern society. There are many types and each has its own unique properties, but all have certain common features which account for their ever-growing use in a wide sphere of applications.

Their advantages are:

Light weight.
Ease and speed of manufacture, which usually offsets the higher cost of materials.
Corrosion resistance.
Good electrical insulation.
Good thermal insulation.
Colourful and attractive appearance.
Pleasant to handle.

On the other hand, there are certain disadvantages in many plastics which offset the above.

Comparative low strength.
Low dimensional stability.
Poor heat resistance.
Generally poor weather resistance.
Relative high cost of material.

Most modern plastics are synthetic resins produced by a polymerisation process. These resins are usually hard, brittle and weak. Their qualities are improved by the addition of chemicals known as *plasticisers*, and other materials known as *fillers*, to give greater mechanical strength.

PLASTICISERS

Plasticisers are usually high-boiling-point liquids with low evaporation rates at room temperature. They must be good solvents for the materials to be plasticised; otherwise a bloom or oily film will appear on the product. This is caused by the plasticiser sweating out.

By carefully selecting plasticisers and using them in the correct proportions, brittle synthetic resins can have their physical characteristics completely changed. Brittleness disappears, and tough pliable materials with lower softening temperatures are produced.

In thermosetting materials, plasticisers are generally used for a different purpose. In hot-setting conditions, the plasticiser acts as a flux, helping the material to flow more easily, and is chemically combined in the plastic of the finished material.

FILLERS

Wood flour, wood pulp, paper, cloth and glass fibre are used as fillers to improve the tensile and impact strength. Mica reinforcement improves the electrical insulation. Graphite reinforcement is used where some self-lubrication is required in moulded bearings. Asbestos fibre or cloth is used to increase the usable temperature range. Glass-fibre reinforcement is used to increase strength and temperature range and to reduce water-absorption rate.

CATALYSTS

These are substances which are added during the polymerisation process to increase the speed of the chemical reaction. They do not chemically combine with the polymer produced; their function is to whip the chemical process into action. Catalysts are sometimes referred to as acclerators. On the other hand, *negative* catalysts slow down chemical reactions which are of an unduly violent nature.

CLASSIFICATION OF PLASTICS

Plastics are broadly classified into two groups:

1. Thermoplastic.
2. Thermosetting.

Thermoplastic materials are hard and rigid at normal temperatures but soften quickly when heated. They can be repeatedly softened and hardened by heating and cooling. It is therefore obvious that for articles required to withstand even moderate heat thermoplastics should not be used.

No chemical change takes place on repeated heating and cooling, so that all clean scrap and reject articles can be ground and reworked, thus making a substantial saving on raw materials.

For setting in the final required shape, thermoplastics need to be cooled after being formed. This is achieved simply by air cooling, immersion in water, or by water-jacketing of the moulds and dies.

Thermoplastic resins are usually supplied by manufacturers in a granular form.

Thermosetting materials differ fundamentally from thermoplastics in that under hot-pressing conditions they become a plastic mass and flow into the shape of the mould, where curing takes place. They are irrevocably set by the continued application of heat. The material undergoes a chemical change, sets rigid and will not become plastic again.

The main thermosetting materials are phenol-formaldehyde, urea-formaldehyde and melamine-formaldehyde resins. All the resins are mixed with fillers, such as wood flour, wood pulp, cotton paper, rag, mica, asbestos and mineral powder, to give improved physical properties. The phenolics are generally manufactured in rich, dark colours, whereas the amino-plastics (urea and melamine) are manufactured in the full colour range from white and pastel shades to black.

Thermosetting plastics are usually supplied as a liquid mixture or a partially cured granular moulding compound.

Table 13.1 gives a selection of the plastics in use today, with brief descriptions and an outline of uses.

TABLE 13.1 THERMOPLASTICS

Material	*Description*	*Some uses*
ABS plastics (copolymers of acrylonitrile, butadene and styrene)	Exceptionally tough and strong. Tensile strength 35–55 MN/m² (5000–8000 lbf/in². Temperature range up to 100°C. Remains tough at low temperature. Resistant to acids, alkalis, salts and, in many cases, petroleum solvents. Good overall electrical properties. Materials are available as granules and powders for injection moulding, extrusion and calendering and as sheets for thermoforming.	Pipes and fittings, appliance wheels, safety helmets, textile bobbins, refrigerator parts, battery cases, water-pump impellors, radio cases, tool handles, toys, TV implosion screens, sun visors, instrument covers, lavatory seats, business machine parts.
Acetal resins (polyformaldehyde)	Feel and look something like nylon. Combines high tensile strength, stiffness and hardness with toughness, good resilience and dimensional stability, retaining a large degree of these properties over wide temperature ranges and differing service conditions. Tensile strength 90 MN/m² (13 000 lbf/in²) at −35°C, 35 MN/m² (5000 lbf/in²) at 120°C. Temperatures up to 150°C can be tolerated for short periods. Low moisture sensitivity. High resistance to petroleum solvents and alkalis, but limited resistance to acids. Good colourability. A versatile material of special interest to engineers, competing favourably for many uses with ferrous metals, brass and die-casting alloys of zinc, aluminium and magnesium. Good electrical insulator. Material available as moulding powders and granules. Suitable for injection moulding, extrusion and blow moulding.	Automobile parts, including speedometer and wind-screen wiper gears, cams, carburettor parts, door and window handles and mechanisms, heater fans; plumbing applications, including taps and pipe fittings, ball valves, valve seats and glands; water-softener cases; washing-machine agitators and bearings; extractor fans; aerosol bottles; switches; petrol tanks; terminal blocks; extrusion dies for macaroni; shoe heels; mechanical toys; instrument bearings; toffee moulds; bottled gas regulator housings; farm equipment; drive sprockets; conveyor links.
Acrylics (polymethyl methacrylate)	A hard, strong, rigid material with a tensile strength of 55–70 MN/m² (8000–10 000 lbf/in²). Excellent optical clarity. and shows good resistance to sharp blows, but scratches easily. Water absorption negligible. Electrical insulation good. Maximum service temperature 90°C. Loses shape in boiling water. Does not become brittle at low temperatures. Resistant to	Telephones; contact lenses; dentures; handles and knobs; automobile industry as rear-light lenses, reflectors, insignia medallions, windows and canopies; electrical industry as TV implosion screens, radiogram and pick-up arms, lighting fittings and meter cases; drawing instruments; clock faces and glasses; pens; disposable hypodermic syringes; display

Material	Description	Some uses
	most chemicals. Attacked by benzene, turpentine, acetone, chloroform and strong oxidising agents. Good colourability. Will pipe light round corners. Materials available in injection and extrusion grades, and as solutions and emulsions for coating compositions, cloth and leather finishes.	signs; traffic signs; aircraft glazing; goggles; sinks and baths; see-through models; architectural murals and panels; refrigerator interiors and ice-cube trays.
Cellulosics	Cellulose in the form of cotton linters is the raw material from which the following common cellulose plastics are made. They all have excellent colourability and toughness. Fabrication is easy. In general, should not be in service above 90°C.	
	1. *Cellulose nitrate* (originally *Celluloid*). Highly flammable, tough, resilient. Available in a range of bright colours; pearl and tortoiseshell effects. Good electrical insulation.	Table-tennis balls; knife handles; cycle mudguards; spectacle frames; drawing instruments.
	2. *Cellulose acetate*. Not so tough, and does not discolour as quickly as Celluloid. Much less flammable but otherwise similar to Celluloid. Injection moulded, sheet and film.	Toys; containers; lampshades; brush handles and combs; spectacle frames; business machine keys; sound tape; buttons; textile bobbins; door handles.
	3. *Cellulose propionate*. Similar to cellulose acetate, with lower absorption and superior heat resistance. Moulding powder for compression and injection moulding and extrusion.	Transistor radio cases; pen and pencil barrels; film spools etc.
Nylon	The mechanical toughness, durability, lightness in weight, good abrasion resistance, low frictional resistance, and ability to be used without external lubrication make nylon an attractive material to the design engineer. Nylons filled with glass fibre give improved mechanical strength, and the addition of graphite and molybdenum disulphide lowers frictional resistance. Resistant to attack from most chemicals, greases and solvents, but attacked by concentrated mineral acids. Good insulating properties. Dimensional stability affected by water absorption. Can often replace traditional materials more cheaply and satisfactorily, and for many applications goes beyond the scope of metals. Can withstand reasonably high temperatures. Nylon yarn can withstand ironing temperatures. Materials available as granules for injection, blow moulding and extrusion. Selection of the right grade of material for design is most important. Nylons can generally be used over a temperature range of 40 to 120°C. Glass-filled grades can withstand much higher temperatures.	Gear assemblies for small mechanical units in farm machinery; vehicle door-locking mechanisms; tubing for high-pressure lubrication; carburettor floats and parts; terminal blocks; cable sheathing; insulating clips; relay bobbins; coil-formers able to withstand high cure temperatures (175°C) for lacquer on windings; power tool housings; conveyor belt rollers and chains; components for food mixers; casing and machine parts for domestic equipment; boat propellors and fittings; helmets; water flasks; pistol grips; rifle stocks; bayonet scabbards; pulleys; transistor cases; handles; flexible couplings; clutch facings; lift guide shoes; filaments for the clothing industry; fishing lines; surgical sutures; ropes; tennis racquets and strings; brush bristles; car-tyre reinforcements.

Material	*Description*	*Some uses*
Fluorocarbons (polytetra-fluoroethylene, PTFE)	PTFE is a tough, flexible, non-resilient, non-toxic thermoplastic with a rather waxy feeling. Ivory white in colour. Mechanical strength comparatively low. Outstanding properties include inertness to most chemicals, with a working temperature range from that of liquid nitrogen (−195°C) to about 250°C. The best of the solid dielectric materials. Not wetted by water, does not absorb water, and has a very low coefficient of friction. In addition, practically no substance will adhere to its surface, making it the best of the non-stick materials. PTFE cannot be processed by conventional moulding techniques because it does not soften like other thermoplastics. It is formed by a method similar to powder metallurgy. The powder is placed in a mould, compacted and heated to 325°C. The process is known as sintering. PTFE is also manufactured in standard forms of rod, tube and sheet for machining. Dispersions of PTFE are used for spraying thin coatings.	Non-lubricating bearings of all types; pump parts; pipes, expansion joints, tank linings and valve diaphragms for the chemical industry; high-temperature electronic parts such as missile radomes; sealing rings and linings for flexible pipes used at high temperatures; coatings for thrust and journal dry bearings; coatings for epoxy resin moulds; non-stick surfaces for food-handling containers; coating rolls for textile and paper industries; insulation for power cables and high-frequency applications; gaskets; packings; printed circuits for ultra-high frequency and miniaturisation which results in increased operating temperatures; razor blade coating.
Polythene or polyethylene (polymer of ethylene gas)	Polythene is a tough, flexible, white translucent material with a waxy feel and a soft surface which is easily marked. It is available in low and high densities, the primary differences being increase in rigidity and in temperature resistance and ability to sustain loads. Excellent dielectric qualities. Excellent resistance to acids, alkalis and solvents, although embrittled by ultraviolet light. Good dimensional stability, and unaffected by water. Approximate service temperature 65°C (low density) and 95°C (high density). Can be self-coloured, has good weatherability, and has become a widely used utility plastic. Originally produced after long research by ICI during 1930s and played an important part during World War II in aiding the rapid progress of radar. Polyethylene is used in a variety of forms—injection and blow moulding, extrusion, sheet, film and coatings.	Packaging (the largest application) of farm produce, dairy products, sweets, frozen foods, nuts, bread, coffee, garments, hardware, consumer durables. toys, paper goods etc.; weather protection in building construction; moulding (second largest application) by both blow and injection methods for industrial and household ware—bowls, buckets, brushes, waste bins, dustbins, baskets, caps, closures, bottles etc; in the electrical industry for cable insulation, house wiring, radio relay, telephone and power cables; pipes for agriculture, building and chemical works; coatings for paper, aluminium foil, fabric and board; films for packaging industry; pipes and fittings; tables, lids, surrounds and agitators for washing machines; safety helmets etc.
Polypropylene (polymer of propylene gas)	Polypropylene first appeared as a new plastic in the late 1950s. Somewhat similar to high-density polythene, it differs in being more rigid. It is able to withstand high working temperatures, up to 150°C under no-load conditions, and is the	Films for packaging industry; pipes and fittings; tables, lids, surrounds and agitators for washing machines; safety helmets; brush tufting for road sweepers; tool handles; toys; tableware; bottles and containers, particularly where sterilisation

Material	Description	Some uses
	lightest thermoplastic commercially available (specific gravity 0·9). First of the polymers where molecular structures are tailored to give required properties. Good electrical and abrasion resistance; resistance to acids, alkalis and salts even at high temperatures. Good surface finish. Containers made of this plastic can be sterilised in boiling water. Can be processed by injection moulding, blow moulding and extrusion. Available as sheet filament and film.	is necessary; moulded chairs; lightweight cases for record players, TV and radio; trim panels for cars; integral hinges on car accelerator pedals; ladies' compacts and containers; automobile wheel arches; filler caps; knobs; grills; heater ducts; crates for milk and beer bottles.
PVC (polyvinyl chloride)	PVC and PVC acetate polymers are a group of materials which can be formulated to provide such a wide combination of properties that it is difficult to give a set of characteristics for the materials. In general, they are tough, strong and resistant to abrasion. They can be produced in both rigid and flexible forms. Colourability is almost unlimited. Exceptional resistance to a wide variety of acids, alkalis and solvents. Excellent insulating properties. Resist water, burn only with difficulty, and are non-toxic. Service temperature range between −55°C and 75°C. The most common PVC copolymers are vinyl chloride/vinyl acetate and vinyl chloride/vinylidene chloride. PVC can be injected, blow moulded and extruded. It is produced in sheet for thermoforming, film, laminates and foam. It can be sprayed or used for dipping when mixed with suitable plasticisers and solvents.	Bottles; trays; ducting; buckets; bowls; decorative and finishing panels for buildings, tunnels, car and coach interiors; aircraft toilet fittings, window surrounds, passenger lights and ventilation panels; lift interiors; motor-cycle windshields; traffic and advertisement signs; indicator panels and signposts; radio and TV parts; hose, tubing and electrical cable covering; refrigerator interiors; leathercloth for upholstery, handbags and luggage; curtaining; rainwear, gloves and protective clothing; gaiters; terminal covers for automobile parts; wire baskets; railing and metal-chair coatings; gramophone records; foams for cushioning.
Polystyrene	Polystyrenes are one of the largest and most widely used families of plastics. The good characteristics of general-purpose styrene are low cost, ability to be crystal clear with a hard glossy surface, excellent dielectric properties, resistance to moisture, and unlimited colourability. On the other hand they have low service temperatures (approximately 65°C) but retain their properties at freezing temperatures. Impact strength is low. Many forms of modified styrene are available, each suited to a particular type of use. The important modified types include: 1. Styrene-acrylonitrile copolymer. Provides resistance to chemicals, oils and petroleum. 2. Materials of varying proportions of styrene and butadene. Improve impact strength; decrease rigidity.	Refrigerator parts; vacuum flask cups; wall tiles; radio and TV cabinets; tough shoe heels; storage boxes; cutlery handles; toys; buttons and buckles; brush backs and combs; clothes hangers; clock cases; packaging for foodstuffs such as jam, butter, cream, eggs; containers for drugs and cosmetics; thin-wall containers for ointments, yoghourt, honey, cream etc.; trays; closures with puncturable thin spots for scouring powders; bathroom and kitchen applications where moist conditions are present; ball pens and fountain pens; as expanded foams for thermal, impact and sound insulation in building panels for houses and factories and for refrigerator and aircraft panels; novelty display items; transit packaging with high shock-absorbing capacity; flotation applications in catamaran hulls, surf boards, buoys; as emulsions, for surface coating of textiles and manufacture of water paint.

Material	Description	Some uses
	3. Polymethylstyrene. Improves heat resistance. 4. Methylstyrene-acrylonitrile copolymer. Improves resistance to chemicals, toughness and resistance to crazing. 5. Glass-reinforced polystyrene. Provides higher strength and durability. Polystyrene is available for injection moulding and extruding, and in sheet for thermoforming and film. Rigid foams are obtainable in preformed blocks and planks. Beads are available for expanding in heated moulds. Rigid foams can easily be carved by wood-working tools, hot wire or soldering iron.	

TABLE 13.2 THERMOSETTING PLASTICS

Material	*Description*	*Some uses*
Phenolics	Phenol formaldehyde is the oldest commercial thermosetting plastic, and is commonly referred to as *Bakelite*, which was the name originally given to it by the inventor. The main features are strength and rigidity (tensile strength 35–70 MN/m², 5000–10 000 lbf/in²). Excellent dielectric resistance; most dark-brown or black electric fittings are of this material. Good heat resistance; service temperatures of 175°C to 200°C are common, even higher in some compounds. Used within 75 mm of the flame in automatic steel cutting and welding machines. Good chemical resistance and dimensional stability. Phenolics are limited in colourability, and are usually moulded in rich dark hues; their natural colour is brown and darkens on ageing. Moulding powders are available for compression or transfer moulding, and liquid resins for laminating, foundry practice, casting and foaming applications. Moulding compounds with a variety of fillers—wood flour, wood pulp, asbestos fibre, mineral powders, cotton flock, mascerated rag, sisal, glass, mica, graphite etc.—are available.	Electrical applications—plugs, switches, ignition equipment, arc-welding torches etc.; buttons; knobs; radio cabinets; saucepan handles; vacuum-cleaner parts; as laminated material reinforced with paper or cloth (commonly known as *Bakelite* or *Tufnol*) for machining into a multitude of forms—gears, aircraft propellers, jigs and dies, clutch and brake linings etc.; liquid resins for adhesives in joinery work, chipboard adhesives, veneering and plywood adhesives, decorative laminates, sandcore bonding and shell moulding.
Melamine formaldehyde	Probably the hardest of all commercial plastics. Similar to urea formaldehyde but harder, with lower water absorption and less tendency to shrink after moulding. Improved heat resistance (service temperature 99°C). Excellent electrical resistance, with even greater arc resistance than urea formaldehyde. Available as moulding materials and laminates. More expensive than phenol formaldehyde and urea formaldehyde.	Electrical equipment; aircraft ignition and distributor parts; handles for electric irons, saucepans and ovens; household ware in constant contact with water, such as cups, saucers, handles and soap dishes; decorative panels (*Formica*, *Warerite* etc.) for counters, tabletops and wall coverings; textile and paper finishes. Hardness, colourability and washability are the primary reasons for the choice of melamine.

Material	Description	Some uses
Urea formaldehyde	Mainly fabricated as mouldings either in compression or transfer moulds. Urea formaldehyde is a colourless resin which can be coloured to almost any colour, with high translucency and light-fastness. Hard and rigid with scratch-resistant finish. The strength and toughness are increased when certain fillers are used. Absorbs water, resulting in a loss of dimensional stability. Excellent electrical insulator with resistance to arcing. Service temperature 75°C. Burns with difficulty. Good resistance to most solvents and common chemicals. Unaffected by petrol, oils, greases, detergents and cleaning fluids.	Electrical fittings—plugs, sockets, ceiling roses, junction boxes, switches and bulb holders; bottle and jar closures made in a wide range of colours and decorative shapes for foods, sauces, detergents, perfumes, cosmetics and spirits; household goods—buttons, lavatory seats, hairdryer housings, food-mixer parts, extractor fans, louvres and crockery; foundry core production; surface coatings for paper to give improved wet strength, and for textiles to give crease resistance; base for enamel stoving.
Polyurethane	One form of polyurethane is used for making bristles, filaments and films. Polyurethanes in the main are used in the form of foams, which may be rigid, semi-rigid or flexible, and can be prepared *in situ* to fill cavities.	Rigid foams are used as thermal insulation, to give lightweight structural strength in sandwich panels, and for buoyancy applications. Semi-rigid foams are used for thermal insulation where a measure of resilience is necessary for energy absorption. Flexible foams are in common use as cushioning for furniture, aircraft and car seats, crash pads and sponges.
Epoxy resins	These are thermosetting resins with the following properties: 1. Excellent adhesion to metals, rubber, glass, china, concrete, stone and many synthetic plastics. 2. Great mechanical strength and toughness, particularly in laminated form. 3. Excellent resistance to chemicals and moisture. 4. Excellent dielectric strength. 5. Negligible shrinkage on setting. 6. Absence of volatiles on setting. These resins are produced by combining two complex organic chemicals into another yet more complex form and, by introducing suitable hardeners in varying amounts, may be made suitable for a wide variety of purposes. By the addition of dyes, pigments, fillers, plasticisers and solvents further variations can be achieved. Epoxy resins are usually prepared as two-component systems in which equal quantities of each component are mixed together. Some forms will set at room temperature but heating accelerates curing. These types must be used within 3 to 4 hours of mixing. Hot curing forms can be held in the mixed condition for long periods because heat is required to produce curing. Epoxy resins are available as adhesives, moulding powders, potting or encapsulation compounds, tooling resins, coatings.	Bonding of metals, glass, plastics, china, wood, ceramics etc.; particular application to glued wooden structures for buildings, bridges, boats, jetties and furniture; impregnating, sealing and encapsulating electrical and electronic components and systems; casting of high-grade insulations for large switch-gear and power-transformer components; casting of architectural decoration and statues; glass-fibre laminates for nuclear, electrical, mechanical and aircraft engineering where strength-to-weight ratio comparable to that of aluminium can be achieved, particularly in corrosive environments and where dynamic loading is involved; transfer and compression moulding of insulation components to dimensionally close limits; construction of tools for forming sheet-metal or plastics patterns, jigs, and fixtures for drilling and checking the accuracy of assembly; anti-corrosion surface coatings (in the form of paint which, due to the absence of volatiles, reduces fire risk and can be applied in confined spaces) for metals, wood, concrete etc.; coating insides of squeeze tubes for cosmetics, pharmaceuticals and food.

Material	Description	Some uses
Polyester resins	The term polyester covers a large number of plastics. Laminating and casting resins are the form most commonly used. As liquids, they are easily handled, and curing either takes place at room temperature or can be accelerated at higher temperatures. They can also be used as adhesives or for surface coating. Resins are produced which vary from colourless to brown. The physical properties of polyester resins vary according to the type of polymer and the reinforcing material. Glass-fibre polyester resins are hard, strong and durable, have good resistance to impact, and are easily repairable. They are attacked by strong acids and alkalis, but have good resistance to most common solvents and to dilute acids and alkalis. They have good weathering properties and are good electrical insulators. Laminating moulding processes are carried out in simple one-part or two-part moulds made from wood, metal, plaster or concrete. Moulding can be achieved with or without pressure, at room temperature or at elevated temperatures. Moulding powders are available and can be used with a variety of fillers, such as mineral powders, glass fibres, mica and wood powder. In general, properties are similar to those of phenolics, but polyester resins have better resistance to moisture. They have the same electrical resistance to tracking as the ureas, coupled with better resistance to dry heat. They are used for electronic and electrical components where dielectric characteristics must be retained after exposure to moisture.	The most common structural applications for glass-laminated polyesters are in the manufacture of car bodies, boat hulls, aircraft parts, chairs, luggage trunks and cases, building panels and partitions, and fishing rods. Moulding compounds produce components of high dimensional accuracy and are used for vacuum tube bases, car ignition parts, power-circuit insulation, computer components, connectors, rotors and switchgear. Mineral-filled putty is used for low-pressure moulding round delicate inserts; typical uses include insulation in the manufacture of capacitors, resistors and coils.
Silicones	Silicones are unique among plastics in that they are only semi-organic. They are polymers based on chains of alternate silicon and oxygen atoms with organic side groups. They possess a unique range of properties, including extraordinary resistance to extremes of heat and cold (heat distortion points in some materials are above 480°C), inertness to most chemicals, excellent water repellency, and high dielectric properties. Mechanical strength is not as great as in some of the other thermosetting plastics, but strength is maintained exceptionally well at elevated temperatures. Silicones are available in many forms. They are produced as liquid laminating and casting resins, as emulsions and bases for paints, and as rubber-like materials called elastomers.	As liquid resins, for electrical and electronic insulation and encapsulation that must operate at high temperatures; insulation of electric motors operating in hot and damp conditions; laminating glass cloth and other textiles for high-temperature electrical use; paint for high-temperature points; water repellents for masonry, glass, ceramics etc. Elastomers are used where extremes of temperature are encountered—for cable insulation and sleeving, grommets, bushes, gaskets, valve diaphragms, roller coatings, oven door seals, aircraft parts where low-temperature flexibility and high-temperature stability are critical, and moulds for plastics and low-melting-point alloys.

PLASTIC FORMING AND FABRICATION TECHNIQUES

The following is a simplified description of the methods of production in the plastics industry.

THERMOPLASTICS

Injection moulding. This method is probably the one with the widest application. Granular moulding powder is fed from a hopper into a heated cylinder, where it is softened and then forced into a single- or multi-cavity split die by a piston operated by hydraulic pressure.

The moulded article has to remain in the die until set, when the die is opened and the moulded article is removed. The piston is then drawn back, recharging the cylinder with fresh material for a new shot, and the whole cycle is repeated.

Injection moulding is used for mass production on long runs; equipment is expensive and tooling costs run to hundreds of pounds for even small dies.

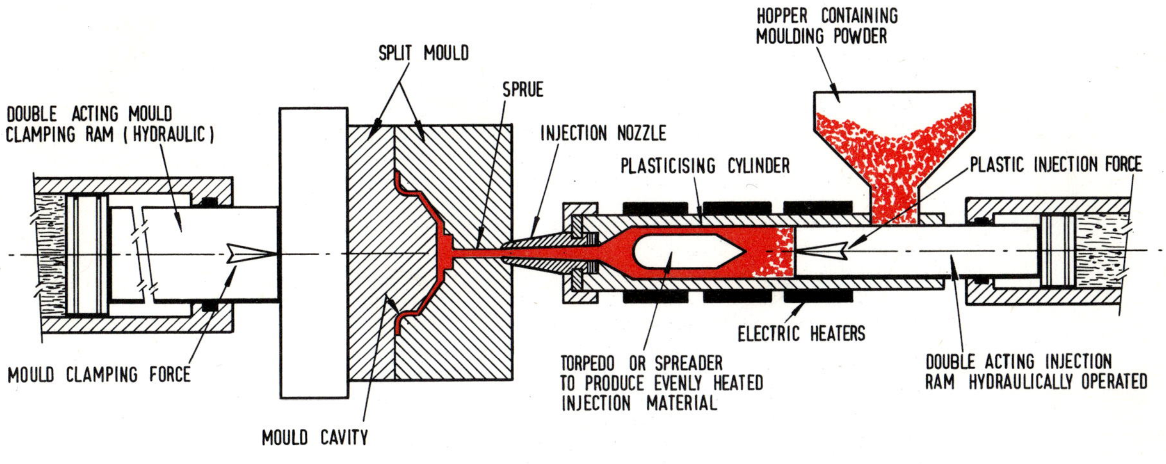

Figure 13.1 Diagram of injection moulding process

Extrusion. (See Figure 13.2). This method is a continuous process. The moulding powder is fed continuously into a revolving screw-feed mechanism operating in a heated chamber; there it is completely plasticised and fed to an outlet die of the cross-section required for the particular extrusion. Wire-covering is achieved by use of a cross-head die.

Wire or cable is fed into the die so that it is located at the centre of the plastic material when it is extruded. Extrusion processes require additional equipment such as cooling baths, coiling and wind-up devices, and take-off conveyors to deal with continuous production.

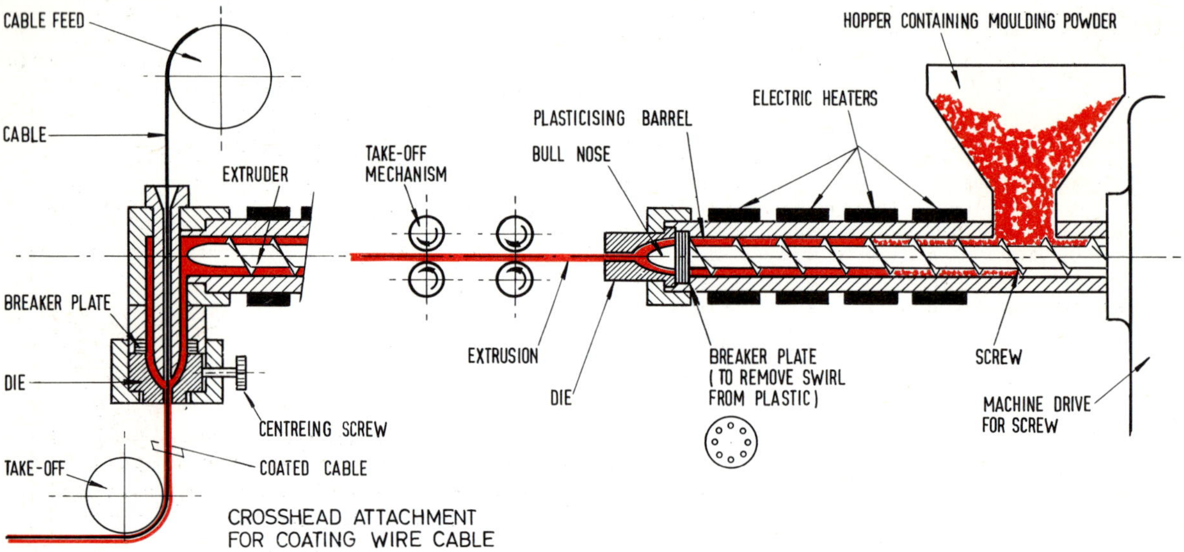

Figure 13.2 Diagram of single-screw extruder

Blow moulding (see Figure 13.3). This is a process involving the extrusion of plasticised tube called a parison. A suitable length is automatically clamped between the halves of a split die and cut off, sealing one or both ends of the tube while the plastic is still hot. Air is introduced through a hollow needle or sizing core, and the tube is blown into the shape of the mould. The method is used for hollow toys and for bottles and containers from a very small size up to 40-gallon drums.

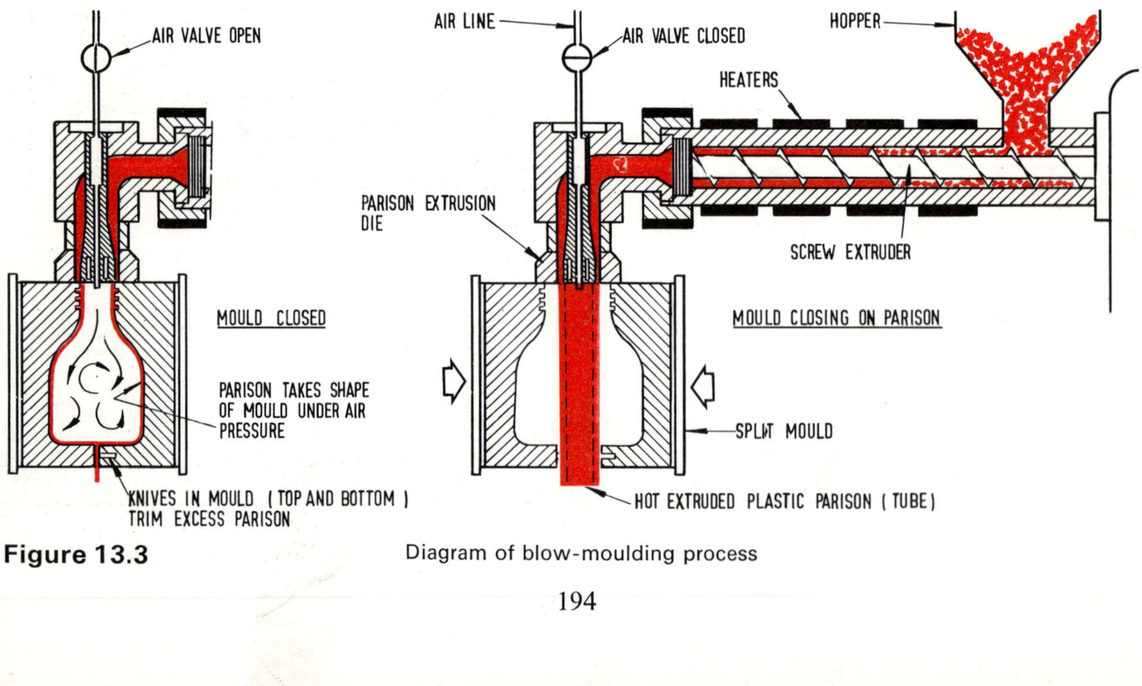

Figure 13.3 Diagram of blow-moulding process

Thermoforming (*see Figure 13.4*). This term covers a number of techniques used to fabricate articles from thermoplastic sheet material. The most used method is *vacuum forming*, a reasonably economical method of forming even short runs of large articles. Heated and softened sheets are drawn into close contact with either male or female mould shapes by evacuating the air between the sheet and the mould. Moulds may be made of such materials as metal, wood, plaster of paris and thermosetting plastic material.

Film casting (*see Figure 13.5*). Film for packaging can be formed by the use of a conventional extruder delivering hot plastic through a slit die on to a water-cooled, highly polished, steel roller that revolves at greater speed than the extrusion rate. This method quickly cools the plastic and reduces the thickness. The plastic passes over further cooling rolls and is finally edge-trimmed and wound on to cores ready for use.

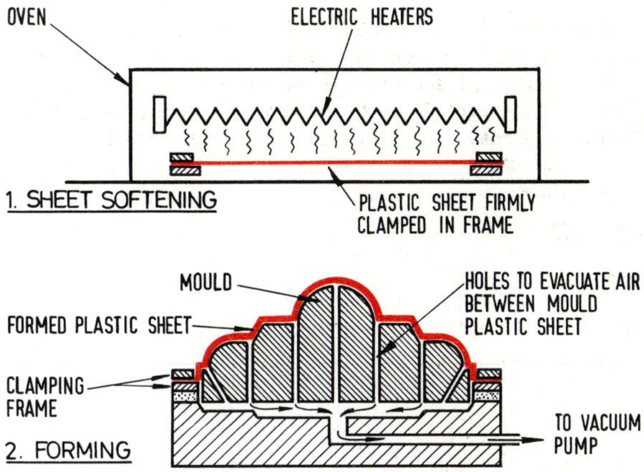

Figure 13.4 Diagram of vacuum-forming process

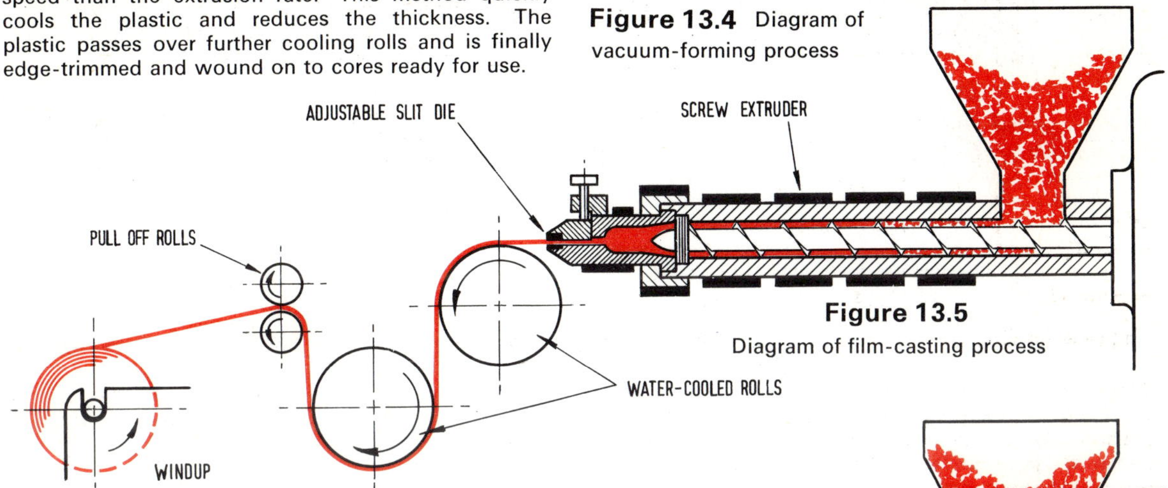

Figure 13.5
Diagram of film-casting process

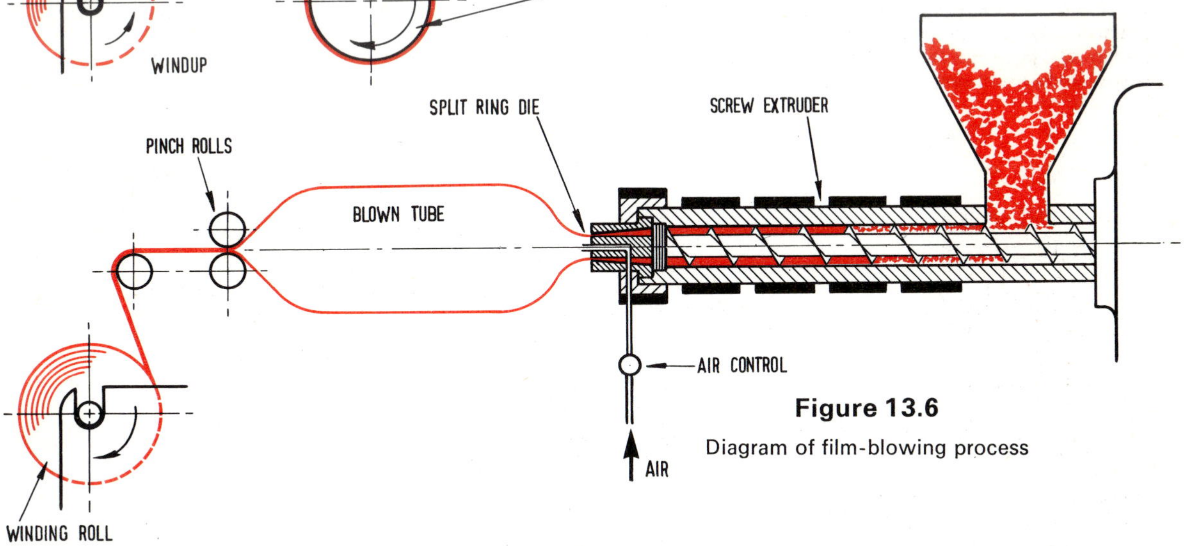

Figure 13.6
Diagram of film-blowing process

Film blowing (*Figure 13.6*) is a method used for the production of polythene packaging film. A thin tube of hot polythene is extruded through a circular slit die on a conventional extruder and air is introduced through the centre of the extrusion to inflate the tube and produce a bubble of controlled size and gauge between the die and a pair of pinch rolls. The film is wound on to a core beyond the pinch rolls as a flattened, continuous tube of film. The edges are sometimes split to reduce the tube to sheet before winding on to cores for use.

Calendering. In this process plasticised PVC is delivered to the heated rollers of a machine called a calender, which is like a huge mangle. The sheet thickness produced depends on the gap setting between the rollers. Patterns can be embossed or printed on the calendered sheets by passing them through suitably engraved or printed rollers. A wide range of upholstery materials and plastic sheet for curtains, tablecloths and plastic clothing is produced by this method.

Dipping process. This method may be carried out as either a hot or a cold process. Pastes of PVC with suitable plasticisers are prepared. Articles such as wire baskets, metal stool legs and railings can be permanently coated by hot dipping. Cold dipping is used to produce rubber protective hoods, bellows, gloves, grommets, teapot spouts, tap swirlers, protective gaiters for dust exclusion, waterproof covers for vehicle ignition components etc., by dipping suitable formers to which the paste adheres. On setting, the articles are stripped off.

Slush moulding. PVC paste is poured into a heated mould which can be split. The paste adheres to the wall of the mould and the excess is poured out. The mould is then heated until gelation is complete. This method is used to produce dolls' heads, hollow flexible toys and overshoes. It can also be used for permanently coating the interiors of hollow metal containers.

Rotational casting. Measured quantities of PVC paste are inserted into closed moulds, which are rotated simultaneously on two axes at right angles as they are passed through hot-air ovens, thus allowing all interior surfaces to be covered. After a suitable curing time, the moulds are removed from the oven, cooled in water, and the mouldings extracted. This method is superseding slush moulding because it offers high output rates.

THERMOSETTING PLASTICS

Compression moulding (see Figure 13.7). Moulding by this method is carried out by heat and pressure in steel moulds which can be split. The moulding compound (powder, flake, pellets, tablet or preform) is placed in the cavity of the heated mould. The mould is closed and the heat and pressure applied, gently at first and then with full pressure. In large mouldings, the mould is sometimes breathed or gassed by momentarily releasing and quickly applying pressure to remove entrapped air or gas given off by the moulding compound. This reduces cure time and the tendency to blistering. The heat and pressure plasticise the compound, forcing it to flow into the mould shape. At the end of the curing time the mould is opened and the hot moulded article is ejected.

This method is sometimes used with thermoplastics, for example, in making gramophone records.

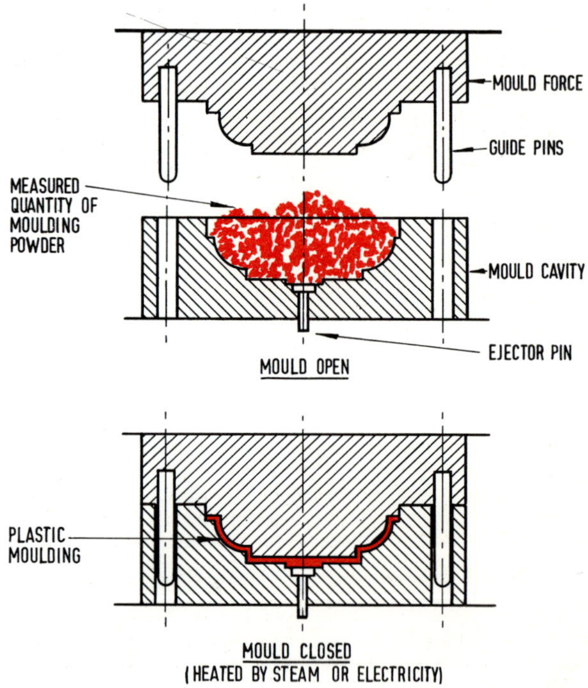

Diagram of compression-moulding process

Figure 13.7

Transfer moulding (*see Figure 13.8*) is similar to compression moulding in that the plastic material is cured in a heated mould under pressure. It differs from compression moulding in that the moulding material is forced by a plunger from a separate charging pot or chamber, through a feed channel with an orifice called a gate, into the heated mould cavity. This method is used when mouldings with sections of widely differing thickness are being produced. It prevents overcuring of thin sections. Precise feeding of material can be arranged to assist moulding with awkward shapes such as small-diameter deep holes. There is also less risk of shifting and damage to small delicate inserts.

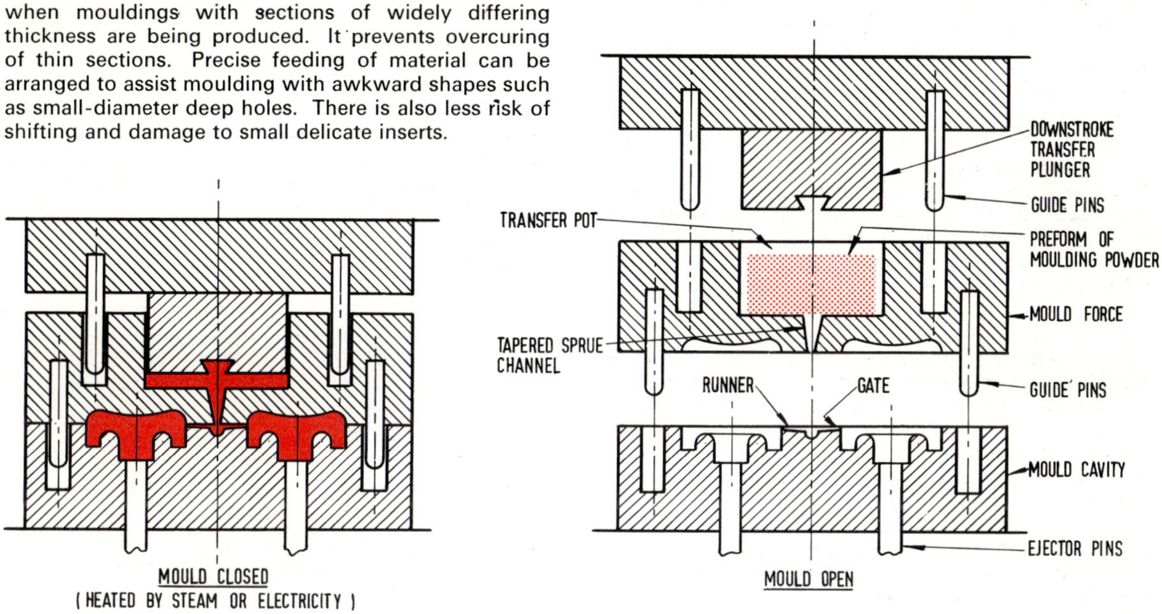

Figure 13.8 Diagram of transfer-moulding process

Matched metal moulding. This method is used for moulding reinforced plastic shapes. Matching male and female moulds with a gap between faces are used. A premoulded mat of reinforcing material is laid in the open mould, and liquid resin is then poured over the mat. The mould is then closed and heated to cure the mat.

Low-pressure laminating. Pressures from 0 to 6·9 MN/m² (1000 lbf/in²) are used. There are a number of techniques in which very large shapes can be produced using mainly glass fibre cloth and epoxy resins. Cheap moulds of either male or female form are made from wood, metal, plaster etc. The reinforcing glass cloth is laid up by hand and impregnated with resin. Pressure is sometimes applied by a rubber bag, and curing may be acclerated by applying heat.

High-pressure laminating. Sheets, rods and tubes known as industrial laminates are manufactured from reinforcing cloths or mats, using a variety of materials pre-impregnated with thermosetting resin which has partially polymerised. The materials are formed in heated presses under pressures exceeding 6·9 MN/m² (1000 lbf/in²).

Casting. No pressure is required. Liquid epoxy resins are poured into moulds in which components are inserted for encapsulation.

Filament winding. Reinforced filament (usually glass) in the form of tape is passed through a resin bath and wound on to a former which is rotated between centres. Spherical and cylindrical shapes can be built up and the resin is then oven-cured.

Iron is the world's most versatile and durable metal. Steel is derived from iron, and the difference between them is briefly discussed below.

Iron contains more carbon than steel and in a different form. Carbon is present in iron to the extent of 3% to 4%, some of which is combined with the iron to form iron carbide (Fe_3C), known as cementite, which is a very hard, brittle compound; the rest of the iron remains in a free state as graphite which, unlike cement, is soft and flaky. The mechanical properties of iron are determined by the relative proportions of graphite to cementite.

Steel is an alloy primarily of iron and carbon. The carbon in steel is always chemically combined with the iron and is present in quantities varying from 0·05% to 1·4%. The proportion of carbon in steel determines its mechanical properties and crystalline structure, and so does the condition in which the carbon is combined. This gives rise to the wide range of steels that are now available. Heat treatment and alloying with other elements are the methods used to control the condition of the carbon, and thereby the mechanical properties. The introduction of small amounts of alloying elements helps to arrange the carbon in the right condition to produce the required properties.

PRODUCTION OF IRON

The raw materials are (1) iron ore, (2) coke and (3) limestone.

In modern plants, the iron ore is preprocessed by crushing the ore and fusing it in gas-fired ovens with coke being haematite, which contains approximately 70% iron. Combined with the iron ores are impurities in the form of alumina, silica, lime, sulphur and phosphorus.

In modern plants, the iron is preprocessed by crushing the ore and fusing it in gas-fired ovens with coke breeze. This is known as sinter. Sinter thus prepared from various grades of ore produces a standard blast-furnace 'burden' with some of the impurities already driven off, thus giving a good iron output per ton of coke used in the furnace.

The iron ore or sinter, together with coke and limestone, are charged in a continuous process into tall cylindrical blast furnaces. A blast of hot air introduced at the bottom of the furnace oxidises the coke, which produces the heat to reduce the iron ore to liquid iron. The iron sinks to the bottom of the furnace, combining with and absorbing carbon on its way down; most of the impurities, including large amounts of sulphur, but not the phosphorus, are removed by the limestone, which is decomposed to form a liquid surface slag. Both the iron and the slag are tapped off at intervals. Nowadays, in integrated steel-making plants, the molten iron is conveyed directly to the steel-making plants. If the iron is not immediately required for steel making it is run off into moulds to form iron lumps weighing about 45 kg (100 lb) each, and called 'pigs' (hence the term pig iron).

Iron suitable for castings is produced by remelting pig iron with coke and limestone in a cupola which resembles a small blast furnace. An air blast through the furnace burns the coke and melts the iron, which collects at the bottom of the furnace, from where it is tapped off. This process provides the means of adjusting the composition and improving the quality of the iron.

CAST IRON

Cast iron is a most useful engineering material and can be obtained in many grades. It is cheap and has a comparatively low melting point, and castings weighing only a few grammes to others weighing many tons are produced from it relatively easily. It is an ideal casting material; it is fluid when molten, due to the high carbon content, and expands slightly on solidification, thus producing castings with clean, sharp profiles.

Cast irons are machinable without difficulty and have high compressive strengths, making them very suitable for the bed-plates of heavy industrial machinery. The tensile strengths of cast irons, however, are not high, varying between 60 and 430 MN/m^2 (4 and 28 $tonf/in^2$) according to the grade.

A disadvantage of cast iron is its lack of ductility, which prevents its being worked by the mechanical methods of forging or rolling, so that it can only be used in the as-cast condition.

The carbon in cast iron is in two forms:

(a) Iron carbide or cementite.
(b) Free carbon.

The iron carbide or cementite is iron combined with carbon, which is extremely hard and brittle, and gives these properties to cast iron. The free carbon is usually in the form of sharp flakes which have the effect of breaking up the metal structure. These discontinuities, being sharp and spiky, give rise to local high stresses within the metal when tensile loads are applied; they thus tend to produce failure at relatively low tensile loading. By the addition of small quantities of magnesium, the carbon flakes change into small spheroidal shapes; this doubles the strength and considerably reduces the brittleness of the material, which is known as spheroidal-graphite cast iron.

Another way of rearranging the carbon in cast-iron castings is by prolonged annealing by the Whiteheart (U.K.) or Blackheart (U.S.A.) methods of producing malleable iron castings. The carbon is reduced from the flake form to clusters of finely divided graphite, known as temper carbon, which is part way to the spheroidal form.

TYPES OF CAST IRON

1. Grey cast iron.
2. White cast iron.
3. Malleable (Whiteheart and Blackheart castings).
4. Alloy cast irons.
5. Spheroidal-graphite cast iron.
6. Meehanite.

Grey cast iron (*BS 1452: 1961*)

This contains between 3 and 4 per cent carbon, of which most is in the free state of carbon flakes, giving the iron a characteristic grey colour and causing it to be weak in tension (tensile strength of 60–125 MN/m², 4–8 tonf/in²).

The high carbon content gives high fluidity when the iron is being cast, and this results in castings with sharp, clean profiles.

Grey cast iron is used where cheapness and accurate shape are more important than the ability to carry load. It has widespread industrial applications, such as cylinder blocks for cars, soil pipes, domestic stove and industrial furnace parts, domestic pots and pans, and ingot moulds.

White cast iron

This contains up to four per cent carbon, most of it being in the combined state as cementite. The fracture appearance is silvery white. It is much stronger and harder than grey cast iron but does not produce such sharp castings; it is also difficult to machine. White cast iron is used where a high degree of abrasion resistance is required, as, for example, in rolls for steel mills and in abrasion-resisting parts for pulverisers and rolls in cement, ceramic and mining industries.

Malleable cast iron

The *Whiteheart process* (BS 309: 1958) was developed in Europe during the 1700s. Briefly, it consists of reducing the carbon content of white cast-iron castings by packing them in iron ore in enclosed boxes and then heat-soaking them at a temperature of about 1000°C for long periods (24 to 100 hours), after which they are cooled slowly in the furnace. The resulting castings have a reduced carbon content, with the free carbon changed from flakes to temper carbon.

The *Blackheart process* (BS 310: 1958) was developed in the U.S.A. and is an annealing method similar to the Whiteheart process. The castings are produced from white cast iron and are packed in sand in closed boxes and furnace annealed, thus reducing the flakey carbon to temper carbon. This process, because of the lower carbon content of the initial castings in white cast iron, is appreciably quicker than the Whiteheart process.

A more recent process has been evolved whereby both Blackheart and Whiteheart malleable iron castings can be produced by heating in a controlled furnace atmosphere without the need to pack the castings in either sand or iron ore.

Heating malleable castings to temperatures above 700°C rearranges the carbon in its cast condition. Care must be taken, therefore, in welding malleable iron parts, to maintain a low welding temperature.

Malleable cast iron has widespread industrial applications for small thin-walled castings such as small pipe fittings, hand tools, electrical fittings, cycle-frame lugs, and parts for motor vehicles and for agricultural and railways equipment, where strength, toughness and cheapness are essential.

Alloy cast iron

This is produced by the addition of nickel and chromium, with small amounts of other elements.

Nickel increases the strength and hardness of cast iron, improving machinability and reducing porosity. Chromium increases the strength and hardness, and improves wear and corrosion resistance; it also produces castings with sharp profiles and gives increased resistance to heat and shock.

Some applications of alloy cast iron are for cylinder blocks, pistons, cam shafts, machine beds and tables, and dies for forging and sheet-metal press work.

Spheroidal-graphite cast iron or ductile iron (*BS 2789: 1961*)

This is a form of cast iron produced since World War II. The addition, under specific conditions of small quantities of magnesium to cast iron reduces the flakey carbon to small spheroids, thus doubling the strength, increasing ductility and resistance to shock, and improving machineability. Spheroidal-graphite castings can be hot- or cold-worked to a limited extent. Welding on these castings can be carried out by normal methods with electric-arc and gas equipment. Gas welding requires magnesium-treated cast-iron filler rods to produce a deposit of spheroidal-graphite cast iron.

Spheroidal-graphite cast irons have wide industrial applications for pipes, gears, dies, machine frames, pump and compressor bodies, valves, components for agricultural machinery, office equipment, and many other items.

Meehanite

This is the registered trade name for a range of high-duty cast irons, the structure of which strongly resembles an 0·85% carbon steel.

It has good damping and fatigue strength, and the ability to withstand dynamic overloading. Although its tensile strength is not comparable to high-grade steels, the qualities mentioned make its use preferable in moving parts such as large crankshafts.

Meehanite can be supplied to specifications for use in the following applications:

General engineering,
Heat resisting,
Wear resisting,
Corrosion resisting.

Meehanite is widely used. Its applications include large crankshafts, heads for diesel engines, lathe spindles, cylinder liners, gearing, flywheels, machine tool tables, headstocks and saddles, rollers for the manufacture of rubber and metal and paper, crane drums, bedplates, pistons, piston rings, pulleys, brake drums and clutch plates for cars, drawing dies, furnace doors, combustion chambers, grates and bars for blast furnaces, glass moulds, brick moulds, sand-blasting machinery, chemical plant and many other items.

PRODUCTION OF STEEL

The term *steel* refers to a metal which is almost pure iron combined with small quantities of carbon, manganese and other elements. Steels have a finer granular structure than iron, which can be seen by comparing a fractured surface of cast iron with one of steel.

Steel-making processes are said to be either *acid* or *basic*. These terms refer to the slags and furnace linings used in the processes to remove the impurities (particularly phosphorus) which are present in cast iron. Acid steels are made from pig iron rich in silicon, and basic steels are produced from pig iron made from low-grade ores, high in phosphorus content, such as are found in western Europe.

Steel is manufactured from pig iron and scrap, and the conversion is a process of refinement by oxidation of the impurities. This may be carried out by any of several methods.

Bessemer process

The Steel Age may be regarded as starting in 1856 with the invention of the Bessemer converter. This made large-scale production of steel possible for the first time. The Bessemer converter consists of a cylindrical steel structure lined with refractory bricks and fitted with a removable base through which an air blast can be introduced. The converter is mounted on two large bearings half-way along its length to permit mechanical rotation. For filling and pouring, the converter is horizontally positioned but during the blow, when oxidation of the impurities takes place, it is positioned vertically. No extra fuel is required in this process; the heat is generated by the chemical reaction due to oxidation of the impurities, this reaction being started by the heat contained in the initial charge and the refractory lining when the blow commences. Conversion by the Bessemer process takes about 20 to 25 minutes; capacities range from 15 to 25 tonnes.

The original Bessemer process was unfortunately only suitable for producing acid steels from ores low in phosphorus and rich in silicon. This produced acid slag charged with silica, and it was found necessary to line the converter with an acid refractory material, namely silica bricks, to prevent chemical attack on the converter.

Thomas process

Bessemer's invention was soon followed by the Thomas process, which was due to the work of two men, Gilchrist and Thomas. This process was similar to Bessemer's but the acid refractory lining was replaced by a basic refractory lining of dolomite. Together with the introduction of lime into the molten pig iron to form a basic slag, this method allowed for the processing of ore of high phosphorus content. The process led to the development of the steel industries of Lorraine, Luxembourg and Germany, where the local ore has a high phosphorus content.

Open-hearth process

In 1867, soon after the Bessemer process came into common use, two brothers, C. W. and F. Siemens, produced a new furnace for steel making. Now known as the open-hearth furnace, this utilised the waste heat from the furnace to preheat the incoming air needed for combustion of the fuel gas in the furnace. Nowadays both gas and liquid fuel are used. The furnace consists of a shallow bath, which is either fixed or can be tilted mechanically, and a back wall; it is completely covered by a low roof. The furnace is made from steel, and lined with refractory brick in order to withstand the tremendous heat generated. There are several doors at the front through which the furnace is charged, worked, mixed and skimmed by the furnacemen.

At each end of the furnace are openings; one is for the passage of fuel and heated air, which burn with the flames directed down on to the surface of the melt, and the other is for the exhausting of the hot burned-fuel gases to atmosphere; these gases pass over the brick regenerators, which absorb a large amount of the waste heat and in turn heat the air coming into the furnace on the reverse blow.

The process takes between 7 and 15 hours, and modern furnaces have capacities up to 300 tonnes. By the methods used, high furnace temperatures are produced which allow the mixing of steel scrap with the charge of pig or molten iron, thus reducing the cost of raw materials. The process is much slower than the Bessemer process, allowing for greater control of the quality of the steel. The furnace construction also makes possible the production of a larger tonnage of steel per charge.

The open-hearth method has been widely adopted in Britain for steel production because it meets the conditions and requirements of home-produced ore, which is of lower phosphorus content than that on the Continent. Thus for some eighty years the pattern of steel making has been little changed, Continental countries favouring the basic Bessemer process and British steel companies concentrating on the open-hearth process.

MODERN DEVELOPMENTS

There are several new oxygen steel-making processes in use. These are known as the LD, LD-AC, Kaldo, Rotor and Ajax processes. All are modern developments of Bessemer's method of oxidation by air blast.

The *LD process* was developed in Austria and named after the towns of its origin, Linz and Donawitz. The process uses a converter basically similar to a Bessemer converter but with a closed bottom. It is held upright throughout the blow and the oxygen is blown on to the molten surface by means of a water-cooled lance inserted vertically through a top opening. For pouring and filling, the converter is mechanically tilted. The process has been found suitable for iron with fairly low phosphorus content (up to about 0·4 %).

The *LD-AC process* is a later development of the LD process. It requires two blows, with the formation of two slags aided by lime injected with the oxygen, the first slag and impurities being poured off between blows. The method makes it possible to process pig iron with high phosphorus content into steel.

The *Kaldo method,* named after its inventor, Professor Kaling, and the Swedish steelworks at Domarvet, uses a converter which rotates on an inclined axis during the blow, while oxygen is directed by a water-cooled lance on to the surface of the molten iron.

The *Rotor process* is of German origin, and again consists of a converter which slowly rotates on a horizontal axis during the blow. Oxygen is introduced through two water-cooled lances, one delivering oxygen below the surface of the molten iron, while the other directs oxygen on to the surface.

The *Ajax system* was first introduced in 1958 by a British steel company after much research and development. It is a variant of the oxygen process adapted to existing open-hearth furnaces of the tilting type, whereby oxygen is injected into the bath by retractable water-cooled lances. The excess oxygen speeds up the oxidation of the impurities, and it has been found that output has been doubled by this method.

The *electric-arc furnace,* where the heat is generated by an arc between consumable graphite electrodes and the metal surface, has until recently been regarded as uneconomic for anything except special alloy steels for which the open-hearth and Bessemer processes are unsuitable. However, with modern techniques using tonnage oxygen, and the fact that the cost of electricity has not risen as fast as that of other fuels, it has been shown that large-scale arc-furnace operations of up to 100-tonne capacity are practical for a wide range of steel.

The molten steel which is made by the above methods, other than continuous casting, is used at once to make castings or is teemed into ingot moulds to produce long tapered rectangular blocks weighing from 4 to 20 tonnes. These ingots are of convenient shape and size for further processing into finished wrought steel for the manufacturing industries.

STEEL CASTINGS (BS 3100: 1967)

BS 3100 covers twenty British standards applying to steel castings in various carbon and alloy steels.

Steel for casting is tapped from the furnace into a ladle from which it is poured into moulds of intricate shapes which would be too expensive and time-consuming to manufacture by other methods. Steel castings are usually large, and are used for such applications as locomotive frames and wheel centres, crank cases for diesel engines, hammer blocks for drop forges, frames and bedplate for rolling machines, rock-crushing machinery, presses, forges, boring machines and other large machine tools. Large steel castings are also used for turbine casings, steam chests, hydraulic cylinders, derrick fittings, manhole covers and similar applications.

Cast steel must not be confused with steel castings. Cast tool steel is a special range of high-carbon steel (0·1 to 1·5 per cent carbon) produced in a purified and refined form in a high-frequency electric crucible furnace for the manufacture of small tools which, when suitably hardened and tempered, will cut other metals. Examples are drills, taps, files and scrapers.

Silver steel (BS 1407: 1959), which is supplied in closely toleranced polished rods, is cast tool steel. There is no silver in its composition—and much romantic conjecture of how it got its name.

CARBON STEELS

Steels are classified according to their carbon content as follows:

Type	*Carbon content* (%)
Low-carbon	0·07 to 0·15
Mild	0·15 to 0·25
Medium-carbon	0·25 to 0·5
High-carbon	0·5 to 1·4

In addition, small quantities of other elements such as manganese, silicon, phosphorus and sulphur may be present.

Low-carbon steel and the low range of mild steel cannot be hardened and tempered by heat treatment. Medium- and high-carbon steel can be satisfactorily hardened by heating and quenching in water.

WROUGHT STEELS

Wrought steels are those which, from the steel ingot form, are worked into shape by rolling into plates, sheet and strip, extruding into bars, rods and sections, drawing into tubes and wire, and forging into shape by mechanical hammers.

BS 970: 1955 (*En series*) *Wrought steels*
This Standard provides a schedule of more than a hundred steels of different compositions suitable for automobile and general engineering purposes.

The series starts with carbon steels covered by the specifications from En 1 to En 9. These steels, although termed carbon steels, contain small percentages of silicon, manganese, phosphorus and sulphur.

The series then continues with low-alloy steels through to high-alloy steels, some of which contain as much as 20 per cent chromium and 10 per cent nickel. Table 14.1 gives a brief outline of the uses of wrought carbon steels in the En series.

BS: 970 Spec. No.	*Carbon content* (%)	*U.T.S.* MN/m² (tonf/in²)	*Remarks and uses*
En 1A En 1B	0·07–0·15 0·07–0·15	355–540 (23–35) 355–420 (23–27)	En 1B contains a higher percentage of sulphur than En 1A. Free-cutting steels intended for the production of small parts in automatics, low-duty nuts, bolts, studs etc.
En 2	0·20	310 (20)	Mild steel for making lightly stressed parts. Easily machinable and can be suitably cold-worked where plate has to be bent and riveted.
En 3 En 3A En 3B	0·25 0·15–0·25 0·25	380–540 (25–35) 430 (28) 430 (28)	Mild steel suitable for general-purpose lightly stressed machined details, such as brackets, plugs, levers, bolts, sockets. Cannot be cold-worked like En 2 and is unsuitable for deep-drawing operations.
En 4 En 4A	0·30 0·30	430–590 (28–38) 500–650 (32–42)	A steel of slightly higher carbon content. Obtainable as bright bars and suitable for machining into small lightly stressed components such as bolts, pins, collars, brackets etc.
En 5 En 5D	0·25–0·35 0·25–0·35	540–620 (35–40) 540–770 (35–50)	En 5 is supplied in the hardened and tempered condition. En 5D supplied in hardened and tempered condition and cold-drawn. Has a slightly higher carbon content than En 4 and is used for machining similar lightly stressed machined details.
En 6	0·4	540–700 (35–45)	Obtainable in the form of bright bars and suitable for machined components and general engineering purposes.
En 7	0·10–0·30	540–620 (35–40)	A steel with a high sulphur content to improve the machining qualities, especially when manufacturing nuts, pins, bolts, etc. En 7 has similar mechanical properties to En 6. It is not recommended for case-hardening.
En 8 En 8M	0·35–0·45 0·35–0·45	540–620 (35–40) 540–700 (35–45)	This is a 4% carbon steel which may be obtained in the normalised cold-drawn or hardened and tempered condition. For the production of machined components which require strength and resistance to wear without high impact strength, such as automobile crankshafts and connecting rods, pins, shafts, levers and small rotating parts. En 8M has a high sulphur content which improves the machining qualities.
En 9	0·50–0·60	700–1000 (45–65)	A carbon steel which can be obtained in the normalised cold-drawn or hardened and tempered conditions. Suitable for cylinders, gears, machine-tool parts, gun parts and wear-resistant parts which do not require to be too deeply hardened. The 0·50–0·08 percentage nickel in the composition actually classifies En 9 as a low-alloy steel rather than a carbon steel.
En 32A En 32B En 32M	0·15 0·10–0·18 0·10–0·18	460–550 (30–36)	This is a carbon case-hardening steel suitable for manufacturing parts which require hard surfaces to resist continuous wear, with tough interiors to resist shock and prevent breakdown. Typical uses are tappet rollers, cams and camshafts, gudgeon pins, gun parts and lightly stressed gears. En 32M has a higher percentage of sulphur, which improves machinability.
En 42 En 43 En 44	0·70–0·85 0·45–0·60 0·90–1·20		These heat-treatable steels are used mainly for the manufacture of vehicle springs in the form of both coil and laminated leaf springs.

TABLE 14.1 Wrought carbon steels. Bars, billets, stampings and light forgings

Iron and steel

Chapter 14

BS 1449: 1962–1967 Steel plate, sheet and strip

This standard is in several parts, is an extension of the En series, and can be regarded as companion to BS 970. It covers plate, sheet and strip made by continuous and hot- and cold-rolling processes in both alloy and carbon steels.

The uses of sheet steel are manifold. The car industry and the manufacture of domestic appliances, such as refrigerators, washing machines, sinks, stoves, electric fires, consume millions of tonnes a year. Other uses include office machinery, cabinets and trays, rolled and welded tube, heating and ventilating ducting, roofing, and sheet-metal tins and closures for the canning industries.

Plates are used in shipbuilding, bridge construction boilermaking and large fabrications for machine tools where castings would be too expensive.

Strip is the common form of supply to press shops to give a continuous feed to automatic, power-operated presses for making the multitude of cheap sheet-metal articles that we use daily, such as clips, handles, brackets, buckles, fasteners, cigarette-lighter parts, pen nibs, ash trays, spoons, keys, toys, cages for ball bearings, pawls, levers and ratchets for typewriters and other office equipment, and many similar items.

Sheet metal is supplied in many finishes, such as black, bright, tinned, galvanised, and coated with tough plastic decorative surfaces which withstand pressing and require no further finishing after forming.

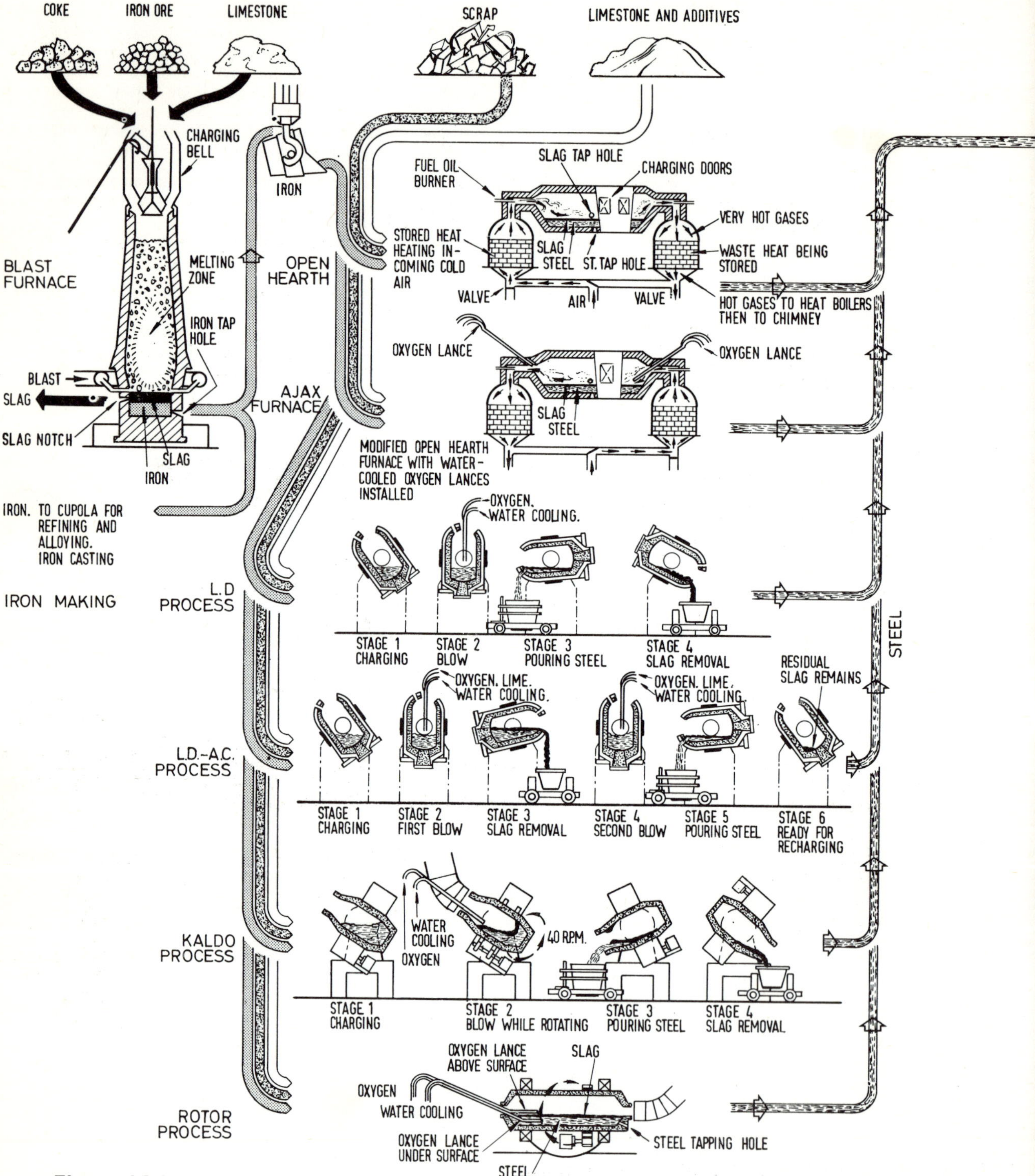

Figure 14.1 SOME METHODS OF MAKING STEEL

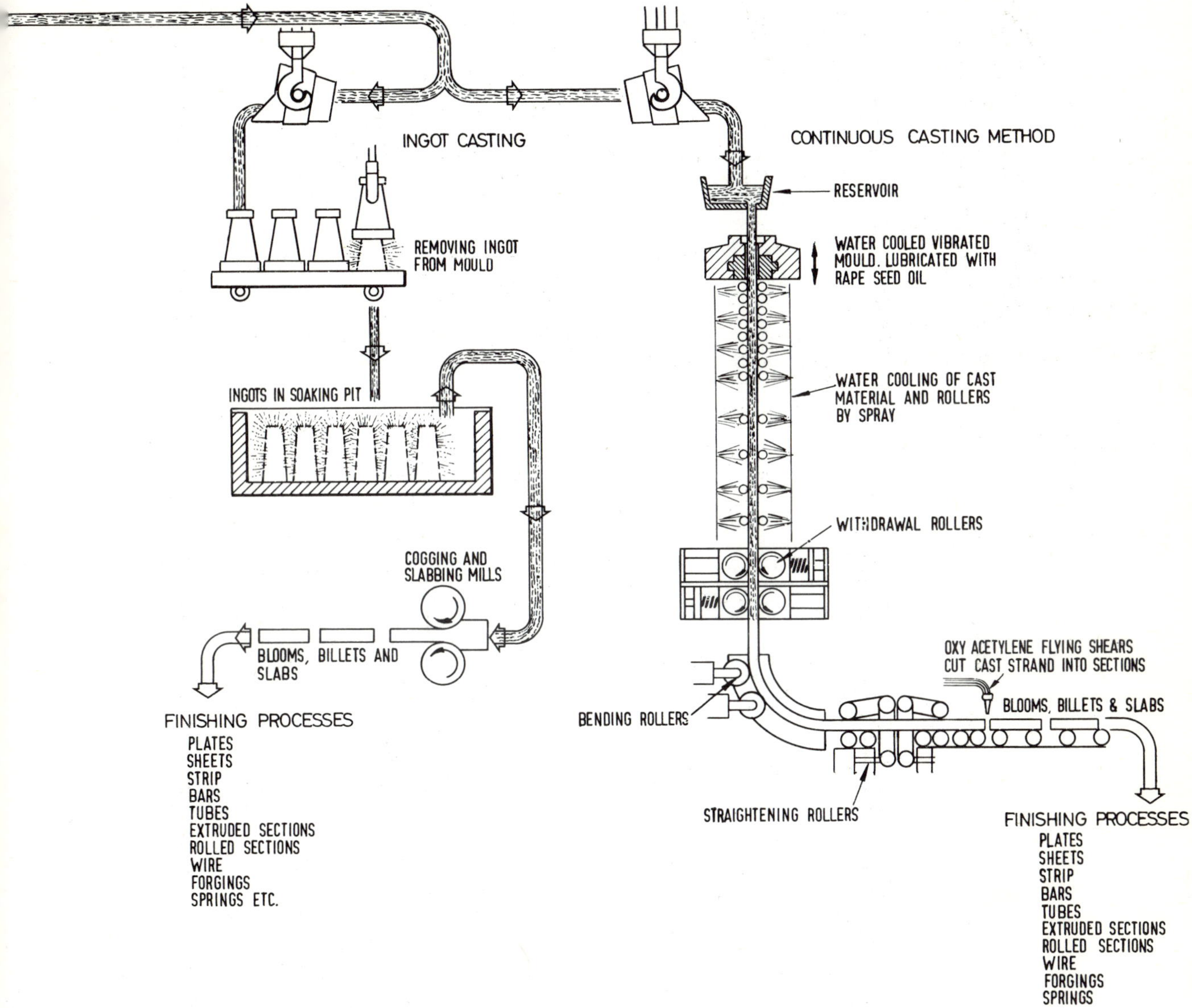

DIAGRAM SHOWING INGOT CASTING AND SLAB MAKING, PRIOR TO FINISHED STEEL PROCESSES

Figure 14.1

Copper and copper alloys

Copper is one of the most useful of the non-ferrous metallic elements. It was discovered and first used by Neolithic man during the late Stone Age (about 8000 B.C.) and was found in a free state in nature. This 'native' copper was used as a substitute for stone for making into hammers, knives and utensils.

Its malleability made it relatively simple to shape with the assistance of heat. Working the material by a rough hand-forging process hardened it, giving a keen edge to knives and swords. Its bright reddish colour and durable qualities made copper a prized possession.

Some 2000 years later it was discovered that copper could be cast—probably following accidental melting in a fire.

Copper was extensively produced in Cyprus about 3000 B.C. It was known as *aes Cyprium* (ore of Cyprus). The word *Cyprium* was corrupted to *cuprum*, from which the English word *copper* is derived.

The modern use of copper extends to practically every sphere of industry, and it is an alloying element in many of the important metal alloys we use today. You may be the proud possessor of a 9-carat gold watch or cigarette case without realising that it is two-thirds copper.

The electrical industries absorb the largest percentage of the pure copper produced. More than a quarter or a million tonnes of copper wire is produced annually for use for the manufacture of overhead power-transmission lines, telegraph and submarine cables, and motor and generator windings, transformer windings, electro-magnetic windings for nuclear power stations, the sheathing of mineral-insulated electric cable, and a multitude of other applications in the manufacture of the electrical equipment which modern living requires.

Copper has a wide variety of other domestic and industrial applications. For example, copper piping is used in most water, gas, heating and sanitation systems, and copper is also used for roof coverings, flashings, cooking utensils, household ornaments and many other items. Copper, as sulphates and oxides, is also an important constituent of paints, fungicides and coloured glassware. Some of the industrial uses of copper are tubes for boilers, refrigerators, oil and water coolers, brewing vats, jam boilers, whisky stills, oil pipes for machinery, instrumentation components, oxygen lances for steel making, gas nozzles for oxy-acetylene cutting equipment etc.

Copper in the form of powder is used to compact into small industrial and domestic components by die-pressing into shape, followed by sintering (a long controlled heat treatment). It is usually alloyed with other metals. Electric-motor brushes are often made from a mixture of carbon and copper powders.

THE OCCURRENCE OF COPPER

Copper deposits are widely distributed over the earth's surface in rocks, soils, oceanic clays, river salts etc. The main areas, which contain about 90 per cent of the world's known reserve, are:

The Rocky Mountains and the Great Basin area of the U.S.A.
The Urals in the U.S.S.R.
Zambia.
The west slope of the Andes in South America.
Central Canada.
The Congo Republic.

From these areas millions of tonnes of copper are extracted every year.

Other smaller but important deposits occur in Japan, Australia, Mexico, Finland, Jugoslavia, Germany, Norway, Sweden, Spain, central Asia and South Africa.

CLASSIFICATION OF COPPER ORES

Copper ores are classified into three groups:

1. *Native*, containing 99·9 per cent copper.
2. *Sulphide ores*, containing 47 to 80 per cent copper.
3. *Oxide ores*, containing 56 to 89 per cent copper.

The copper-bearing ores are quite complex and often contain other metallic elements such as silver, gold, cobalt, nickel, zinc, lead and platinum, together with an enormous amount of rocky waste (*gangue*) which has to be handled and disposed of. Although the sulphide and oxide ores contain a high percentage of copper, they are finely dispersed through the mined deposits, from which, even in the richest deposits, only about 4 per cent results as copper.

BRIEF DESCRIPTION OF THE REFINING PROCESS

The method of extracting copper varies according to the type of ore mined but, in general, the process is:

1. The ore is crushed in cone mills and finally ground in ball mills to a fine powder.
2. The next step is a flotation process. The fine powder is vigorously agitated in flotation tanks containing water mixed with suitable wetting agents, causing the valuable copper-containing dust to float to the surface in a froth which overflows from the tank, in a continuous stream, to a tank where the excess moisture is removed and the concentrate is thickened and filtered. The concentrate is now ready for the next process of smelting, and is passed to charging bins, where it is mixed with suitable fluxes.

   The precipitated waste, in the form of fine mud known as tailing, is run off from the flotation plants into settling tanks, and finally deposited as large mounds of greyish dust. Further copper is extracted from this waste by leaching. The waste is sprayed with acidified water, which percolates through the mass into large settling tanks from which copper is precipitated and recovered.
3. The wet concentrate may pass either direct to a reverberatory furnace or through a complicated

roasting process that drives off some of the sulphur content when this is necessary. In any case it finally reaches the reverberatory furnace where, with suitable fluxes to form a slag, it is smelted to a crude melt called the *matte*. This matte is tapped from the reverberatory furnace and charged into a converter similar to a Bessemer converter used for steel making. Air is blown through the melt, oxidising out iron, which enters the slag and is poured off, and oxidising the sulphur, which is blown out as sulphur dioxide gas. The copper is then poured into cakes of 'blister' copper which may be as much as 99 per cent pure. The name *blister copper* is descriptive of the rough blistery surface of the cast cakes of copper.

4. The blister copper is further refined by one of two processes:
   (a) fire-refining in a reverberatory furnace, or
   (b) electrolytic refining.

Fire refining

The blister copper is melted in a reverberatory furnace, where the remaining impurities are slagged off; these impurities include iron, antimony and arsenic. To remove the remaining oxygen from the melt, it is poled with green wood, and the surface of the melt is covered with coke of low sulphur content.

During this process, small test 'buttons' are taken from the melt and the percentage of oxygen remaining is determined by the amount of surface sink of these buttons. This is continued at intervals until no sinking occurs, which indicates that the melt now contains between 0·04 and 0·08 per cent oxygen. This is called *tough pitch* copper (*tough* from the toughness brought about by the process itself, and *pitch* from the surface level of the test buttons, i.e. high pitch signifies low oxygen content).

Fire-refined copper includes a high-conductivity grade and is suitable for most purposes. It is cast into shape as bars, billets etc. for further processing by rolling, drawing or extruding.

Electrolytic refining

This process is used to produce copper of very high purity and to recover other precious metals known to be present in the blister copper. However, it is not always necessary to use this process to produce copper of high quality; some ores yield remarkably pure blister copper not contaminated by other metallic elements.

The blister copper is remelted in an anode furnace, from which it is cast into thick slabs to form anodes for processing in a refining tank. This tank is constructed of wood or concrete with a lead lining, and is filled with a solution of dilute sulphuric acid and dissolved copper sulphate to form the electrolyte, which is heated and slowly circulated. The thick blister-copper anodes, about 30 or 40 to a tank, are hung in the electrolyte and connected to the positive bus-bar of a direct-current supply. Thin starting sheets of pure copper to form the cathodes are connected to the negative bus-bar of the direct-current supply and are lowered into the bath to interleave with the blister-copper anodes.

The electric current produces a plating effect, removing copper from the anodes and depositing it on the thin cathode starting sheets. These starting sheets grow from their original weight of about 5 kg (10 lb) to some 140 kg (300 lb) in about 14 days. They are then removed, melted, and cast into shapes suitable for forming by further industrial processes into wire, bar, sheets etc. Electrolytic copper is oxygen-free, having a purity of 99·95 per cent. It also has high electrical conductivity properties, second only to pure silver.
In this process of electrolysis, the impurities sink to the bottom of the tank to form sludge, from which precious and industrially important metals are recovered.

COPPER ALLOYS

Although millions of tonnes of pure copper are used in engineering where the conduction of electricity and heat are the important factors of design, copper in the pure state is too weak for many engineering purposes. However, some thirty other metallic elements can be alloyed with copper to improve its physical properties and so widen its use. The chief alloying metals for general commercial applications are *tin, zinc, aluminium, lead* and *nickel*.

The three main copper alloys which are widely used for industrial and domestic purposes are:

1. Brass (copper–zinc alloys)
2. Bronze (copper–tin alloys)
3. Aluminium bronze (copper–aluminium alloys).

Brasses

These include a wide range of copper–zinc alloys containing from 55 to 95 per cent copper. The alloys with the highest percentages of copper (from 80 per cent upwards) are known as *gilding metals* because of their colour. Other romantic names have been given to some of these alloys in the past—*Princes' metal, red brass, pinchbeck, tombac* etc. Some of these alloys and names are still used but are outside the scope of British Standards.

Brass alloys show marked resistance to both atmospheric and marine corrosion and are used in engineering applications where they are exposed to adverse conditions and where continuous mechanical functioning depends on their non-rusting characteristic.

One of the advantages of brass is the ease with which it can be joined. Depending on the mechanical application, it can be soft-soldered with tin-based solder, hard-soldered with silver solder, brazed or welded.

Brass is obtained in the form of sheet, strip, rods, sections, tubes, hot stampings and castings. The brasses fall into two main groups:

1. Cold-working brass which contains up to 39 per cent zinc.
2. Hot-working brass which contains from 39 to 45 per cent zinc.

With a zinc content of more than 45 per cent, brass becomes brittle and has little commercial value.

Copper and copper alloys

Chapter 15

GENERAL PROPERTIES OF COPPER

1. Ductility.
2. Malleability.
3. High electrical conductivity (97·5 per cent and second only to silver, taken as 100 per cent). It should be noted that small quantities of impurities will drastically reduce conductivity; for example 0·04 per cent phosphorus will reduce electrical conductivity to about 75 per cent of that for pure copper, and 0·1 per cent arsenic will have the same effect.
4. High thermal conductivity.
5. Good resistance to corrosion.
6. Will alloy well with some 30 other metallic elements to produce useful engineering alloys.
7. Unsuitable for producing sound castings free from 'blow holes' due to the absorption of oxygen and other gases during melting and casting. Under controlled conditions however, copper castings can be satisfactorily produced.
8. Specific gravity 8·7 (steel 7·8).
9. Melting points 1083°C.

BRITISH STANDARD GRADINGS OF COPPER

Table 15.1 lists the British Standard Specifications referring to copper in its commercial unworked forms. These specifications are all published in one volume dated 1964.

TABLE 15.1 GRADES OF RAW COPPER: TYPICAL APPLICATIONS

BS No.	*Grade*	*Minimum % copper*	*Typical applications*
1861	Oxygen-free high-conductivity copper	99·95	Uses involving the conduction of heat and electricity; welding and brazing; exposure at high temperatures.
1954	Ditto for special applications		
1035	Cathode copper	99·90	For the production of high-grade alloys and high-conductivity castings. Used to produce coppers to BS 1036 and BS 1861 specifications.
1036	Electrolytic tough pitch high-conductivity copper	99·90	Conductors, fabricated electrical components, high-grade alloys, and high-conductivity castings.
1037	Fire-refined tough pitch high-conductivity copper		
1038	Tough pitch copper	99·85	General engineering, chemical engineering, building, and various other applications where the highest conductivity is not essential.
1039	Tough pitch copper	99·75	Used mainly for production of other alloys.
1040	Tough pitch copper	99·50	
1172	Phosphorus de-oxidised non-arsenical copper for general purposes	99·85	Chemical engineering, manufacture of tubing, and general engineering applications where welding and brazing are necessary.
1173	Tough pitch arsenical copper (0·3% to 0·5% arsenic, 0·1% oxygen, no phosphorus)	99·20	Not recommended for welding applications. Used in situations of medium high temperatures.
1174	Phosphorus de-oxidised arsenical copper (0·3% to 0·5% arsenic, no oxygen, 0·15% to 0·08% phosphorus)	99·20	For general and chemical engineering where brazing and welding are involved; for situations exposed to medium high temperatures; for tubing and boiler plates for steam locomotives.

TABLE 15.2 SOME AVAILABLE BRASSES

Description	*Nearest BS No.*	*Composition (%)*	*Remarks and uses*
Cap copper		97 copper 3 zinc	Ammunition caps
Gilding metal	711 712 713	80 copper 20 zinc 85 copper 15 zinc 90 copper 10 zinc	Architectural metal work, sections formed from strip, cheap imitation jewellery, bullet cases, miscellaneous applications requiring brazing.
Cartridge brass 65/35 brass Basis brass	267 266 265	70 copper 30 zinc 65 copper 35 zinc 63 copper 37 zinc	Cold-working alloys which can be used for deep drawing and press work. Used for cartridge cases, deep-drawn shells, car radiator shells and articles of similar unsymmetrical shape.
Clock brass	2785	59 copper 39 zinc 2 zinc	Used for plates; gears and wheels in clock and instrument manufacture.
'Yellow' or 'Muntz' metal	1541 149	60 copper 40 zinc	High resistance to corrosion. Difficult to machine but improved by addition of lead. Used to manufacture sheet, bars, extrusions, hot stampings, bolts, spindles, pump components, regulator bodies, condenser plates.
Free-machining brass	249	58 copper 39 zinc An increase to 61% copper with a decrease to 36% zinc increases impact strength.	Supplied in rods and sections. Suitable for high-speed turning and milling of engineering components.
Brass for castings	1400		BS 1400 is a schedule which covers a number of casting copper alloys, the copper content of the brass alloys varying between 55 and 85 per cent, and the remainder being zinc with small quantities of other materials added. The uses for brass castings cover a considerable range from cheap inexpensive castings, such as general-purpose boiler fittings, to high-tensile brasses having strength and toughness combined with excellent corrosion resistance.

Bronzes

These are alloys of mainly copper and tin, with small amounts of lead, zinc and phosphorus. Bronzes fall naturally into two main classes:

1. *Wrought alloys* containing up to 10 per cent tin are used for the manufacture of sheet, strip and wire which are cold-rolled or drawn. Cold working will increase the temper of the alloy, and spring temper can be achieved in alloys having from 8 to 10 per cent tin.
2. *Cast alloys* for industrial use are more complex and contain up to 14 per cent tin. In practice, bronzes are not straight copper–tin alloys but contain alloying constituents of nickel, lead, phosphorus and zinc. Bronze alloys for casting bells contain up to 20 per cent tin; such alloys are brittle, but give a particularly sonorous note when struck. Gun metals are casting alloys of tin bronze containing zinc.

TABLE 15.3 BRONZES

Description	*Composition* (%)	*Nearest BS No.*	*Remarks and uses*
Bronze coinage alloys	95·5 copper 3·0 tin 1·5 zinc		Used for making copper coins in the U.K.
Phosphor-bronze (ordinary quality)	94·5 copper 5·25 tin 0·1 phosphorus	407/2 369 384	For the manufacture of sheet, strip, rod, wire and tubes. Phosphor-bronzes have good elastic properties combined with resistance to corrosion and corrosion fatigue. This accounts for their use as springs and instrument components. In rod form, they are favoured for engineering components subject to friction, tubes used in condensers, piping for fuel systems, and Bourdon gauges.
Conductivity bronze	1·0 tin 0·5 cadmium remainder copper		Used for telephone and trolley wire.
Leaded bronzes	70–85 copper 5–10 tin 5–25 lead	Covered by BS 1400	A series of sand-cast leaded bronzes of progressively increasing lead content; with increase of lead the alloys become more plastic, and this makes them suitable as bearings materials, particularly where the shaft material is soft steel.
Phosphor-bronzes	5–20 tin 0·1–1·5 phosphorus remainder copper	Covered by BS 1400	Alloys with the higher percentage of copper are used for bearings and gear wheels (particularly worm gears). Other uses are for springs and general-purpose sand castings.
Gun metal	Up to 88 copper, 8–10 tin, 2·4 zinc	Covered by BS 1400	Good casting qualities. The tin imparts fluidity and makes gun metal preferable to brass. Widely used for pumps, valves, miscellaneous castings and statues. Has good anti-corrosion properties and is used where pressure tightness in castings is required. Admiralty gun metal (88% copper, 10% tin, 2% zinc) was formerly the standard composition for marine purposes. Other alloys are now frequently used. Addition of 1% to 5% nickel improves strength and hardness.

Aluminium bronzes

These are copper alloys containing up to 14 per cent aluminium, although commercial alloys usually contain not more than 10 per cent. Iron, manganese, nickel and other elements are also added, and the manufacture of the alloys for commercial use is by no means simple.

Aluminium bronzes have many good features in respect of mechanical strength and resistance to creep, oxidation at high temperatures, corrosion and wear; their pleasing golden colour also makes them suitable for decorative applications. They are often used in cheap jewellery to simulate 18-carat gold.

Aluminium–bronze alloys can be cast and wrought by both hot and cold processes. Mechanical strength increases in the conventional manner by working, and in certain alloys the mechanical strength can be modified by quenching and tempering techniques similar to those used for steel.

TABLE 15.4 ALUMINIUM BRONZES

Description	*Composition* (%)	*Nearest BS No.*	*Remarks and uses*
Cold-worked aluminium bronze	4–7 aluminium, up to 4 magnesium or nickel, remainder copper.		Manufactured in cold-rolled sheet, strip and drawn tubes. Excellent corrosion resistance. Resistant to oxidation at high temperatures. Used for tubes in heat exchangers.
Casting aluminium bronze	9–12 aluminium; varying small quantities of iron, nickel, and manganese; remainder copper.	Covered by BS 1400	Castings may be either sand- or die-cast. Good resistance to wear, and will support heavy compressive loads in service. Used as stern bearings for marine propeller shafts, and for gears, cylinder heads, chains in acid pickling tanks, pump bodies. Many marine applications because of excellent resistance to sea-water corrosion.

Aluminium is the most modern of the common engineering metals. Although it exists in abundance in the earth's crust (an estimated 8 per cent by weight), it was not isolated until 1825 by the scientist Hans Christian Oersted. Another fifty years passed before the dynamo had reached a stage of development where high-amperage currents were commercially available to enable aluminium to be produced on an economic scale. In 1886, two scientists, Charles Martin Hall in America and Paul Heroult in France, discovered almost simultaneously the electrolytic method of extracting aluminium which forms the basis of production in the present aluminium industry.

PRODUCTION

Aluminium, owing to its tendency to chemical activity, is never found in the natural metallic form but combined in soils, clays and rocks as chemically complex alumino-silicates.

The rich ore from which aluminium is extracted is bauxite. This is found in large deposits in many parts of the world—north and south America, Jamaica, France, Germany, Italy, Jugoslavia, Russia and many other countries.

Unlike the other common engineering metals (iron, copper, zinc, lead etc.), aluminium cannot be extracted by direct smelting of the aluminium ores because of its affinity to oxygen. If smelting of aluminium ore were attempted, the other metallic impurities such as iron, silicon and titanium would more readily reduce to the metallic state, and the result would be an alloy too impure to be of any use. Also, at the high temperature necessary for smelting, the greater part of the aluminium would vaporise and be lost in the flue gases.

One method of production from the crude bauxite is first to produce a purified alumina (Al_2O_3) and then extract the aluminium by electrolysis. The ore is mixed with sodium carbonate and roasted to produce sodium aluminate, which is dissolved in water through which carbon dioxide is blown, producing a precipitate of aluminium hydroxide. The precipitate is then roasted in ovens, and results in alumina from which the principal impurities have been removed.

The purified alumina (Al_2O_3) is dissolved with molten fluxes of cryolite and fluorspar and electrolised in a carbon-lined steel bath or cell. The carbon lining acts as the cathode (negative pole) of the cell while large carbon block dipping into the molten contents of the bath form the anode (positive pole).

Most commercial cells for the production of aluminium use currents of 50 000 to 100 000 amperes, under the influence of which the oxygen in the alumina is driven off and burnt at the anode, while the aluminium, which is heavier than the flux, sinks to the bottom of the bath, where it is siphoned off at intervals and cast into ingots weighing about 25 kg each. By this method aluminium can be produced in a range varying from 99·99 per cent (super purity) to 99 per cent (minimum purity).

The production of every tonne of aluminium requires 4 tonnes of bauxite, three-quarters of a tonne of carbon anode, approximately 20 000 kW h of electricity, and smaller quantities of several other materials.

The baths, which are 6 metres or more long, are usually arranged in lines coupled together electrically in series, as each cell requires only some 5 to 6 volts. They are housed in suitably large buildings with the necessary large direct-current generators and suitable fume-extraction plant. From time to time the baths are stirred, and additional alumina and flux added as the aluminium is siphoned.

Figure 16.1 Electrolytic-cell production of aluminium

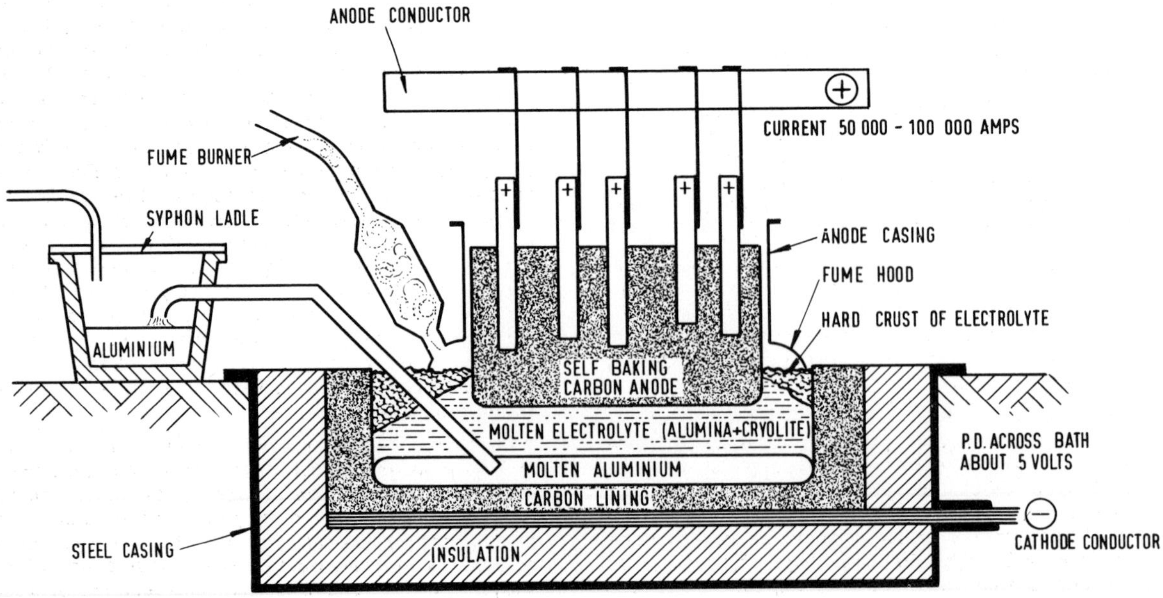

Aluminium and aluminium alloys

Chapter 16

PROPERTIES

In its commercially pure state, aluminium is a soft, silvery, ductile metal. Its characteristic properties are lightness (being less than one-third the weight of steel), good electrical and thermal conductivity, and affinity to oxygen. Size for size, aluminium has an electrical conductivity equivalent to three-fifths that of copper, but weight for weight it is far superior. The affinity of pure aluminium to oxygen readily provides it with a tenacious oxide film that protects it from further corrosive attack. Because of this property, aluminium is used for vessels and plant in the processing of foods and chemicals, and aluminium foil is widely used as a wrapping for foodstuffs, confectionery, cigarettes and medicines.

Aluminium in its pure state is too low in strength to be of real interest as a structural material. However, alloying it with small amounts of other elements—the most important being copper, magnesium, silicon, manganese, zinc, nickel and chromium—has a remarkable effect on its properties. The alloys produced are of considerably increased strength, are capable of heat treatment, and work-harden more readily. Which elements are added, and in what quantities, depends not only on the mechanical properties required in the finished product, but also on the method to be used in manufacture.

There are a bewildering number of aluminium alloys. Each manufacturer of aluminium produces a considerable number of different alloys, in many instances duplicating each other's efforts and creating confusion by giving different trade names and specification serial numbers to identical alloys. However, the independent British Standards Institution and the Directorate of Technical Development of the British Ministry of Supply, with the co-operation of the industry, have produced standards that are recognised at a national level. For marine applications, Lloyds specify their own standards in respect of both chemical composition and mechanical requirements.

To bring the multitude of available alloys into some sort of order, it is convenient to arrange them in four main classes as shown in Figure 16.2. From this figure it can be seen that the chief distinction is between casting and wrought alloys, which are further subdivided into those that are heat-treatable and those that are not.

Casting alloys, which are poured when molten into sand moulds or permanent steel moulds of intricate shape, and allowed to cool into the shape of the mould, have compositions largely dictated by the requirements of this method of fabrication, and differ considerably from the wrought alloys.

Wrought alloys, having been cast into ingot form, are mechanically worked to produce the finished shape that is required. Rolling produces plates, sheets, and foil down to 0·005 mm (0·0002 in) thick. Extrusion, where the metal is forced through steel dies, forms continuous lengths of structural and decorative sections. Forging makes strong, intricate shapes by forcing the alloy into a steel die under the impact of a power

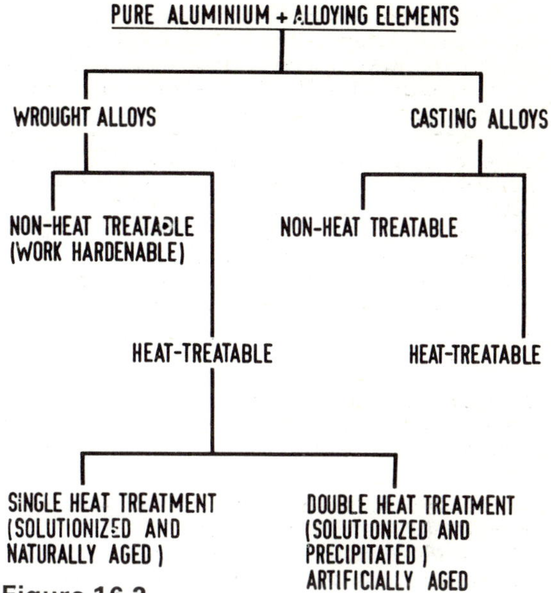

Figure 16.2

hammer, and seamless tubes and containers are produced by drawing. These alloys can be severely distorted without tearing or cracking, and some of them, without further heat treatment, improve in mechanical properties by work-hardening in the forming process.

HEAT TREATMENT

This term, when applied to suitable heat-treatable aluminium alloys, primarily denotes *solution* treatment whereby considerable hardening and strengthening can be induced in the material while retaining a useful amount of ductility. *Solution* treatment does not refer to liquids in salt baths or quenching tanks. It is a metallurgical term to describe the movement of the atoms which occurs in an alloy in the solid metal condition. All metals are made up of small crystals or grains. With some alloys, as the crystals slowly form from the molten state, the formation of the crystal centre differs from the boundary layer which is the last to solidify. The crystals formed have centres rich in one metal, while the boundary layers are rich in another of the alloying metals. When heated to a certain temperature below the melting point, the alloy remains solid but the atoms in the boundary layers move into the metal in the crystal centre, producing what is known as a *solid solution* and at the same time producing other chemical effects. Rapid quenching when a *solid solution* has been achieved traps the atoms in the *solid-solution* condition.

Aluminium alloys are heated to temperatures of about 500°C for several hours, then rapidly cooled by quenching in oil or water. After quenching, the metal remains soft for a period, when forming may be carried out, and then gradually hardens or 'ages', the process being complete after a few days at room temperature. Most alloys, however, require double heat-treatment so that a greater degree of hardening can be achieved; the second treatment is carried out by reheating the solution-treated material to temperatures between 100°C and 200°C for a suitable period. This process is referred to as *artificial ageing* or, in metallurgical terms, as *precipitation treatment*.

CASTING ALLOYS (NOT HEAT-TREATED)

Aluminium–silicon alloys (9–13 per cent silicon).
Aluminium–copper alloys (up to 10 per cent copper).
Aluminium–magnesium–manganese alloys.

Tensile strengths between 125 and 185 MN/m^2 (8 and 12 $tonf/in^2$).

Pure aluminium, like most pure metals, solidifies sharply from the molten state at one particular temperature. For aluminium this is 660°C. Because of this, it can only with great difficulty be made into commercial castings; it is almost impossible to feed molten metal into the casting during solidification to take up the shrinkage that occurs.

Alloys, on the other hand, solidify over a temperature range. Crystals of aluminium begin to form in the molten alloy, producing a mushy stage during which the shrinkage takes place gradually and is progressively replaced by molten metal from a feed reservoir. Copper in aluminium alloys decreases the shrinkage which takes place, while silicon counteracts the shrinkage by actually expanding on solidification.

Alloys in this class are used when strength is not the most important factor. They are good general-purpose materials, having fluidity in casting and producing rigid products with good resistance to corrosion.

The most widely used alloys in this class are the *aluminium–silicon* alloys containing between 9 and 13 per cent silicon, with small amounts of other elements. One much-used alloy in this range is specified in BS 1490 as LM6 (10% to 13% silicon); it is suitable for both sand- and die-casting of large, intricate, thin-walled castings, such as housings, gear boxes, radiators, crank cases and sumps for general engineering, aircraft and marine applications. The silicon in the alloy produces a high resistance to corrosion, which is of extreme importance for aircraft engineering, where metal is reduced to the necessary minimum, and for marine engineering, where material is exposed to highly corrosive conditions.

Aluminium–copper alloys with varying amounts of silicon and other elements have good foundry qualities, machine well, and produce good rigid castings; they are used for applications where they will not suffer mechanical shock. Aluminium–copper alloys were the earliest casting alloys used, but since World War II they have become obsolescent due to the superior qualities and corrosion resistance of the aluminium–silicon alloys. Aluminium–copper cannot be anodised and the best method of preventing corrosion is painting, specially when used out of doors.

Aluminium–magnesium–manganese alloys are used where corrosion resistance is of first importance, this resistance making them very suitable for marine use. They can be sand cast or gravity die cast, producing rigid castings suitable for moderately stressed components, and will receive a very high polish.

CASTING ALLOYS (HEAT-TREATED)

The strongest aluminium castings are produced from the heat-treated alloys. Alloys in this group contain, in addition to silicon, other elements which act as hardening agents. Many of these alloys are of the 4% copper group, containing smaller amounts of magnesium. Some also contain up to 2% nickel, which acts as another hardening agent.

One of the best known alloys in this group is Y alloy, containing approximately 4% copper, 2% nickel, 1·5% magnesium, and smaller amounts of other elements. At elevated temperatures it has superior strength, hardness and wearing qualities which, as a casting alloy, make it suited for the manufacture of pistons and cylinder heads for high-duty internal-combustion engines for aircraft and marine applications. While fundamentally it is a casting alloy, Y alloy can be supplied in the wrought condition as rod, bar, sheet and strip.

Heat treatment of castings by solution treatment followed by precipitation produces tensile strengths in sand castings of between 230 and 280 MN/m^2 (15 and 18 $tonf/in^2$); with chill castings made in permanent iron moulds, tensile strengths of between 260 and 340 MN/m^2 (17 and 22 $tonf/in^2$) can be achieved.

WROUGHT ALLOYS—NOT HEAT-TREATED (WORK-HARDENED)

Aluminium of purity above 99%.
Aluminium alloys including 1% to 1·25% manganese.
Aluminium alloys including 2% to 9% manganese.

Aluminium in its purest form is the most suitable for manufacturing wrought articles. This is because its softness and ductility allows it to be mechanically formed into sheets, forgings, extrusions etc. with comparative ease. As small amounts of alloying elements are added, the material becomes less ductile, harder to work, and tends to crack or tear. To overcome this, annealing of the alloy is necessary between the forming stages.

The necessary requirements for this class of alloy are adequate strength and rigidity, with good corrosion resistance in the finished work-hardened state. The desired mechanical properties are produced by the cold working given to the product after the last annealing operation. Aluminium-alloy plate, sheet strip, tube and wire are supplied in a range of tempers designated as soft, quarter hard, half hard, three-quarter hard, and full hard. It must be realised that when the material has been worked to the desired thickness no further increase in strength can be obtained without making it undersize by further working.

Aluminium of the highest purity (99·99%) is used for making foil for electrolytic capacitors. It is also used for wrapping foodstuffs and dairy products, confectionery, cigarettes and medicines, and for the capping of milk bottles.

Aluminium of between 99·5% and 99·8% purity is used for producing electrical cables for carrying high-tension currents, particularly on the grid system. It is also used in the manufacture of the so-called collapsible tubes for packing toothpaste, medicines, cosmetics, greases etc.
Aluminium of 99·25% purity and alloys containing manganese are available as sheet, plate, extrusions, forgings, tube and wire. These are used for architectural and transport applications such as building and structural panels (in both flat and corrugated forms), window mullions, flashing, decorative assemblies and general sheet-metal work.

Alloys containing magnesium are very ductile when soft, and easily weldable, but work-harden very quickly. They have a high corrosion resistance and are suitable for shipbuilding applications where they are exposed to a corrosive marine atmosphere.

Aluminium alloys in the non-heat-treated group are used extensively for cooking vessels, where their high thermal conductivity spreads the heat, eliminating hot spots which would burn the food; also, the products of any corrosion are colourless and non-toxic. For similar reasons these alloys are widely used in the brewing industry for brewing kettles, fermenters, storage tanks and barrels, and in the chemical industry for piping, vats, reaction vessels and subsidiary equipment.

It should be noted that, although aluminium has excellent resistance to many corrosive chemicals, alkalis attack it very vigorously and quickly dissolve it, irrespective of the purity.

WROUGHT ALLOYS (HEAT-TREATABLE)

Alloys containing approximately 4% copper, with smaller amounts of silicon and magnesium.
Alloys containing no copper, with silicon and magnesium up to 2%.
Alloys containing varying amounts of copper, below 4%, with silicon and magnesium.

As in casting, the heat-treatable wrought alloys are the stronger alloys, and are used for structural purposes and highly stressed parts in aircraft engineering, such as forgings for undercarriages, engine mountings, wing spars, attachment brackets and propellers. They are also used in structural members for all forms of transport vehicles, ladders, scaffolding and tubular furniture.

In sheet form they are used for structural panelling, for aircraft frameworks, and for stressed skin construction of fuselages and wings.

In general, heat-treatable alloys contain more alloying elements than the non-heat-treatable ones, copper forming the basis of many of the precipitation-hardening alloys and the other alloying elements being magnesium, silicon, manganese and zinc.

Duralumin, although originally the trade name of the first aluminium alloy in which the phenomenon of hardening by heat-treatment was observed, is now used as a general trade term for a group of alloys containing approximately 4% copper.

The alloys of highest strength contain copper, magnesium and zinc; after heat-treatment strengths of over 540 MN/m² (35 tonf/in²) result.

The corrosion resistance of this group of alloys is not as good as that of the non-heat-treatable ones, and alloys in the form of rolled sheet are often protected by coating the surfaces with a thin skin of pure aluminium. The resulting composite clad-sheet is a high-strength material suitably protected with an integrally bonded skin of high corrosion resistance. Duralumin coated with aluminium as above is sold under the trade name *Alclad*.

BRITISH STANDARDS RELATING TO ALUMINIUM ALLOY

As previously mentioned, there appears to be a bewildering number of alloys, but when dealing with British Standards they can be roughly divided into three groups:

1. Casting alloys (BS 1490: 1963)
2. Wrought alloys (BS 1470–1477: 1963)
3. Aluminium for electrical purposes (BS 215: Parts 1 and 2: 1956; BS 2897: 1957; BS 2627: 1961; BS 3242: 1960)

Casting alloys (BS 1490)

The alloys covered by this standard are numbered 0 to 24 and are prefixed with the letters LM, e.g. LM 0, LM 4, LM 6, LM 21. When the material is thus referred to without a suffix, it signifies that it is in ingot form; other conditions of the material are indicated by the following suffixes:

M	As cast
P	Precipitation treated
W	Solution treated
WP	Solutionised and precipitation treated

For example, LM 6 M would be quoted for an intricate casting having thin sections and where ductility and corrosion resistance are required. The alloy would be in the as-cast condition, with no heat treatment.

LM 14 WP would be quoted for high-duty castings such as pistons and cylinder heads to operate at high temperatures; after casting, the parts would be given full heat treatment to increase tensile strengths.

Wrought aluminium alloys (*BS 1470–1477*)

BS 1470	Sheet and strip
BS 1471	Drawn tube
BS 1472	Forgings
BS 1473	Rivet, bolt and screw stock
BS 1474	Extruded round tube and hollow sections
BS 1475	Wire
BS 1476	Bar and sections
BS 1777	Plate

The wrought alloy specifications are arranged in numerical order corresponding move or less to increasing strength and hardness. No. 1 is used for pure aluminium and graded 1, 1A, 1B, 1C in descending order of purity. The wrought alloys start with the non-heat-treatable alloys (2 to 8) following by the heat-treatable alloys (9 to 30); several numbers are missing from the series. The form in which the material referred to in the above standards is available, is indicated by a prefix letter as follows:

S	Sheet and strip
C	Clad sheet and strip
B	Bolt and screw stock
V	Extruded tube and hollow sections
G	Wire
T	Drawn tube
F	Forgings
R	Rivet stock
E	Bar and sections
P	Plate
PC	Clad plate

Non-heat-treatable alloys are indicated by prefixing the above form letters with the letter N. These alloys are supplied in a range of tempers which is produced by cold working or annealing. The tempers are indicated by the following letters.

O	Annealed
OD	Annealed and lightly drawn
M	As manufactured forged, rolled, drawn etc.
$\frac{1}{4}$H, $\frac{1}{2}$H, $\frac{3}{4}$H, H	Various degrees of temper produced by cold working in the case of sheet, strip, plate, rivet stock and drawn tube.

For example, NS 3—$\frac{3}{4}$H would denote non-heat-treatable alloy No. 3 in the form of sheet or strip, of temper three-quarters hard.

Heat-treatable alloys are prefixed with the letter H, and the condition of supply is indicated by suffix symbols as follows:

O	Annealed
M	As manufactured—rolled, drawn, forged etc., with no subsequent heat treatment
OD	Annealed and lightly drawn
W	Solution treated only and will respond to precipitation treatment
T	Solution treated and requires no precipitation treatment
WP	Solution and precipitation treated
P	Precipitation treated only

For example, HE 30 WP denotes heat-treatable wrought alloy No. 30 in the form of bar which has been fully heat-treated (solutionised and precipitated).

Aluminium for electrical purposes

BS 215 (*Parts 1 and 2*) lays down a standard for hard-drawn aluminium conductors and for hard-drawn steel-cored aluminium conductors for overhead power transmission (grid-system cables).

BS 2627 specifies requirements for round aluminium wire for use as electrical conductors for purposes other than overhead transmission lines.

BS 2897 specifies requirements for wrought aluminium drawn strip of various tempers for use as electrical conductors.

BS 3242 specifies requirements and dimensions for heat-treated aluminium–magnesium–silicon alloy for use as overhead line conductors.

Some other British Standards

BS 3313. Aluminium capping, foil and strip for milk and cream bottles.

BS 1683. Coated aluminium foil for wrapping cheese.

TIN

Tin is a soft ductile white metal with a melting point of 232°C; it is readily extruded and worked by drawing, stamping and spinning. It does not work-harden appreciably during these processes. However, if pure tin is exposed to low temperatures for long periods, it loses its ductility, becomes brittle, and can be easily crushed to a grey powder.

Most of the tin produced annually is used for plating or in the production of alloys. Tinplate is a common material in everyday use. It is made by rolling mild steel into sheet or strip form, to a thickness of 0·25 mm (0·010 in.) and then plating it with a coating of tin 0·0025 mm (0·0001 in.) thick by either hot dipping or an electrodeposition process. The tin coating protects the mild steel from corrosion and, both metals being non-toxic, tinplate is a safe material for the packaging of a wide range of processed foods, drink and other merchandise. Tinplate, being ductile, is easily formed into shape and is suitable for soldering at high speeds. The tin coating imparts a surface which can be readily painted, lacquered or enamelled.

Tin is a main constituent in some of the white-metal or high-tin Babbit bearing alloys used for crankshaft bearings in large engines. These alloys may contain as much as 90% tin alloyed with antimony and copper. In the intermediate and lead-based alloys, tin is always present in varying percentages.

Alloys of tin and lead, sometimes with silver and antimony, are used extensively as solders for joints between other metals. The range of solders contain between 62% and 12% tin, the remainder being mainly lead.

Tin is used to alloy with copper to produce an extensive range of bronzes and gunmetals. It is also added to brass (copper–zinc alloys) in small quantities to improve corrosion resistance.

Tin is alloyed with bismuth, lead and cadmium to produce low-melting-point fusible alloys, some of which can be melted at temperatures much below that of boiling water. These alloys find important applications in automatic safety devices used for fire alarms and fire sprinklers. They are also used to provide links to melt at predetermined temperatures to operate fire doors and valves in fuel-storage systems, to guard electric water-heater installations against overheating, and many other similar applications. They can be used to prevent crushing during pipe bending, and then by low heat (often by immersion in hot water) can be melted out. Vacuum-tight joints in glass apparatus or between glass and metal components can be successfully made with readily fusible alloys. Fusible alloys are widely used to make dispensable foundry patterns and assist in intricate core making, being melted out on baking and recovered for re-use.

Other applications for fusible alloys are in heat-treatment baths, press tools, assembly jigs and fixtures, dies for plastic moulding, positioning magnets in electric meters and generators, and many similar uses.

TABLE 17.1 TIN–LEAD SOLDERS

Type and use	*Composition percentage*				*Solidification range (0°C)*
	Tin	*Lead*	*Antimony*	*Silver*	
Tinman's solder	62	38			instantaneously at 183
Coarse Tinman's solder	50	50			220–183
Plumber's solder	33	67			260–183
Plumber's solder for wiping joints	31	67	2		235–188
General-purpose solder	43·5	55	1·5		220–188
Substitute general-purpose solder	30	69·35		0·65	250–180
Substitute plumber's solder	18	80·8	0·8	0·4	270–180
Solder for iron and steel	12	80	8		250–243

LEAD

Lead is the softest of the common metals and can be easily worked at normal temperatures, with simple hand tools and without periodic softening by annealing. It is, when fractured, a silvery white metal, but contact with the atmosphere tarnishes it by formation of an oxide film. Subsequently, a film of lead carbonate covers the surface by reaction with the carbon dioxide in the air, forming a protective skin. In clean surroundings it takes on a pleasing silvery-grey appearance.

Lead in its pure state is a heavy metal. It is highly resistant to corrosion attacks, as can be seen by the many lead roofs of buildings such as St Paul's Cathedral and Westminster Abbey, which have stood the corrosive atmosphere of London for centuries. Lead plays a modern role as radiation shields in nuclear reactors, and in hospitals it is used to protect operators from the harmful effects of irradiation through constant usage of X-ray equipment.

It is estimated that lead is present in the earth's crust to the extent of about 0·002%. The ore from which it is obtained is often quite complex, containing gold, silver, copper, tin, zinc, antimony, arsenic, bismuth and sulphur, as well as a large quantity of earthy waste which must be handled and disposed of. Often only 3% to 4% of the mined ore results as pure lead.

The yearly world consumption of lead is over 3 million tonnes, of which a large proportion is reclaimed scrap. The largest outlet for the metal is for cable manufacture, where its outstanding resistance to corrosion over a very wide range of service conditions make it an ideal material for sheathing. Lead can be applied to the cable in continuous lengths of seamless, joint-free sheathing; its ductility then allows it to withstand the repeated coiling and straightening which occurs during the manufacturing process, and also the bending which occurs as it is winched along the cable run during installation.

Another large consumer of lead is the electric-battery industry. The electric battery, in its many forms, has both its negative and its positive plates made as lead grids. As every car has a battery, and there are more than 16 million registered vehicles on the roads of Britain alone, can be seen that thousands of tonnes of lead are required each year by the automobile industry for this use alone.

Other uses of lead are for weathering, flashings and plumbing in the building industry. In the chemical industry, it is used for vats and piping for the production of corrosive chemicals, especially sulphuric acid. It is also used in machine bearings, solders, foils for packaging and damp protection, type metals for printing, shot for ammunition, collapsible tubes and many other industrial and manufacturing processes. It is alloyed with steel to improve machinability and is added in various forms to paints, petrol, inks, insecticides, plastics, matches and explosives.

CADMIUM

This is a silvery-white metal which does not occur naturally but only in a combined state. It is closely related to zinc, and a large percentage of cadmium production is a by-product of the production of zinc, being recovered from the fumes released during the sintering of zinc concentrates.

More than 65% of the annual production of cadmium is used for the electroplating of iron, steel, copper, brass and other alloys as a protection against corrosion.

Cadmium is used as one of the plates in nickel-cadmium storage batteries. These batteries have long life and, because of their small bulk and light weight, are particularly useful in aircraft. They can be stored in the discharged condition without deterioration.

Cadmium is also used in the making of pigments for paints, enamels and lacquers. The best yellow and red colours of automobile finishes can usually be credited to pigments produced from cadmium sulphides and sulphoselenides.

Because of its high neutron-absorption characteristics, cadmium is used in the control rods of some nuclear reactors. It is also added, in quantities of up to 1%, to pure copper to produce hard-drawn high-conductivity wire for overhead power-transmission cables.

Other non-ferrous metals

Chapter 17

ZINC

Zinc is a bluish-white metal which, in its pure form, is structurally weak. Zinc-bearing ores are widely distributed throughout the world, Australia, north and south America, north and central Africa, Japan, the U.S.S.R., Scandinavia, Germany, Poland, Spain and Italy being the most important producing countries. The extraction of zinc from ore is complicated, and several thermal processes as well as electrolytic refining are in current use.

Zinc's usefulness in engineering is due to its extremely corrosion-resistant qualities, and 40% of some 3 million tonnes produced yearly is used in some form of protective coating for iron and steel. It gives ideal protection in structural engineering where maintenance by painting would be costly and often dangerous, for example, small slender broadcasting towers and aerial masts, electric grid pylons, bridges, large barns, aircraft hangars, fencing, barriers and other similar applications.

Brass and the making of zinc-based diecasting alloys account for some 20% of the annual production, while zinc sheeting accounts for a further 10%. Zinc in various forms is used in many manufacturing and industrial processes, some of which are paint making, rubber processing, match heads, cosmetics, pottery glazing, linoleum, paper products and opaque plastics.

Zinc coatings

There are five principal methods of zinc coating iron and steel products—hot dip galvanising, zinc spraying, zinc plating, sheradising and applying zinc-rich paints.

Hot dip galvanising. Components are degreased, pickled in an acid bath to remove scale etc., dipped in a flux solution, and finally immersed in a bath of molten zinc, where reaction takes place to produce a chemically bonded coating.

Zinc spraying. A special pistol, fed with zinc wire or powder, sprays atomised particles of molten zinc on to surfaces of articles prepared by grit blasting. This process is most suitable for large structures which, due to their size, cannot be dipped in a hot galvanising bath, or would suffer distortion due to the heat. Where a thicker coating of zinc is required, spraying is by far the most economical method.

Zinc plating (*electrogalvanising*). This is a zinc-coating treatment which is used where small components cannot accept a thick, rough or uneven finish, or withstand the heat of hot dip galvanising. The process gives a fine smooth finish which can be controlled in thickness; the method is suitable for accurately machined components and instrument parts, nuts, bolts and similar fasteners.

The parts are thoroughly degreased, pickled and washed before going into the electrolyte of the plating tank. In large plants this process is often automated and used for the continuous plating of wire and strip.

Zinc plating is usually followed by a *passivation* process consisting of passing through a bath containing chromic acid, which adds a film which remains passive (not active) in many environments, especially damp atmosphere, and so increases the corrosion resistance.

Sheradising. This is a zinc-coating process where small components, after degreasing, pickling, washing and drying, are placed in a barrel containing very fine zinc dust and are then heated to a temperature just below the melting point of zinc. The barrel is rotated during the process to give an even coating of zinc to components of complicated and awkward shape.

Zinc-rich paints. These paints contain over 90% zinc dust, and are applied by the normal method of brushing. The bond between the zinc and the coated steel is physical, not chemical as in hot dip galvanising.

Surfaces should be clean and dry before application, preferably grit-blasted. Painting with zinc-rich paints should be regarded as only a short-term protection from corrosion. These paints are often used as priming coats before conventional painting.

Zinc-based die casting

Die-casting metal alloy containing 4% aluminium, 0·04% magnesium and the remainder pure zinc (BS 1004 : 1955) is used in the production of a wide range of mass-produced articles. All kinds of everyday articles such as door knobs, locks, window catches, brackets, toys, small tool handles, small gear-box casings, bedplates for office machinery, side plates, mechanical parts and covers, carburettor parts and other articles of complicated shape are produced by die casting.

In most cases these zinc-base die castings are unrecognisable as such when finished by chromium plating, enamelling or plastic coating.

MAGNESIUM

Magnesium is a silvery-white metal which melts at 650°C. It is the sixth most abundant metal in the earth's crust. It does not appear in the pure metallic form but as oxides and carbonates from which it is produced by complicated processes, one being electrolytic and another a vacuum reduction method. It is also produced from seawater, and it is interesting to note that a cubic mile of seawater represents 5¼ million tonnes of magnesium metal.

Magnesium is a modern engineering metal. It was first isolated in 1808 by Sir Humphrey Davy, and it was not until 1886 that commercial production was achieved by the Germans, who began to develop alloys for industrial purposes. Before World War II it was already in use for aero engines and for a wide range of civilian products, but it was the impact of the War which created the tremendous demand for magnesium, and much research was carried out into the production of this metal.

Pure magnesium is used for flares, photographic flash bulbs, incendiary bombs and bullets, and also in various chemical processes.

As an engineering material it is weak in the pure form but, by additions of aluminium, zinc, manganese, zirconium, rare earth metals, silver etc., useful magnesium alloys of considerable strength and lightness are produced. Magnesium alloys are approximately two-thirds the weight of aluminium and one-quarter the weight of steel. The ultimate tensile strength (U.T.S.) of magnesium varies between 125 and 150 MN/m^2 (8 and 16 tonf/in^2) for castings, and for wrought alloys can be as high as 340 MN/m^2 (22 tonf/in^2).

Magnesium alloys can be obtained in the form of both sand and die castings, as forgings and extruded sections, and as rolled sheet and plate from 0·45 mm (26 s.w.g.) up to 75 mm (3 in) thick. Magnesium is a very fast machining metal, can be argon arc-welded, and has excellent forming qualities at elevated temperatures, with no spring-back.

For transportation, its lightness gives the advantage of increasing pay loads and thus reducing freight charges. When used in mechanisms, the driving power is reduced, and also the inertia of the moving parts. Magnesium is a metal of high strength/weight ratio which offers designers the means of achieving highly efficient products; combined with its excellent damping qualities, it offers quieter and smoother running in high-speed mechanisms.

Typical uses of magnesium alloys are for aero-engine and automobile crankcases and gearboxes, landing wheels for aircraft and road wheels (particularly for racing cars), air-frame and automobile-chassis components, spacecraft structures, portable tools, business machines, lawn mowers, ladders, warehouse and hospital trolleys, artificial limbs, operating-theatre equipment, television cameras and their mobile platforms, components for textile machinery, and many other applications where the strength and lightness of magnesium reduces human fatigue, the power necessary for propulsion, and the inertia of fast-moving and rotating parts.

BEARINGS

In all mechanisms there is relative movement. Whether it is sliding or rotational movement there is always friction because of the nature of the surfaces in contact, and energy has to be wasted in overcoming the frictional forces. Most of the wasted energy appears again at the bearing surfaces as heat and, in heavily-loaded fast-moving parts where the contact area is small, this can very quickly cause high temperatures. Local melting and fusing together of the moving parts may then take place.

The surface of a metal can be made to flow even by hand-rubbing the surface with a cloth, as when producing a high polish. This is a process commonly used by metallurgists when preparing specimens for microscopic viewing. It can be appreciated from this that, unless some steps were taken to prevent the melting of mating parts, machinery would soon sieze up.

A number of methods are used to overcome friction, which is always present and always in opposition to movement:

1. Lubrication reduces friction by keeping surfaces apart, while also cooling and maintaining the cleanliness of the rolling or sliding elements.
2. Bearing devices and materials are selected which, because of their mechanical and physical properties, reduce frictional resistance and accommodate varying operational conditions.
3. Mating parts are produced with fine finishes to closely machined limits.

REQUIREMENTS OF A PLAIN METAL BEARING

The essential qualities of a plain bearing metal are that it must be hard enough not to wear away in service, and tough and ductile enough to withstand the mechanical shock to which it may be subjected.

It should have a high thermal conductivity to conduct the heat away from the bearing surface, and a low coefficient of sliding friction.

Another desirable property of a bearing alloy is that it should have a melting point lower than that of its mating shaft. If a breakdown in lubrication occurs, causing overheating, the bearing material will then run, and may prevent seizure and severe mechanical damage.

BALL AND ROLLER BEARINGS

In most machines not normally subject to shock, ball, roller or needle bearings are used in preference to the sliding movement in plain bearings, mainly for the following reasons:

1. Because the action is rolling instead of sliding, there is less loss of power; the coefficient of friction is considerably lower.
2. Ball and roller bearings are more compact than plain bearings, and needle bearings even more so. This allows lighter and less bulky equipment to be designed, with subsequent better use of available material and cheapening of the product.
3. Bearings can be sealed for life against all foreign matter, and the wear on the hardened surfaces in contact is practically negligible, thus eliminating costly maintenance.
4. Standard manufactured ranges of bearings are available to suit a comprehensive range of loadings and conditions. This gives designers a wide choice, with saving of time and an assured operating performance.
5. Unit ball and roller bearings simplify production assembly because fitting and scraping of bearings is not necessary.
6. The design of ball and roller bearings practically eliminates the danger of overheating.

Ball and roller bearings are usually made from carbon chrome steel of compositions which are often the manufacturer's secret. In general the balls and races are surface-hardened by case-hardening or by a self-hardening heat treatment, leaving the cores softer and tougher. The balls and races are ground to extremely fine limits, and complete bearings are graded into four BSI categories of diametral clearance; these categories are designated 0, 00, 000 and 0000, the clearance increasing with the number of 0s. These markings can be seen etched on the outside of the ball races. Races used in general engineering are usually of the 00 category.

Cages which separate the rolling elements in ball and roller bearings are made from yellow anti-friction bearing metal, pressed steel stampings or moulded plastic, according to the size and the application for which the bearing is intended.

LINEAR MOTION BEARINGS

These anti-friction bearings, using balls or rollers, are used in engineering where linear motion, as distinct from rotary motion, is required. This type of motion occurs in machine tools, office machinery, textile and food machinery, aircraft controls and instrument linkages, where machine elements such as shafts and carriages have to move to and fro.

In earlier machines, parts moving linearly were supported on plain bearings such as shafts sliding direct in close-fitting plain bushes, carriages on wheels running on rails, or just sliding on accurately machined cast-iron guides. These methods produced high frictional resistance, and such arrangements are often too cumbersome for modern machinery, which is becoming increasingly controlled by servomechanisms. High starting friction in moving parts, commonly called 'stiction' by designers, necessitates excessive pulling and pushing forces to get parts moving; this produces over-run which, when monitored by the servomechanism, causes the machine to 'hunt' (to move backwards and forwards about a mean position). It is like trying to line up the slide in a tight slide rule.

Linear-motion anti-friction bearings effectively reduce this frictional resistance. This is not only an advantage in the design and construction of automatically controlled machines, but it also reduces the power required to operate machines, whether it be muscle or mechanical or electrical power.

Figure 18.1 Plain bearings

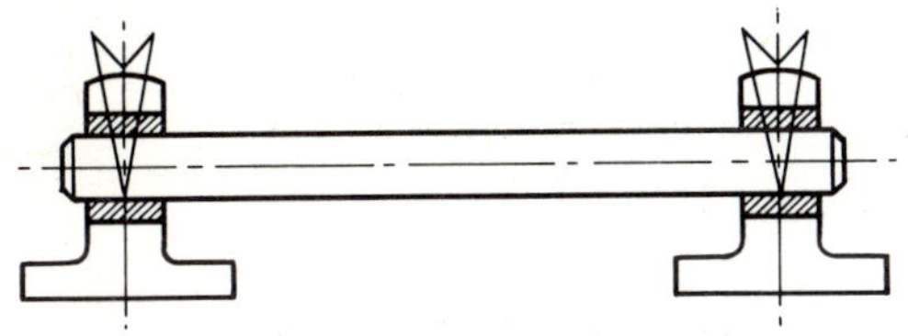

Bearings supporting radial loads

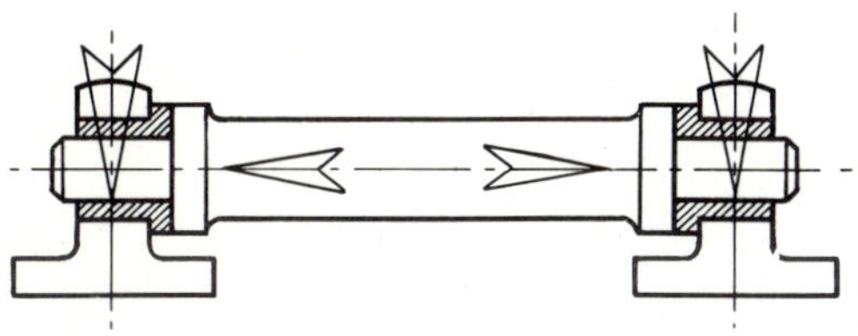

Bearings supporting radial and axial loads

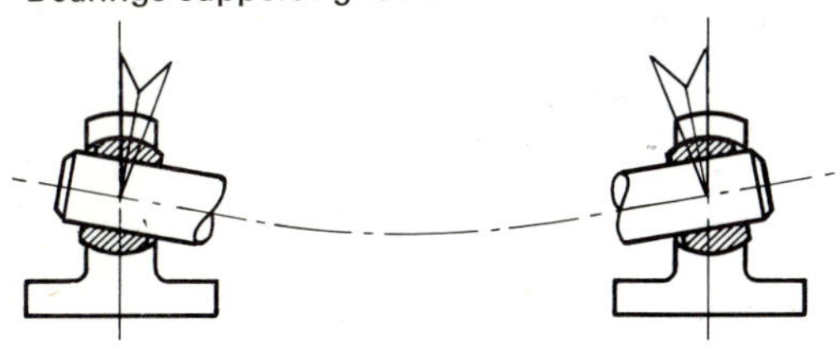

Bearings supporting loads due to misalignment

PLAIN BEARINGS

Bronzes

Bearing bronzes are either phosphor-bronze or gunmetal. Where bearing qualities are of paramount importance, phosphor bronze is used; where the bearing is intricate in shape, gunmetal is more often used to assist the foundry in casting. Gunmetal is bronze with a small zinc content replacing the phosphorus in phosphor bronze.

Bearing bronzes are copper–tin alloys containing 5% to 20% tin, a small amount of phosphorus (usually less than 1%), and the remainder copper.

Leaded bronzes having up to 20% lead are sometimes used where there is doubt about the efficiency of the lubrication system and the bearing alignment, but the addition of lead to the alloy reduces the loading capacity and the strength of the bearing.

Lead–bronze bearing containing up to 30% lead are often used as connecting-rod bearings in piston aero-engines. They have been found superior to white-metal bearings in this particular application where high speeds and comparatively heavy loads are encountered.

Their high thermal conductivity is an added advantage. Aluminium bronzes are sometimes used for heavy-duty bearings where high strength and heavy duty are required in a corrosive environment.

LOADING CONDITIONS

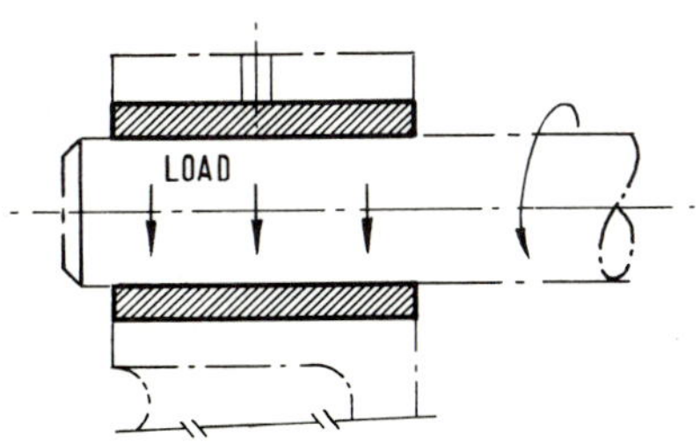

Plain journal bearings:
Material: phosphor bronze
Babbit
porous bronze
plastic

White metal or Babbit bearing metals

These are a large group of alloys varying in composition from approximately 90% tin and no lead to alloys containing 80% lead with only small amounts of tin. Antimony is present in all these alloys as a hardening element, both pure lead and tin being soft metals.

Babbit metals are easily cast and can be bonded without difficulty to steel or bronze shells used as backing members. They all have excellent conformability and embeddability and are nearly always used in conjunction with pressure lubrication. They operate extremely well in contact with soft steel shafts, and are practically 100% corrosion resistant. However, under heavy loads they tend to spread, and at elevated temperatures their strength diminishes. Oil starvation for only a short while will produce enough heat to cause the bearing to melt and run.

Porous bronze bearings

Bearings of this type are not cast or machined from bar, but are made by powder metallurgy. Powdered metal is compressed under heavy loads and then subjected to a high-temperature heating process which fuses it into a strong, solid material resembling alloy bronze. Between the particles, however, there are small gaps which are oil-impregnated by a boiling-in-oil process. These bearings contain as much as 30% oil by volume, which gives boundary-layer lubrication in service. The bearings are usually fitted in machinery, both industrial and domestic, where limited or no maintenance is required or expected during the life of the machine. Similar bearings using solid lubricants such as carbon instead of oil are manufactured.

Figure 18.1 (*cont.*)

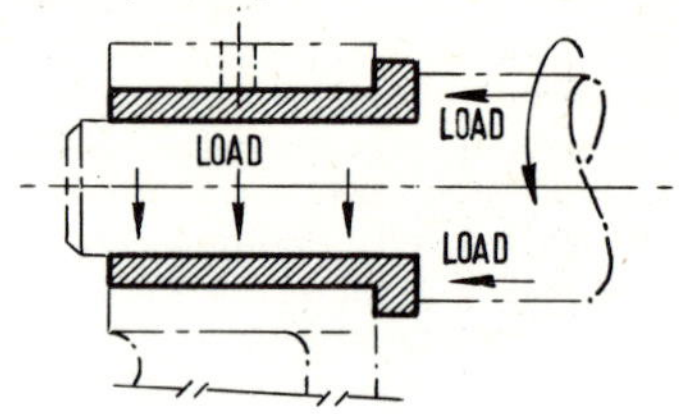

Plain journal and thrust bearings:
Material: phosphor bronze
Babbit
porous bronze
plastic

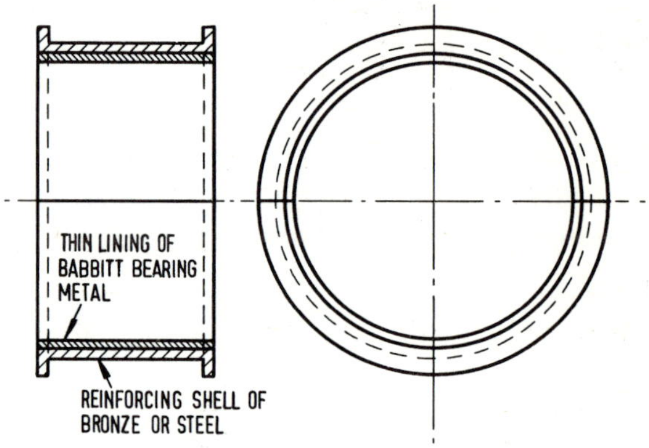

Split thin-shell plain bearings

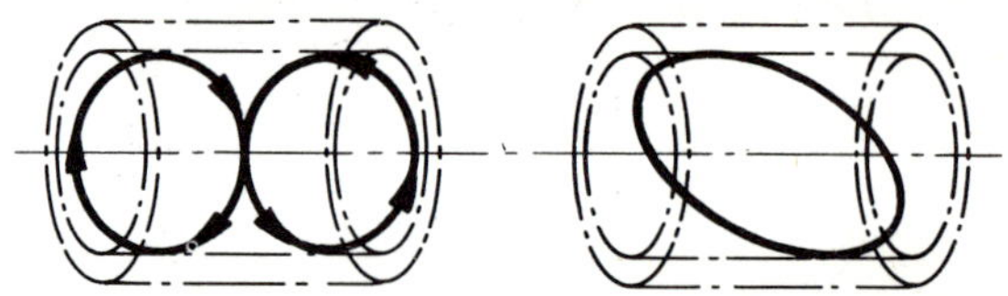

Figure of 8 *Elliptical*

For grease applications where lubrication is infrequent; facilitates quick distribution of lubricant.

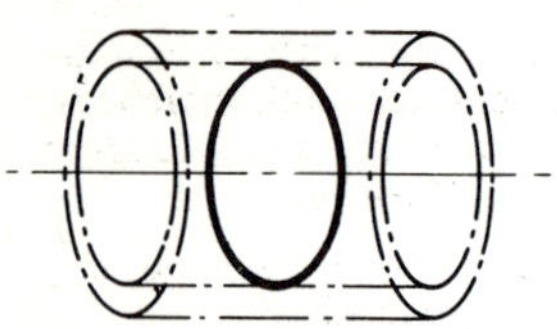

Annular

Pressure oil is supplied to annulus through crankpin, and distributed circumferentially and axially through bearing.

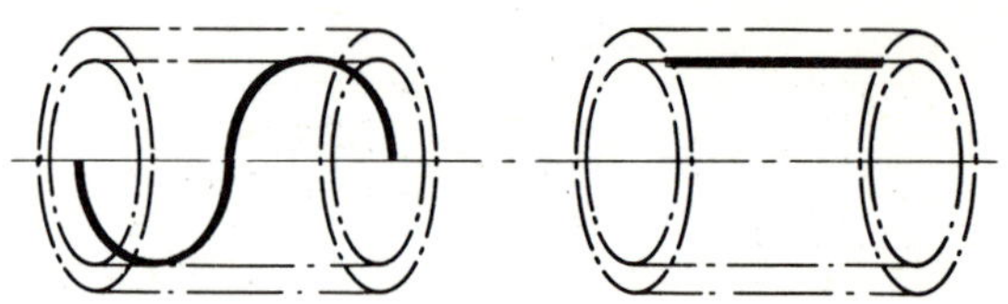

Helical

Used to distribute oil or grease axially.

Axial

Mainly used for simple gravity oil feed to distribute oil axially. Groove must not be located on load-carrying surface.

Lubricant distribution grooves used in plain bearings

Figure 18.1 (*cont.*)

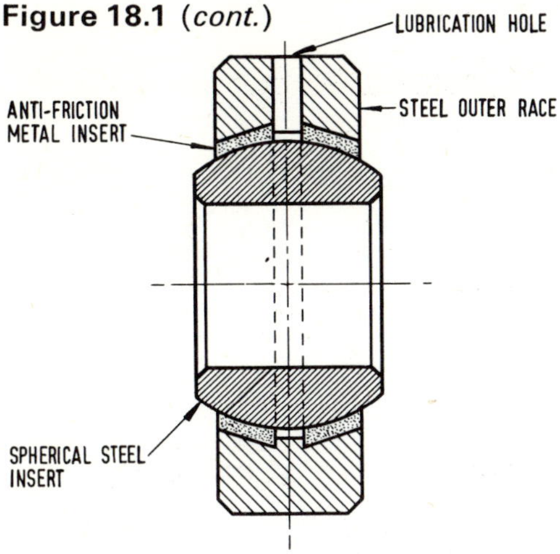

Patented spherical-type plain bearings by Rose Forgove Ltd

Plastic bearings

These offer a new range of materials to the design engineer and, because of their cheapness and long life in service, are being increasingly used in engineering. Their low friction, cleanliness, silent operation, and corrosion resistance are features which are in many circumstances highly desirable. Examples of suitable applications are in pharmaceutical-packing and food-processing machinery, washing and sewing machines, tyepwriters, pumps, valves, car parts and linkages, electrical equipment, and a host of medical and domestic appliances.

Certain of the thermosetting-type plastics, suitably filled, are used as bearings for medium and heavily loaded plant such as rolling mills, textile machinery, chemical plant, paper mills etc., with considerable increase in the life of the plant and economies in production. The bearings are often dry-running or water-lubricated.

Figure 18.2 (a-d) ANTI-FRICTION BEARINGS (Journal bearings which support radial loads only during shaft rotation)

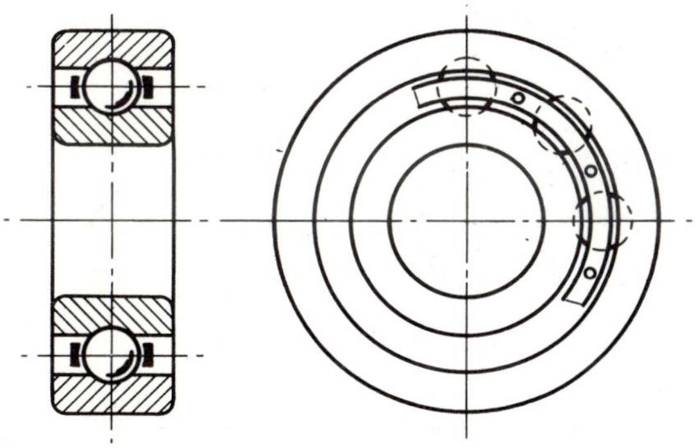

a

Ball journal bearings

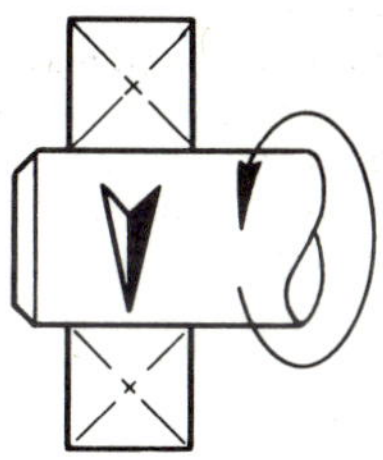

Loading Diagram

Journal bearings (*Figure 18.2*)

Illustration **a** shows a single-row ball journal bearing, the balls running in hardened grooved raceways or tracks. Some bearings have extra-deep grooves which offer a degree of sideways location to take a small amount of sideways thrust load. The width of an anti-friction bearing is usually less than shaft diameter, whereas a plain bearing to support the same loading can be as much as four times the diameter.

Bearings sometimes have a double row of balls, but this does not double the load-carrying capacity.

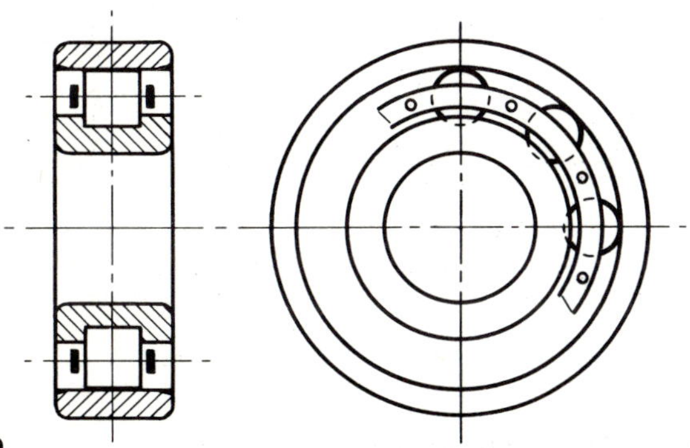

b

Roller journal bearings

Illustration **b** shows a roller journal bearing which has a loading capacity greatly in excess of ball bearings of the same size. The separating cage keeps the axes of the roller parallel with the shaft axis; this type of bearing is satisfactory for high speeds and for very hard and continuous service. It is often cheaper than a ball bearing of similar capacity.

Illustrations **c** and **d** show two types of needle roller. These will adequately support journal loads, and can be supplied without internal races to run direct on to hardened and ground steel shafts. Needle bearings allow more compact design and are extensively used where excess weight is undesirable, as in aircraft engineering, automotive gear boxes, transmission shafts etc. Needle roller bearings have a slightly higher coefficient of friction than roller bearings because of the greater number of rubbing surfaces. Roller bearings allow for higher rotational speeds than needle roller bearings. However, needle bearings of type **d** are usually far cheaper.

Figure 18.2 (*cont.*)

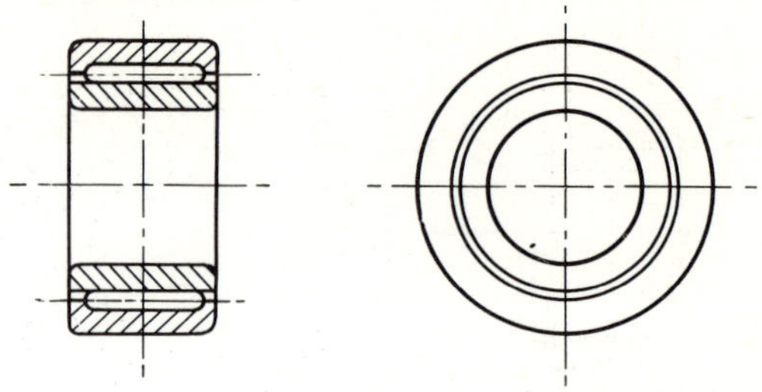

c

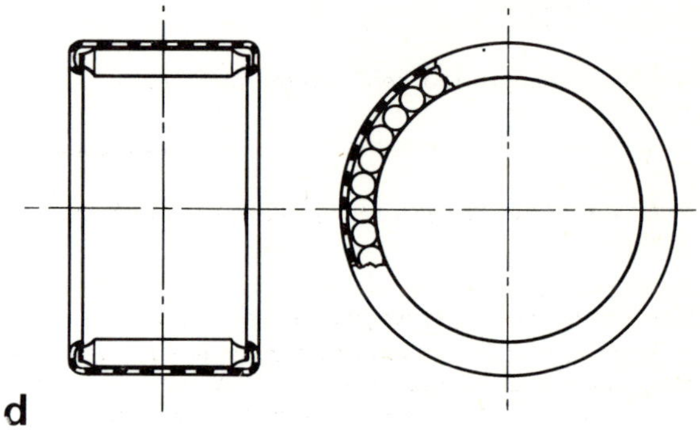

d

Needle journal bearings

Figure 18.3 (e-g) ANTI-FRICTION BEARINGS

(Journal bearings for supporting radial loads while accommodating malalignment often caused by shaft bending while rotating.)

Self-aligning journal bearings

Where there is malalignment of bearing axes due to deflection of long shafts, or in connections where movement is not rotational in a single plane, self-aligning bearings must be fitted.

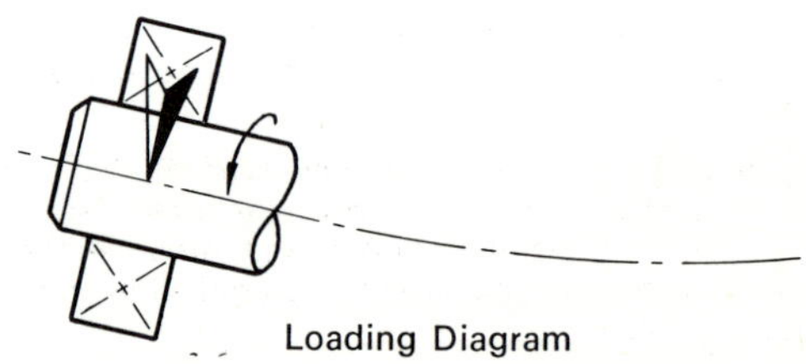

Loading Diagram

Illustration **e** shows a double-row self-aligning ball bearing with a spherical outer race. This arrangement offers an inexpensive solution to malalignment, but should not be used if there is appreciable end thrust to be carried. Also, although a large number of balls can be accommodated, the outer race does not give sufficient support to allow heavy journal loads to be carried. This bearing cannot be fitted with protective end covers.

Illustrations **f** and **g** show single-row ball and single-row roller self-aligning journal bearings. Bearings of this construction can carry loads equal to rigid bearings of the same size, as well as accommodate considerable shaft malalignment.

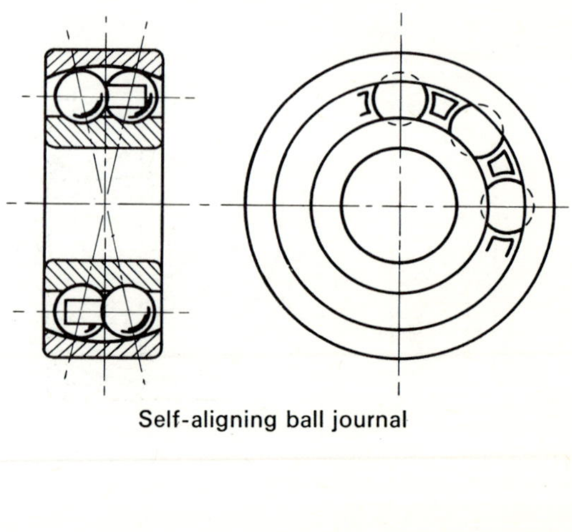

e

Self-aligning ball journal

Figure 18.3 (*cont.*)

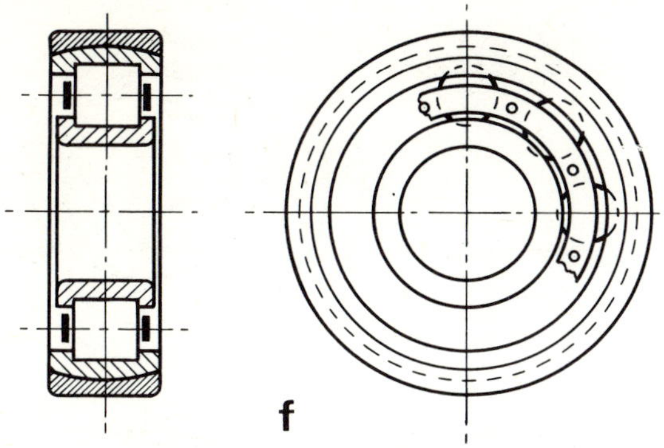

f

Single-row self-aligning ball journal

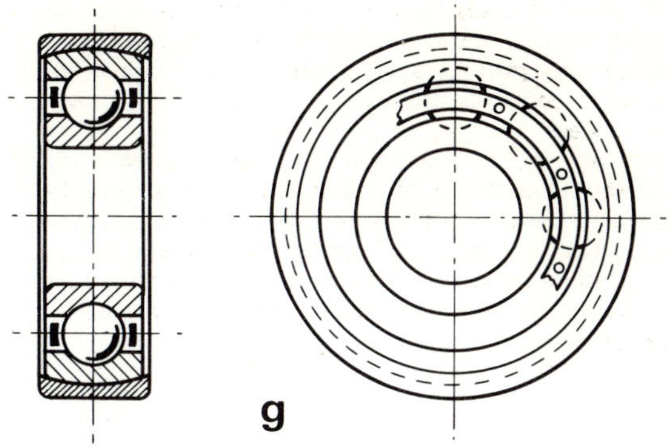

g

Single-row self-aligning roller journal

Figure 18.4 (h-j) ANTI-FRICTION BEARINGS (Bearings capable of supporting both radial and axial loads during shaft rotation.)

Bearings taking combined journal and thrust loads

Illustration **h** shows an *angular contact bearing* with an outer race machined to take end thrust when securely fixed in a housing. These bearings can be fitted singly when thrust is in one direction, or in opposed pairs, suitably clamped to prevent endwise movement, to take thrust in either direction.

Illustration **j** shows a pair of *taper roller bearings.* These bearings are very rugged and are capable of taking heavy radial and thrust loads. By varying the angle of slope of the bearings in manufacture, the ratio of radial to thrust loads can be varied to suit design requirements. When the angle is small the load carried is mainly radial; when it is steep the load carried is mainly thrust.

This type of bearing is commonly used in the front wheels of cars and in agricultural machinery, aircraft landing wheels, helicopter rotors, and other applications where extreme shock loading is applied in several directions at once.

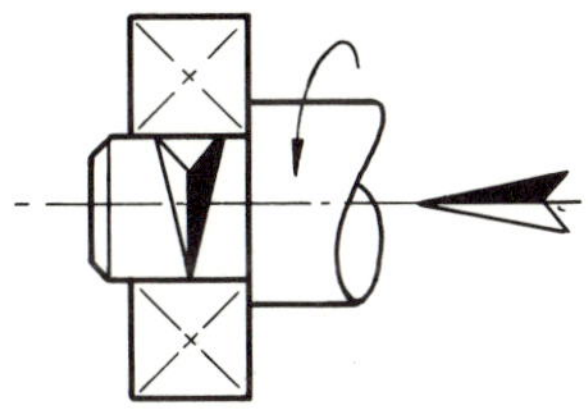

Loading Diagram

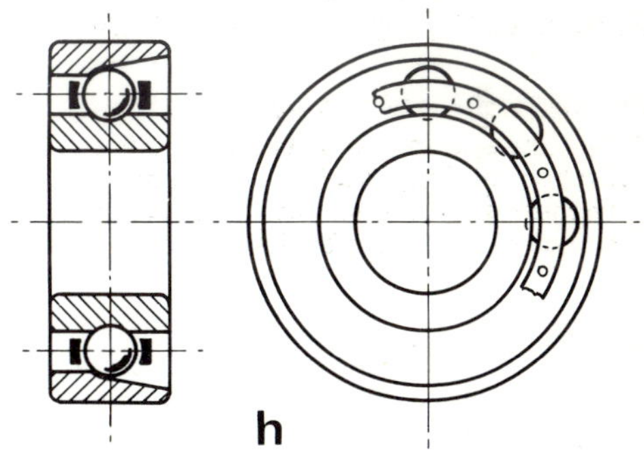

h

Angular-contact ball bearing

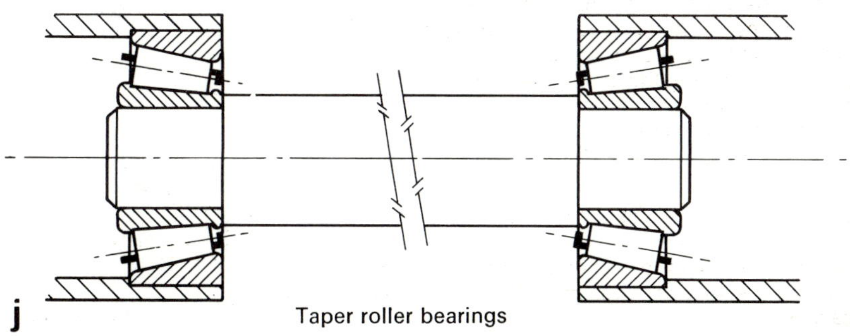

j

Taper roller bearings

Figure 18.4 (*cont.*)

Fig. **k** shows a DUPLEX TWO DIRECTION THRUST AND JOURNAL BEARING. This single bearing will take combined radial, and thrust load. It should only be used when the thrust loads are greater than the radial load. The outer race is split circumferentially and is similar to two angular contact bearings in one.

Fig. **l** shows a patented needle bearing which combines the ability to take both radial and axial thrust loads. Where space is limited, needle bearings are often the only solution to difficult loading problems.

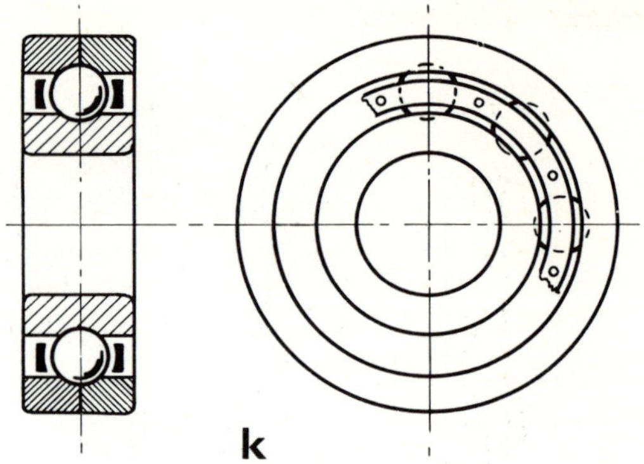

k
Duplex ball bearing

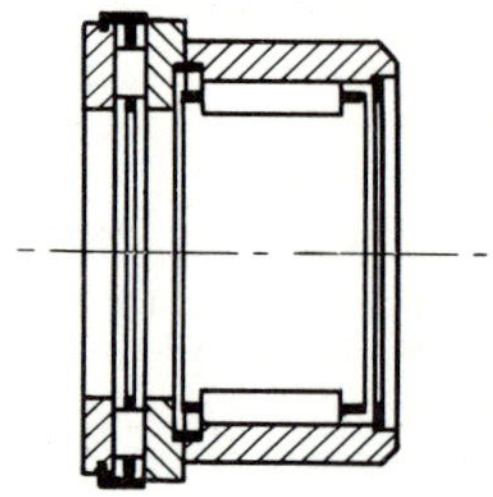

l
Patented Nadella journal and thrust needle bearing

Figure 18.5 (m-n) ANTI-FRICTION BEARINGS (Bearings for supporting axial thrust loads only during shaft rotation.)

Thrust bearings supporting axial loads only (*Figure 18.6*)
Illustration **m** shows a single-thrust bearing which is capable of taking large end thrusts only. This type of bearing is often used with the shaft in a vertical position when support of heavy loads is required. A journal bearing of either ball or roller type must be mounted alongside to resist radial loading and support the shaft axially. Thrust races of this type can be obtained with spherical seating rings to be used with self-aligning journal bearings. It is most important to protect these bearings from dirt and moisture.

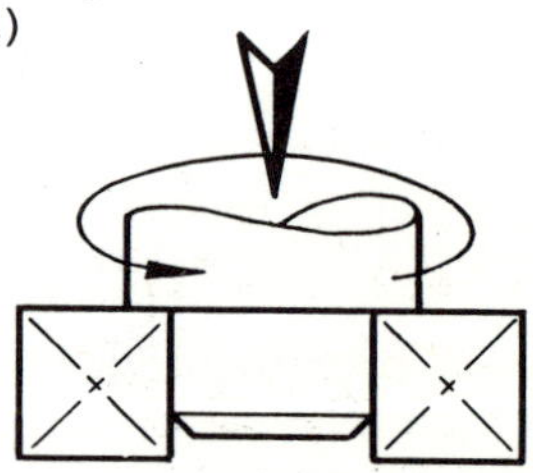

Loading Diagram

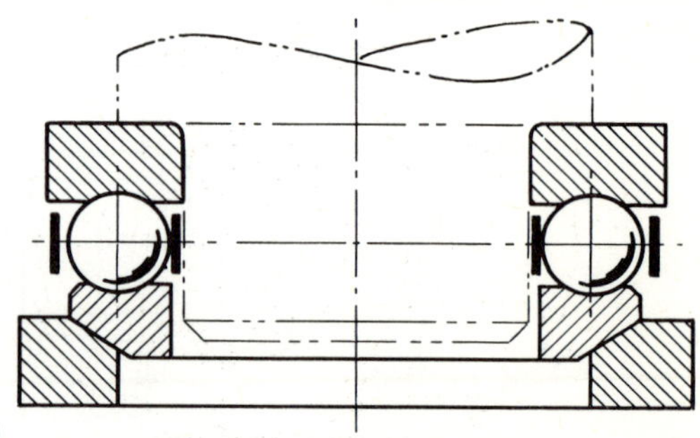

m
Single-row ball thrust bearings

Figure 18.5 (*cont*).

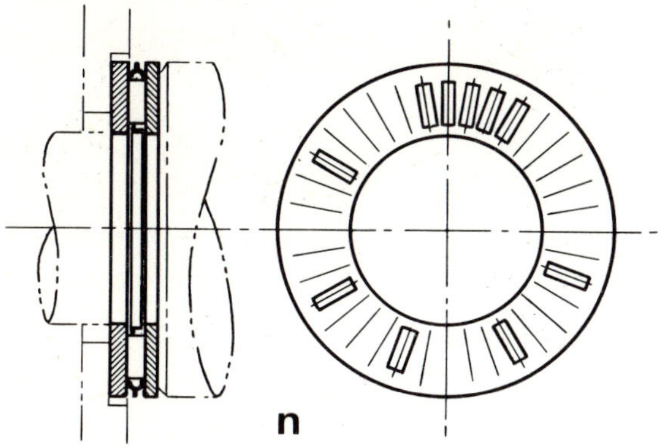

Needle thrust bearings

Illustration **n** shows a needle-thrust race which can be used between rotating hardened-steel faces, or be supplied with hardened-steel raceways of varying thickness. Low cost and compact design commend this type of bearing for mass-produced mechanisms, and it is used in large quantities in the automotive industry. The preferred lubricant for this type of bearing is an ample supply of oil. Grease is suitable as long as a prepacked supply for the life of the bearing can be satisfactorily arranged.

Figure 18.6 (p-r) ANTI-FRICTION LINEAR-MOTION BEARINGS (Bearings for supporting loads at right angles to the direction of motion.)

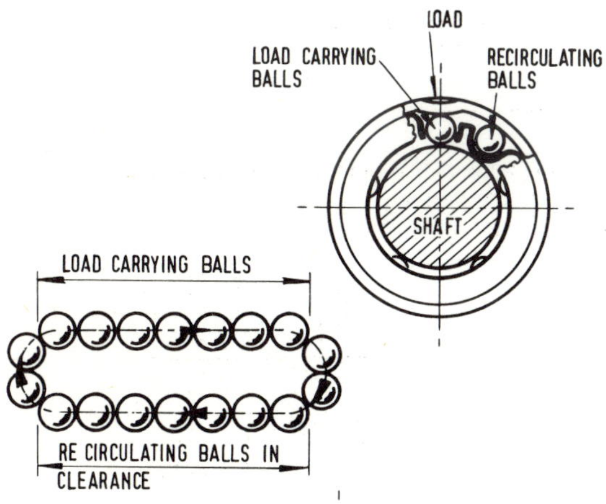

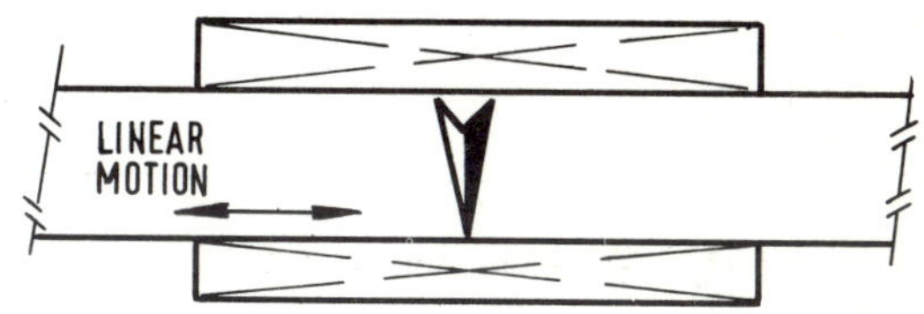

Loading Diagram

Bearings supporting linear motion (*Figure 18.6*)

Illustration **p** shows a patented ball bushing which consists of a strong casing containing three or more recirculating ball circuits. These bearings allow precise linear motion only and are used in aircraft controls, machine tools, office equipment, textile and food machinery, missiles, and many other applications where anti-friction linear motion is required for shafts.

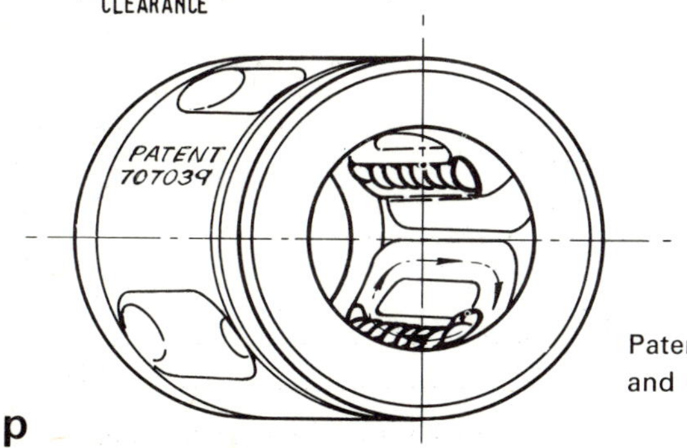

p

Patented recirculating ball bushing by Ransome and Marles.

Illustration **q** shows a flat needle track which, when sandwiched between two sliding machine elements (particularly if of a heavy nature), reduces the sliding friction to a minimum by changing it into a rolling action. These needle tracks can be hooked together to produce raceways of any desired length.

These bearings must be kept clean, especially when operating in gritty or abrasive environments; foreign bodies rolling into tracks produce surface breakdown.

Figure 18.6 (*cont.*)

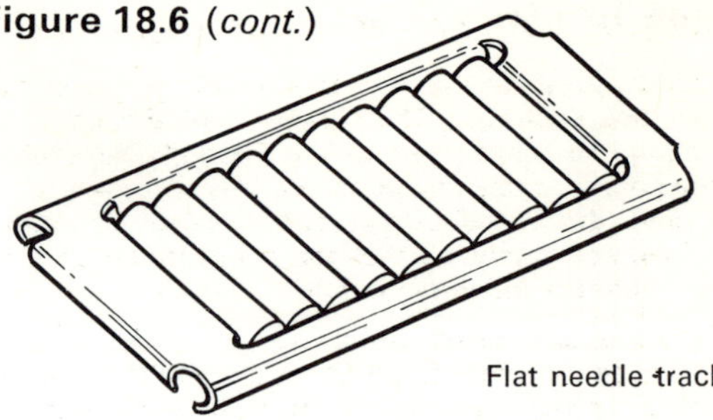

Flat needle track

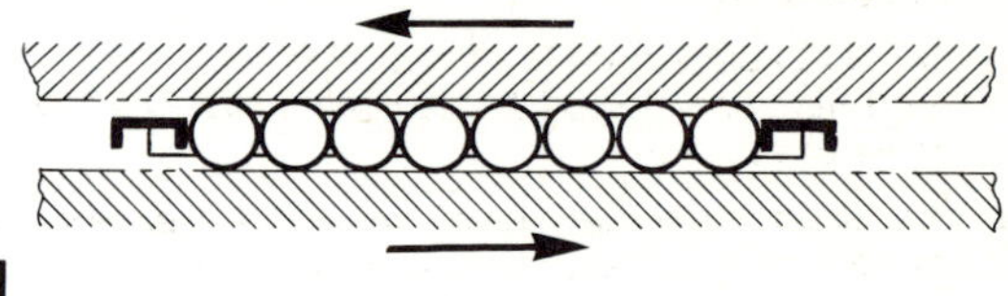

q

Section thro' flat needle track between sliding elements

Illustration **r** shows the patented Dexter slide rail bearing successfully used in tape-controlled machine tools and in atomic reactor plants where easy unlubricated sliding motion is required. These bearings have been developed to provide a really efficient mechanism for slide movements in machinery. Characteristics include:

1. The modular design and construction of this bearing system allows for any length of travel using common basic components.
2. The hardened and ground trackways require little skill to adjust.
3. In reasonable quantities, grinding dust, wood dust, wood chippings, swarf etc. do not impair movement.
4. The bearings have great load-carrying capacity, and machine tables mounted on them cannot lift.

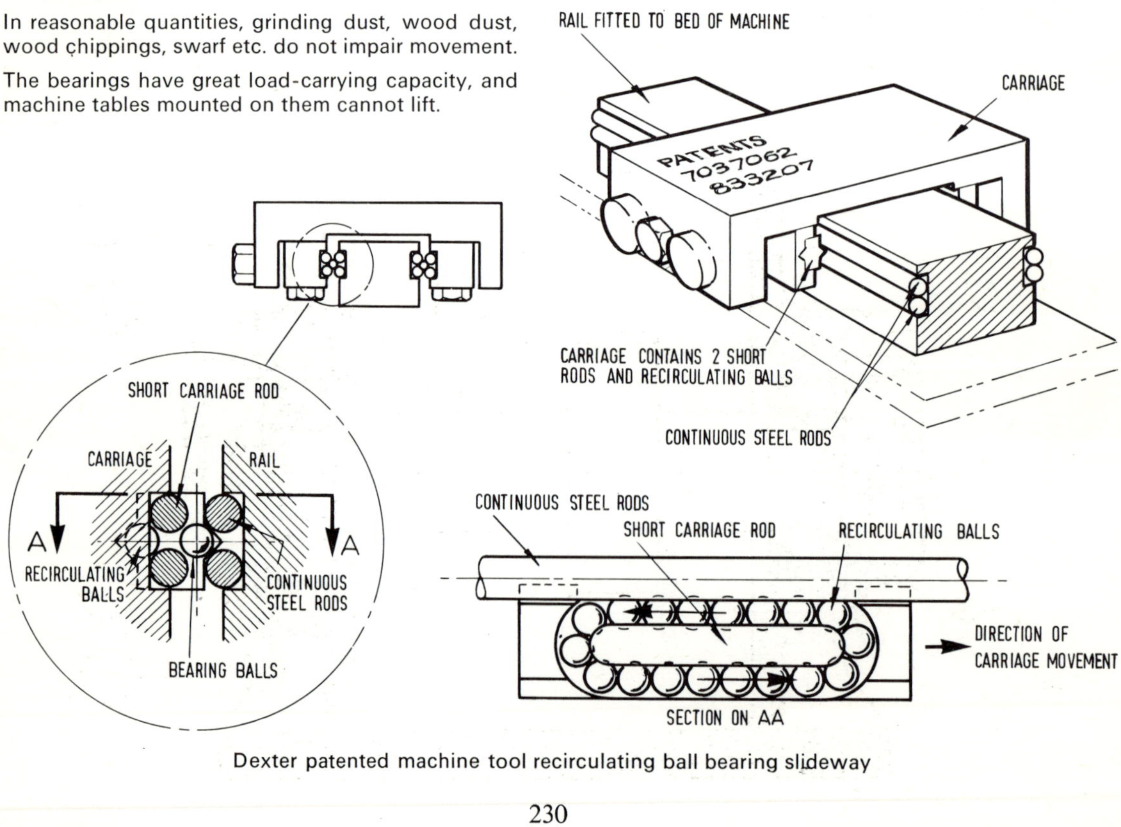

r

Dexter patented machine tool recirculating ball bearing slideway

FITTING OF JOURNAL BEARINGS (Figure 18.7)

The drawing shows a gear shaft mounted between two journal ball bearings. The diameters of the shaft for mounting the inner races of the bearings have been machined accurately and concentrically about the shaft axis, and the bores in the casing have also been machined accurately, parallel and concentric with each other. Note the following points:

1. Both the inner races, which are the revolving races, are made an interference fit on the shaft to prevent 'creep' or slow rotation of the races relative to the shaft. They are also clamped firmly endwise by a washer and nut fitted to the stud machined integral with the shaft.

   The abutment shoulder of the shaft in contact with the inner race should be at least twice the corner radius of the races.

2. The outer stationary race, which is carrying the smallest journal load, is made a nice push fit, and is clamped between the shoulder machined in the bore of the casing and the end cover. To take up endplay of the outer race, shims as necessary may be fitted. This bearing now locates the shaft axially within the casing and will take small amounts of end thrust.

3. The other outer stationary race is also made a nice push fit, but is not prevented from moving endwise within its housing. During the first revolutions after assembly, this race will move to a position allowing the inner and outer races to take up their correct positions relative to each other, thus eliminating sideways stresses introduced when fitting the bearing.

   If all the races are rigidly clamped in fixed positions, without any provision for allowing a stationary race to move endwise, enormous sideways pressures can be permanently applied to the balls, and this will result in early breakdown.
   Races should never be secured by fixing devices such as keys, grub screws or pins; distortion of the races by such methods will also cause early breakdown due to overloading.

Other points to note on drawing (*see BS 308, page 22, para. h*)

4. The shaft and the nuts and washers securing the bearings are *not sectioned* but shown as outside views, although the section plane passes longitudinally through them.

5. Ribs in the plane of the paper with the section plane passing through them are also not *sectioned.*

Because the various parts of the ball race are referred to in this description, a pictorial representation of the rings is shown, and not the recommended convention shown in BS 308, page 24.

Figure 18.7

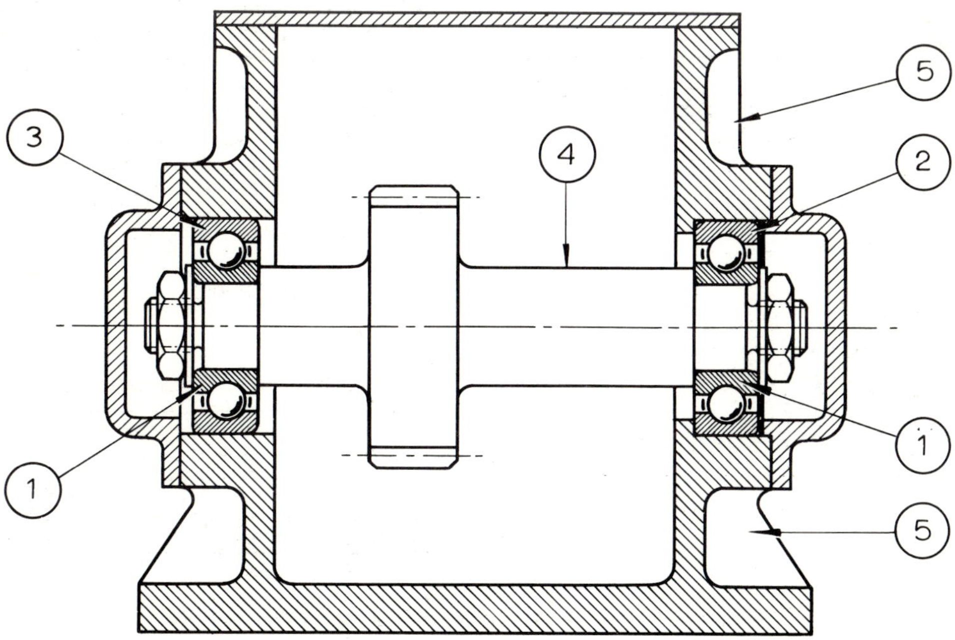

LUBRICATION

Oil and grease are the common forms of lubricant used in machines. Grease is produced in many grades for varying conditions and, because it is more viscous than oil, is used in situations where oil cannot be retained and would tend to run away. Various additives are compounded in greases to improve their qualities; for example, lithium compounds give added resistance to water and operate at high temperatures, and molybdenum disulphide covers metal parts with a tenacious film for a long period and resists the detergent effect of lubricating oil.

It is important to note that ball or roller bearings do not require a large amount of grease. A small amount of lubricant well distributed over the running surface will provide satisfactory lubrication for a long period. A large excess of grease in a bearing will cause overheating owing to the churning which takes place between bearing and grease. The correct grease content of a bearing is between a quarter and a half full.

Oil lubrication

There are two basic systems:

1. Splash lubrication.
2. Pressure lubrication.

In some pressure lubrication systems, a measure of splash lubrication is also employed to lubricate parts where it is difficult to produce oil pressure, and also to provide cooling.

Splash lubrication. This consists of a reservoir of oil into which moving parts dip on rotation and carry the oil round, thus lubricating the complete mechanism. It is usually sufficient for the lower gear teeth to dip into the reservoir, and the lowest ball or roller of a bearing. Excess of oil will cause churning and foaming, introducing oil drag and overheating. Oil splashed around the inside of the housing or casing is collected in troughs in the casing to direct it, by oilways, ducts and passages, to various parts requiring lubrication. It finally returns to the reservoir by gravity, through draining holes or ducts.

Pressure lubrication. This is often used where both lubrication and cooling are required, such as in internal-combustion engines. The system requires a pump to circulate the oil and create a pressure, a large quantity of oil, a reservoir for the oil, a cooling arrangement and a filter.

Lubrication of a motor car. (Fig. 18.8).

The modern car engine carries the bulk of the oil supply in a pan or sump below the engine. The sump is situated in the air stream beneath the car, thus collecting and cooling the oil before recirculation through the engine.

A pump draws the oil from the sump through a metal mesh filter. The oil then passes, under pressure of more than 350–550 kM/m^2 (50–80 lbf/in^2) through a replaceable felt or paper filter which removes from the oil the very fine abrasive particles such as metal particles released by wear and hard gritty particles formed by the heat of combustion. The oil then passes under pressure to a gallery pipe or duct to lubricate the main crankshaft bearings.

The big-end bearings of the connecting rod are lubricated by pressure-fed oil from the main bearings, through oilways drilled in the crankshaft. After passing through the big-end bearings, oil is flung out through bleed holes drilled in the upper side of the big-end, and from the sides of the big-end bearings, to form a cloud of oily spray which cools the cylinder walls and pistons and lubricates these hot, fast-moving sliding components. The oil then drops back into the sump for cooling and recirculation.

Pressure oil is led off the gallery pipe, or from a pressure chamber, to lubricate the bearings of the camshaft. Pressure oil is also led off through internal oilways or an external pipe to the hollow rocker shaft usually situated on top of the engine. Holes drilled radially along the rocker shaft allow lubrication of the bearings of the rocker arms. Oil exudes from these bearings and the movement produces oil splash and mist which lubricates the valve guides, springs and tappet pads. The oil from the valve gear drains back to the sump by way of the push-rod housings, lubricating the cams on the camshaft on its way.

The gearbox and the differential in the back axle are closed units with the gears and moving parts running in oil baths. The bearings of the water pump and fan are often sealed for life with lubricant, or are occasionally lubricated with grease applied with a grease gun.

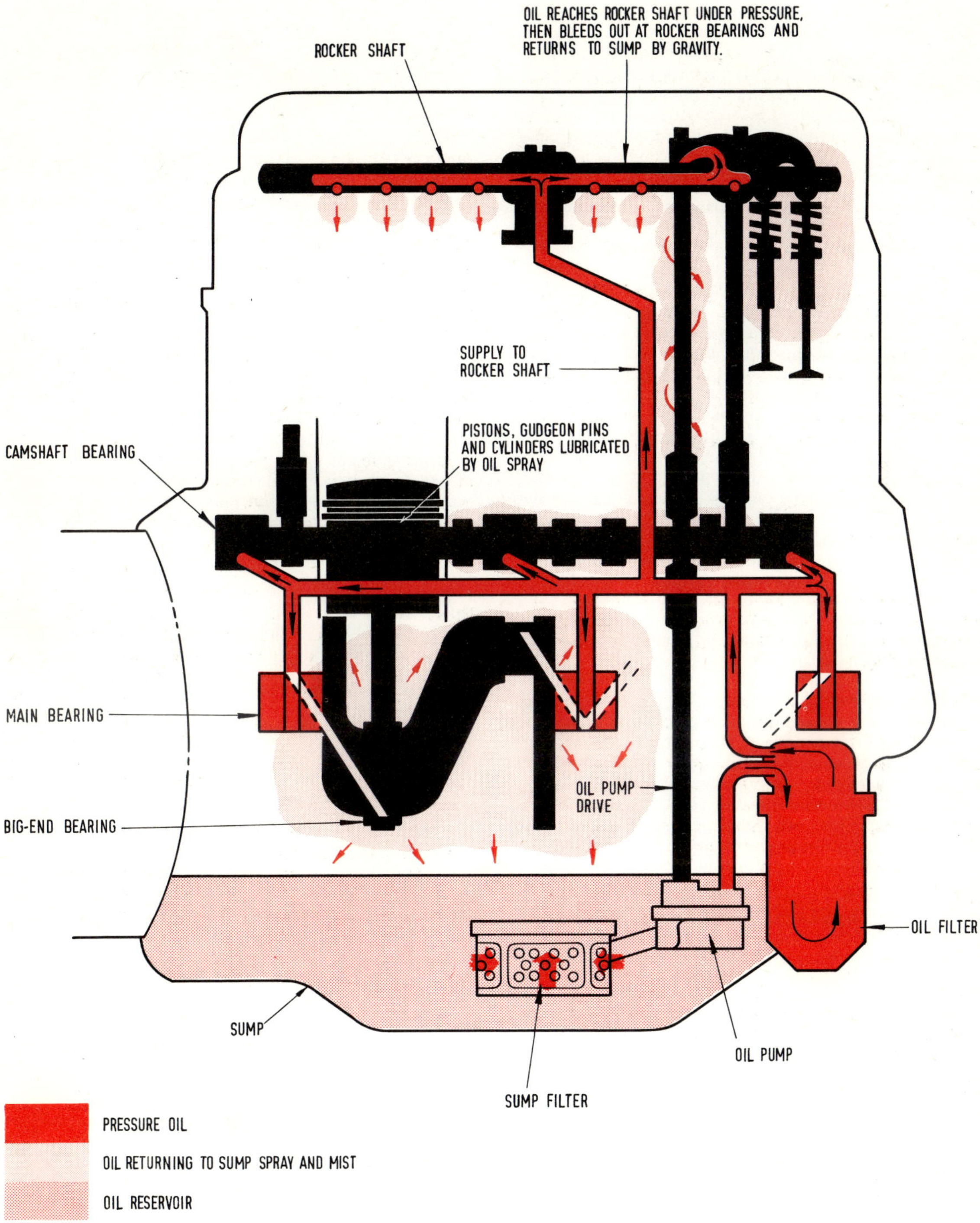

Figure 18.8 Lubrication of a motor car engine

Chapter 19

Insulating materials

Electrical insulation is material that does not conduct electricity and is used to hold electrical conductors in position, separating them from each other and from their surroundings. Insulation thus forms a barrier between energised parts of an electric circuit and confines the flow of an electric current to wires or other conducting paths as desired. Efficient insulation of electric circuits is essential for safe and successful operation of all electrical apparatus.

Various types of material are used as insulation, the selection depending on the specific requirements for each application. The copper conductors used in the electrical wiring of homes are insulated from each other and from the building by rubber or plastic. These materials are low in cost, have a long life, and are sufficiently flexible to withstand necessary bending. Extension cords of heating appliances are usually insulated with asbestos—a material capable of withstanding high temperature without deterioration. Overhead power lines are supported on porcelain insulators which are unaffected by exposure to weather. Paper is used to insulate small wires of telephone cables and high-voltage power cables. In the latter example, the paper is impregnated with insulating oil to improve insulation. Large electric generators and motors which operate at high voltages and temperature are insulated with mica. In some applications, solid insulation is used in conjunction with liquid or gaseous insulation; for example, in high-voltage power transformers solid insulation provides mechanical rigidity while insulating oil increases insulation strength and also serves as a coolant for the structure.

Insulating materials are subject to many deteriorating influences, the most important being moisture absorption and chemical change. Porous materials, such as cotton and paper, must be effectively sealed against moisture, usually by impregnation with insulating material or by immersion in oil. Silicone varnishes are used extensively because they form an effective barrier against moisture and can withstand fairly high temperatures.

Insulating materials deteriorate under influences which change their chemical structure. Chemical changes which progress very slowly at room temperatures are, in some cases, greatly accelerated at higher temperatures.

Most insulation will noticeably deteriorate if operated at high temperatures, and for this reason many types of electrical apparatus have specified maximum operating temperatures. Because temperature is so important a factor in the life of insulation materials, a classification has been drawn up by the International Electrotechnical Commission to established temperature limits for equipment containing various types of insulation. These classifications are now incorporated in BS 2757 : 1966.

Class	*Highest temperature of operation* (°C)
Y (formerly O)	90
A	105
E	120
B	130
F	155
H	180
C	no limit selected

Class Y: cotton, silk, paper and similar organic materials when *neither* impregnated *nor* immersed in oil.

Class A: cotton, silk, paper and similar organic materials when impregnated or immersed in oil; also the enamel of enamelled wire.

Class E: insulation consisting of materials or combinations of materials which by experience or tests can be shown to operate at temperatures up to 120°C.

Class B: mica and asbestos and similar inorganic materials in built-up form combined with binding cement.

Class F: insulating materials or combinations of mica, glass fibre, asbestos etc. with suitable cements or impregnants; also materials, not necessarily inorganic which by experience or test can be shown to operate up to 155°C.

Class H: mica, asbestos, fibre glass and similar inorganic materials in built-up forms, using silicone cements compounds or materials with equivalent properties; also silicone compounds in rubbery or resinous form.

Class C: mica, glass, porcelain, quartz and similar materials.

Paper

As an insulator in impregnated form, paper is cheap and as yet still supreme where reliable service at the highest working stresses is required. Its chief applications are in high voltage—power cable, power transformers and capacitors.

Hydrocarbon insulating oil

This is derived from natural petroleum and is used as an impregnant. For a.c. power duties at high electric stresses, oil is used in power transformers, cables, capacitors and circuit breakers, where it acts not only as a dielectric saturant but also as a heat-transfer medium. Waxes and resins derived from oil are used in d.c. capacitors, solid cables and joint sealing.

Rubber (natural and synthetic)

In many applications, the usefulness of a dielectric depends on a combination of dielectric and mechanical properties and chemical resistance. The use of rubber in such applications is extensive. The most valuable properties of rubber insulations are, in addition to dielectric properties, considerable toughness, ability to sustain large deformations without rupture or permanent distortion, great flexibility over a wide temperature range, and outstanding resistance to abrasion and wear. Synthetic rubbers have some additional properties, in particular high resistance to oil. Rubber is used for all types of cable and insulated protective mouldings.

Glass

This is an excellent dielectric material with widespread applications in electrical engineering. It is used for envelopes for electric light bulbs, radio valves, mercury awitches, X-ray tubes, cathode tubes etc. Glass is slso used to form seals in electrical components, particularly with metals.

Mica

This is a natural occurring mineral. It is expensive because of handling, selection and grading costs, but even in this synthetic age it is still holding its position as an important dielectric for high temperatures. It can be divided into thin, mechanically strong films, but lacks flexibility and cannot be moulded. It is used in electric generators for insulation of commutator segments, and in electric irons, strip heaters for ovens etc.

Plastics

These are a modern and important group of dielectric materials widely used in the electrical industries. They combine excellent mechanical strength, thermal resistance, and resistance to chemical attack. Some of the widely available forms are Bakelite, Tufnol, PVC, polystyrene, Perspex, polypropylene, PTFE, and nylon. They can be obtained in extruded and sheet form and can be moulded to suit specific requirements.

Epoxy resins

These are a combination of resin plus an accelerator or hardener; the most commonly known is Araldite. They have excellent adhesion to metals, glass, ceramics and plastics, good mechanical and insulating properties, and resistance to heat, water and chemicals. The epoxy resins are suitable for casting and are used in encapsulation (potting) of electrical components.

Silicones

These are organic compounds of silicon which, when compounded with other materials, have superior physical properties at high temperatures. They are resistant to embrittlement at low temperatures and to oxidation at high temperatures. They are water-repellant and chemically inert. They are added to rubbers, resins, waxes, oils, greases, impregnating varnishes and glass laminates.

Ceramic

This is a material formed into its final shape by processes which use refined, powdered, naturally occurring minerals as raw material. The shape is then formed into a monolithic body by high-temperature heat treatment. Ceramics are usually glazed with a thin layer of glass to give a highly polished, hard surface. The best known ceramic is porcelain, but others based on barium titanate, magnesium silicates, aluminium silicates and calcium silicates are in wide use. Ceramics are impervious to moisture, are resistant to atmospheric attack, and have a clean surface for cleaning and preventing tracking. Common uses of ceramics are in high-voltage power transformers and switch-gear bushings, spark-plug insulation, and high-capacitance components for low-voltage transistor circuits.

Machine drawing

Chapter 20

In engineering, drawings are the means of communication between designers and all the departments which are involved in manufacturing, marketing, installing and maintaining the product. The drawings provided to make a complex machine or structure often run into many thousands, so it can be appreciated that drawings cannot be made in isolation on just any pieces of paper but have to be systematised, using a numbering code for cross-reference and preparing the drawings on standard-sized sheets for economic photocopying and storage.

Many large engineering organisations have individual numbering systems for referencing drawings, the systems being specially suited to the products manufactured. Some reference systems have evolved over many years, while others have been designed and marketed by companies specialising in large-scale indexing and referencing.

However, in the making of drawings for any engineering project, all design offices follow a common pattern. A complete set of drawings for any project has a master drawing called a *General Arrangement* which is broken down by stages into more informative drawings, the final stage of the breakdown being drawings which describe each individual part in detail. This pattern, which is like a family tree, is shown in the following diagram.

Figure 20.1

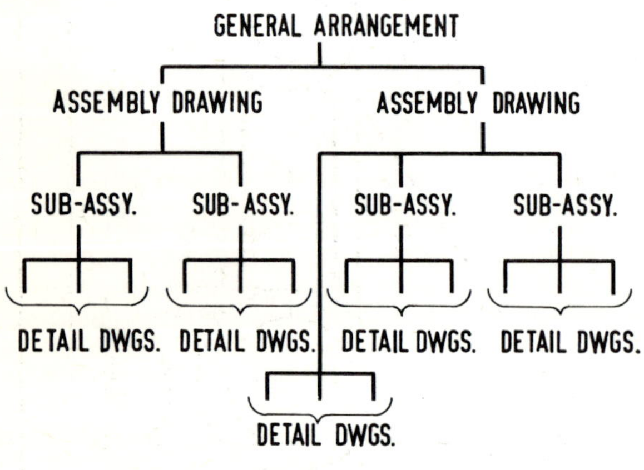

Diagram showing breakdown of a set of engineering drawings required for the manufacture of a product

DETAIL DRAWINGS

These are small drawings containing all the information necessary for manufacturing a single part. They enable the attention of a craftsman to be concentrated on one individual part during manufacture, so that a large workforce can be employed on the many component parts that go to make a complex engineering product. It is the responsibility of the drawing office to make certain that instructions given on each detail drawing are correct, so that when parts are assembled they fit together precisely. Hence, in drawing office work a great deal of time and effort is spent cross-referencing and checking to make certain that the drawings issued for manufacture are accurate; to manufacture large quantities of parts that did not fit together would quickly spell ruin to any industrial organisation.

SUB-ASSEMBLIES

These are drawings which show an assembly of several of the small parts for which the detail drawings were prepared; they show how these small parts are to be assembled to make a larger part of the overall design. These sub-assemblies become the units to be used in the next step of assembling the whole design; they cannot usually be used by themselves but are still building components for the complete machine. For example, the hub of a bicycle wheel could be drawn as a sub-assembly drawing. It would show a hub shell, spindle, cones and ball bearings assembled together correctly. A hub cannot be used by itself, but for manufacturing purposes it makes a compact unit for specialist production.

ASSEMBLY DRAWINGS

These drawings are similar to sub-assembly drawings, but take the construction process a stage further. Many small detail parts brought together to form a large unit could be called an assembly drawing. More usually, a number of sub-assemblies with a few separate details and the necessary fasteners are brought together to produce a large construction unit.

GENERAL ARRANGEMENT

This drawing is the master drawing which shows the complete machine. Sometimes, if the design is large, it is hardly more than an outline picture, but it serves to bring the various assembly drawings together, and call up the fasteners necessary at the final assembly stage.

The following exercises are examples of detail drawing and simple assembly. Draw them to recommendations and conventions shown in BS 308 as though they were required for actual manufacture, giving special attention to line quality, dimensioning and printing.

Exercise 20.1

Draw the two views of the swivel bracket in third-angle projection as shown, using a scale of 1 : 1. Using these views, produce an isometric drawing viewed in the direction of arrow A. Do not show hidden detail. Do not use an isometric scale.

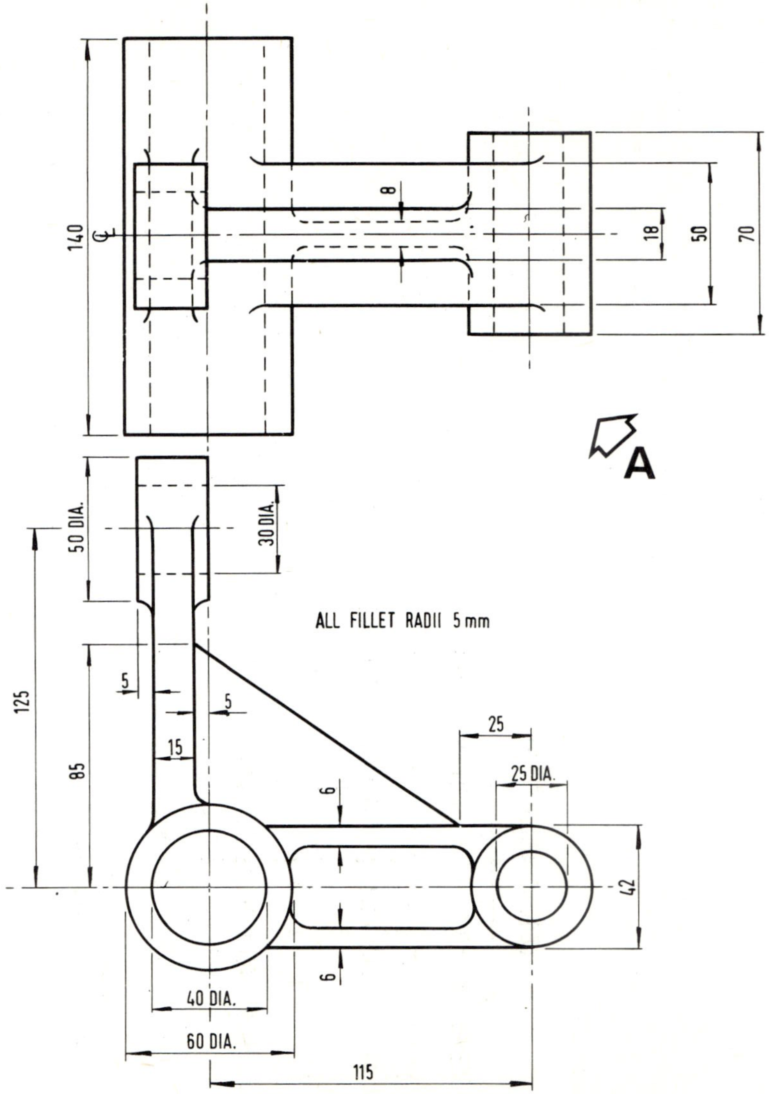

Exercise 20.2

The drawing shows two views of a swivel bracket for use on a 4-wheeled bogie suspension. It was cast in meehanite and finished by machining the bores and faces marked M/C.

1. Copy view B and add a revolved section on the horizontal plane.
2. Copy view C without hidden detail.
3. Draw a section to the right of view C, looking on AA.
4. Draw a view as seen in the direction of arrow D, and show bores as hidden detail.
5. Dimension bore diameters and bore centres only.

The quality of lines, dimensions and printing should conform to BS 308.
Denote machined faces with machining symbols.
Add a title block showing title, scale, material and date.

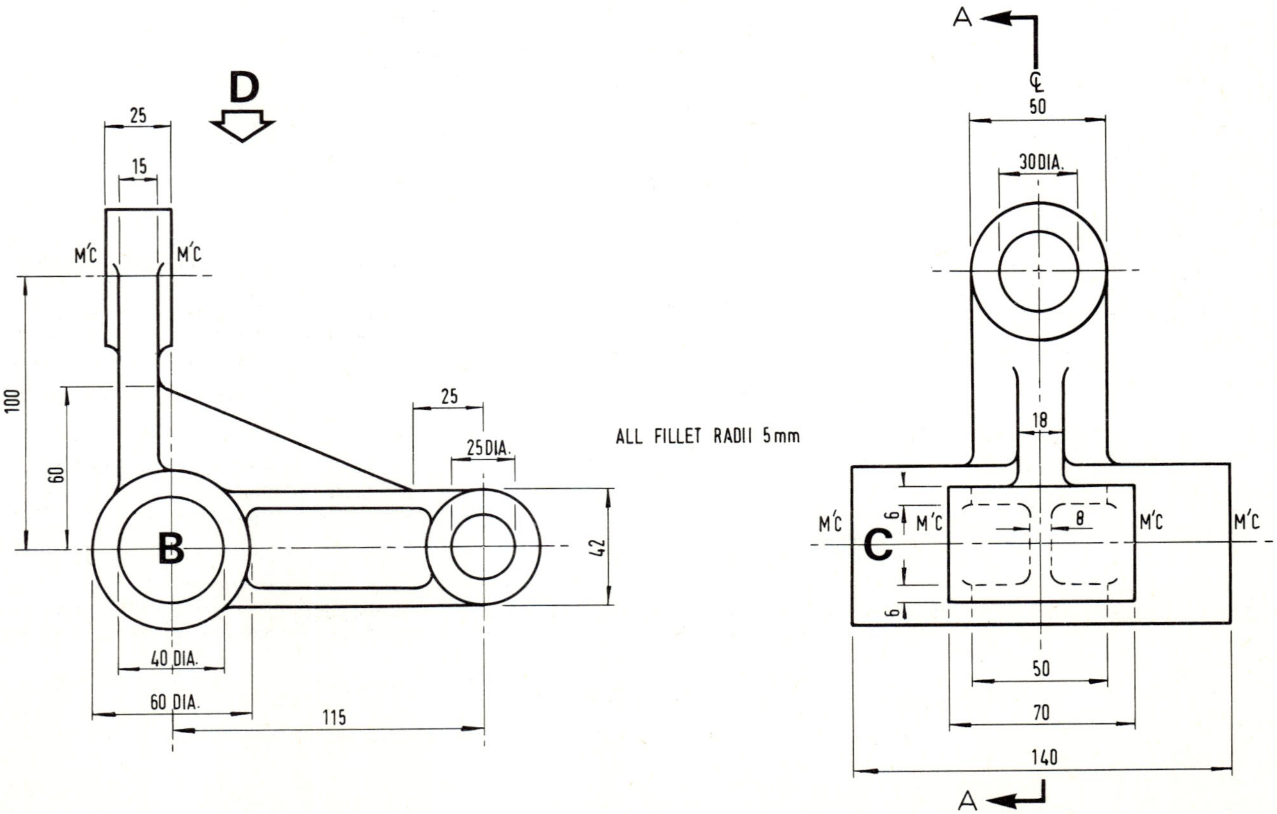

Exercise 20.3

The layout, which is not to scale, shows left-hand support bracket for a hydraulic motor. The bracket is pressure die casting in aluminium LM 24.

Using a scale of 1 : 1 and leaving adequate space between views for dimensions, draw the following views in third-angle projection:

1. A view in the direction of arrow A.
2. A sectional view as indicated by section line BB to conform with recommendations in BS 308, page 22, clause 8, para. h.
3. A view in the direction of arrow C.
4. Add a removed sectional view.

Fully dimension for manufacture, showing toleranced dimensions for machined features and bores. Add machining symbols to surfaces marked M/C. Add a title, scale and date, and state material and method of manufacture, in a prepared title block.

Above the title block add a note as follows:

NOTE: REMOVE ALL BURRS AND SHARP EDGES.

Quality of line, dimensions and printing should conform to BS 308. All dimensions are in millimetres.

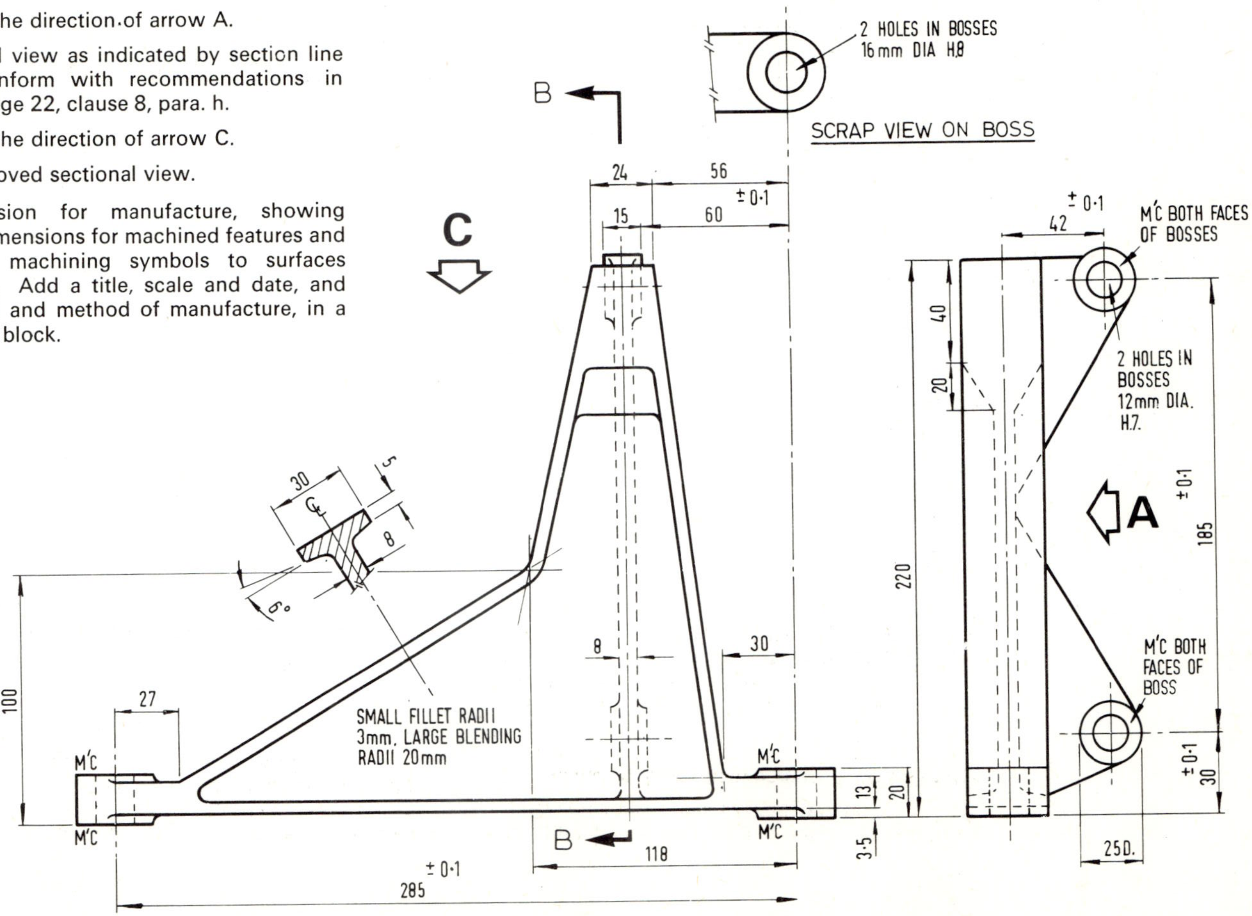

Exercise 20.4

The drawing shows the layout of a bicycle pedal to be injection moulded in acetal resin for 'snap fit' assembly to the pedal spindle.

1. Using a scale of 2 : 1, complete the half-sectioned view marked C.
2. Draw a view in the direction of arrow B.
3. Draw a sectional view on AA projected from view C showing the pedal spindle in chain-dot outline.

Fully dimension for manufacture, showing toleranced dimensions for bearing bores. Add title, scale and date, and state material for manufacture.
Quality of line, dimensions and printing to conform to BS 308. Third-angle projection to be used. All dimensions are in millimetres.

DESIGN INFORMATION FOR A PLASTIC BICYCLE PEDAL

Material–acetal resin–injection moulded

Bearings should have clearance of not less than 0·10 mm, and not greater than 0·20 mm on diameter.

Internal radii 1·5 mm

General wall thickness 3 mm

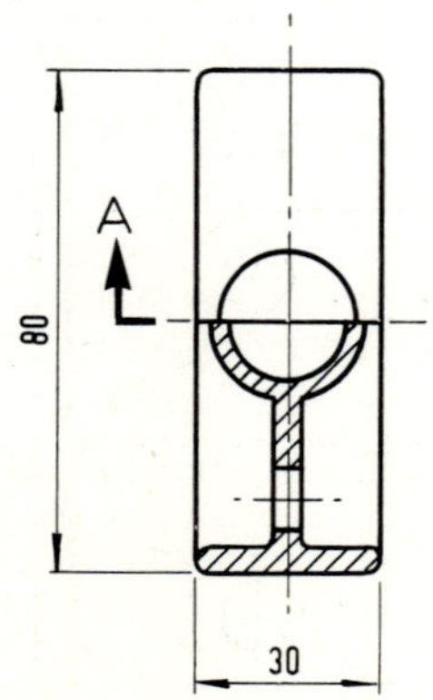

SECTION X

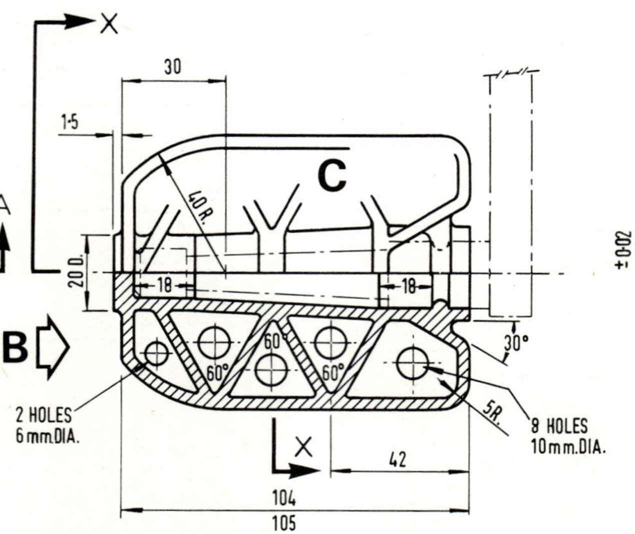

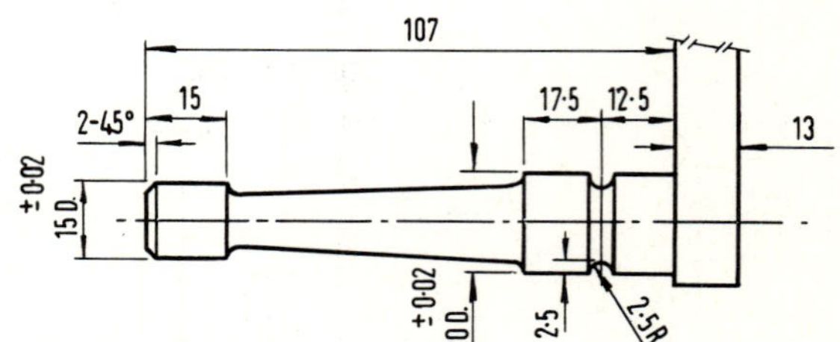

DIMENSIONS OF PEDAL SPINDLE

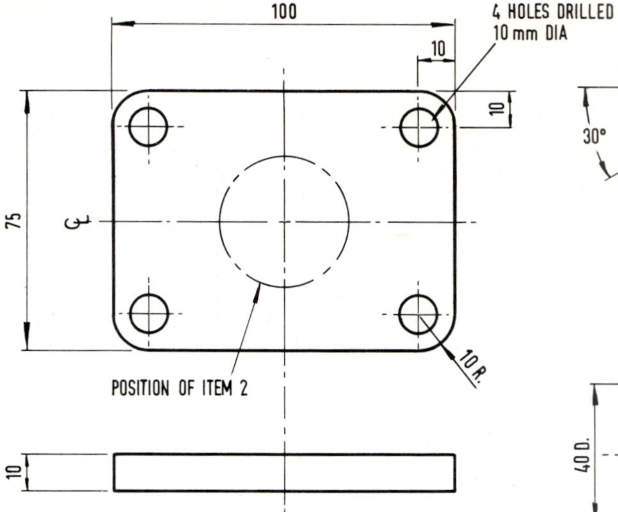

BASE PLATE (ITEM 1)

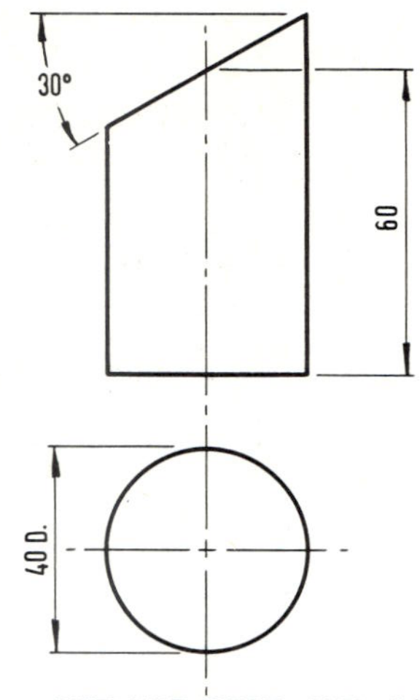

DISTANCE PIECE (ITEM 2)

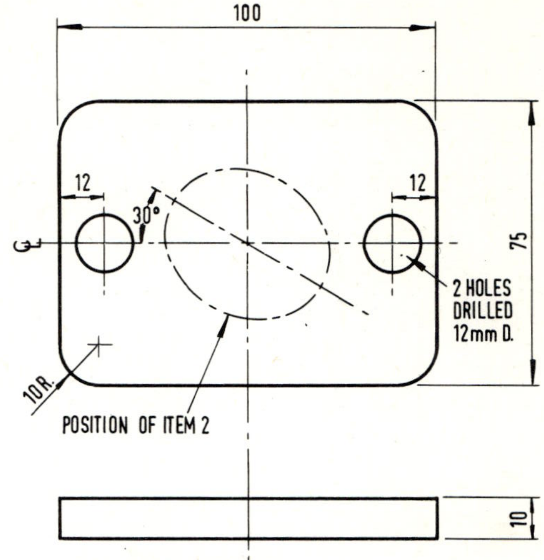

TOP PLATE (ITEM 3)

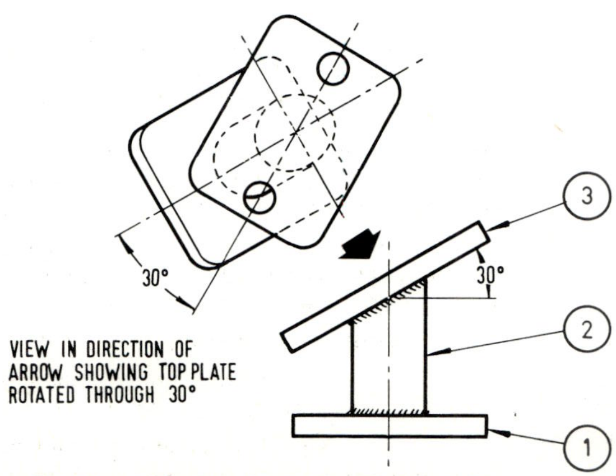

WELDED ASSEMBLY OF DISTANCE BRACKET

Exercise 20.5

A welded steel distance bracket is made from the three detailed items shown. The top plate is rotated through 30° on the sloping face of the distance piece before welding, as shown in the welded assembly.

Draw three principal views of the welded bracket (i.e. two elevations and a plan) and any auxiliary views necessary to assist projection.

Fully dimension the welded bracket for production.

Note: Dimensions may only be applied to views where true lengths and shapes are shown.

Draw in third-angle projection using a scale of 1 : 1. All dimensions are in millimetres.

Exercise 20.6

The sketches show the design for a finned, spigot-faced, flanged outlet for hot exhaust gases of a diesel engine. It is to be cast in gunmetal, and the bore and the mounting faces are to be machined.
Draw the following views in third-angle projection, using a scale of 1 : 1.

1. A view in the direction of arrow B.
2. A half-sectional view taken on AA.
3. A scrap view to show the true shape of a fin.

Dimension the following features:

Outside flange diameter and overall height.
Machined mounting face and bosses.
Inside of bore and thread details.
Add a suitable title block showing:

Title
Drawn by
Scale
Date
Material
Drawing No.

Line quality, dimensioning and printing to conform to Recommendations laid down in BS 308. Machining symbols to be added to all machined surfaces. All dimensions are in millimetres.

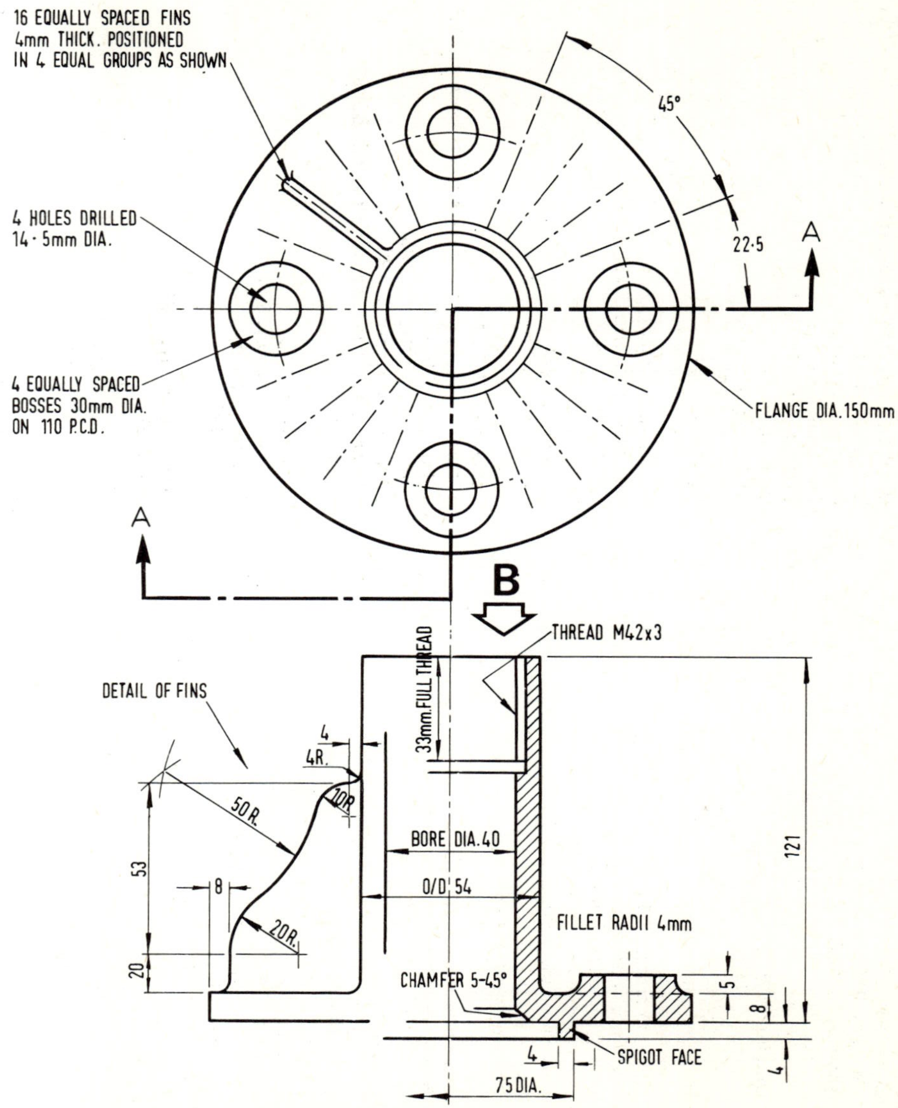

Index

Ajax system, 201
Aluminium, 212
  Alloys, 213–216
  Alclad, 215
  Casting, 213
  Duralumin, 215
  Extrusion, 213
  Forging, 213
  Heat-treatment, 213
  Rolling, 213
Aluminium bronze, 211
Angles (accurate drawing), 35
Archimedian spiral, 46
Assembly drawing, 169, 171
  Balloon referencing, 169, 171
  Parts list, 169, 171
  Title block, 169, 171

Babbitt bearing alloys, 218
Bakelite, 184
Bauxite, 212
Bearings, 222
  Angular contact, 227
  Ball and roller, 222
  Duplex, 228
  Fitting, 231
  Journal, 225
  Linear motion, 223, 229, 230
  Needle, 225
  Plain, 222, 223
  Roller, 225
  Taper, 227
  Thrust, 228
Bessemer, 200
Bisecting an angle, 34
Bisecting a line, 32
Blast furnace, 198
Blow moulding (plastic), 194
Bolts, 92–98
Brass, 207
Bronze, 210

Cadmium, 219
Casein (plastic), 184
Calendering (plastic), 196
Cast iron, 198, 199
Cast steel, 201
Celluloid, 184
Ceramic, 7
Circular arcs (joining), 54, 55
Circle (finding centre), 35
Compasses, 4
Conversion tables
  Inches to millimetres, 28
  Fractions to millimetres, 30
Copper, 206, 207
Cycloid, 41

Development (sheet metal), 108, 126–143
Die casting, 220
Dimensioning, 12, 16, 17
Dipping (plastic), 196

Dividing a line into equal parts, 33, 34
Drawing boards, 4
Drawing instruments, 7
Drills, 26, 156, 157
Drill bushes, 160, 161
Drill holes, 26
Drill sizes, 31
Ductility, 179
Duralumin, 215

Electric-arc furnace, 201
Electrical insulation, 234
Ellipse, 41, 42, 43
Epicycloid, 48
Epoxy resin, 235
Extrusion (plastic), 194

Fasteners, 92–107
Fasteners (draughtsman's method of drawing), 92
Fatigue testing, 183
Film blowing (plastic), 195
Film casting (plastic), 195
Fits and clearance (holes and shafts), 80–87
French curve, 7

Galvanising, 220
Geometry (plane), 32
Geometry (solid), 108
Glass, 235

Hardness testing, 181
Helix, 93
Hexagon, 44
Holes and shafts, 80
Hypocycloid, 48

Impact testing, 182
Imperial units, 24
Injection moulding, 193
Insulating materials, 234, 235
Interchangeability, 80, 156
Interpenetration, 108
  Circular bend and straight outlet, 120
  Cone and tube, 112
  Hemisphere and square prism, 118
  Intersecting pipes at 90°, 109
  Intersecting pipes at 60°, 109
  Milled flat on round shank, 116
  Round pipe with square outlet, 114
  Tetrahedron and cylinder, 122
  Triangular prism and cylinder, 110
Iron and steel production, 198
Isometric drawing, 144–148
Involute, 46

Jigs and fixtures, 156
  Basic design features, 168
  Button locators, 162
  Clamping devices, 165–168
  Freedoms of movement, 162
  Locators, 162
  Materials, 169

Jigs and fixtures—*cont.*
Peg locators, 162
Post locators, 164
Vee locators, 164
Jigs and fixtures, 169
Channel type drill jig, 173
Plate type drill jig, 171
Pot type drill jig, 177
String milling fixture, 175

Kaldo process, 201
Keys and keyways:
Feather, 104
Rectangular, 104
Round, 104
Square, 104
Woodruff, 104

Laminating (plastic), 197
LD process, 200
LD–AC process, 201
Lead, 219
Lines used in engineering drawing, 10
Load/extension diagram, 181
Locking devices (screw threads), 102, 103
Low melting point fusible alloys, 218
Lubrication, 222, 232

Magnesium, 221
Malleability, 179
Materials
Mechanical properties, 179
Physical properties, 179
Selection for design, 178
Testing, 180–183
Meehanite, 199
Metric units, 24
Mica, 235
Micrometer, 80
Milling:
Cutters, 158–160
Gang, 158
Straddle, 158
String, 158

Newall, J. W., 80
Nuts, 92–98
Nuts
lock, 102
self-locking, 103
slotted, 102
Non-destructive testing
X-ray, 183
Magnetic dust, 183

Oblique projection, 149, 150
Open-hearth process, 200
Orthographic projection
1st angle, 8, 62
3rd angle, 9, 64

Paper, 235
Parabola, 49, 50
Parallel lines, 33
Pencils, 4
Perpendicular lines, erecting, 32
Pig iron, 198
Pictorial drawing, 144–152
Plastics
Catalysts, 185
Celluloid, 184
Fillers, 185
Insulation, 235
Parkesine, 184
Plasticisers, 185
Plastic forming and moulding techniques
Blow moulding, 194
Calendering, 196
Casting, 197
Compression moulding, 196
Dipping, 196
Extrusion, 194
Filament winding, 197
Film blowing, 195
Film casting, 195
High and low pressure laminating, 197
Injection moulding, 193
Matched metal moulding, 197
Rotational casting, 196
Slush moulding, 196
Thermoforming, 195
Transfer moulding, 197
Plastic moulding compounds thermoplastic
ABS, 186
Acetal resin, 186
Acrylics, 186
Cellulosics, 187
Fluorocarbons, 188
Nylon, 187
Polypropylene, 188
Polystyrene, 190
Polythene, 188
P.V.C., 189
Plastic moulding compounds thermosetting
Epoxy resins, 191
Melamine, 190
Phenolics, 190
Polyester resins, 192
Polyurethane, 191
Silicones, 192
Urea, 191
Printing, 10
Projection, 5, 8, 9, 26, 66–77
Pythagoras, 38, 40

Rabatment, 124
Reamers, 157
Rivets, 92, 105–107
Rivets, blind, 106
Riveted joints, 107
Rockwell hardness, 180–182
Rotational casting, 196
Rotor process, 201
Rubber, 235
Ruler, 5

Index

Screws, Threads, 26, 29
Acme, 101
American, 94
BA, 94–96
BSF, 94–96
BSW, 94–96
Buttress, 101
Definition, 93
Drawing, 92
ISO—inch (unified), 94, 95, 97
ISO—metric, 94, 95, 97
Lead, 93
Locking devices, 102
Multiple, 93
Pitch, 93
RH and LH thread, 93
Socket head, 100
Square, 101
Sectional views, 12–15
Selection of fits (shafts and holes), 83
Set squares, 5
Sheet metal development, 108, 126–143
Cone obliquely truncated, 134–137
Hopper, 142
Oblique cone, 138
Round pipe, cut at angle, 128
Round pipe, 'T' junction, 130
Round pipe, offset 'T' junction, 132
Square to round transformation piece, 140
Silicones, 235
Sketching, 151
Solder, 218
Steel
Ajax system, 201
Bessemer, 200
Castings, 201
Electric arc furnace, 201
Kaldo, 201
LD process, 200
LD–AC process, 201
Open hearth, 200
Plate and sheet, 203
Rotor process, 201
Sinter, 198
Silver steel, 201
Slag, 198
Thomas process, 200
Stress
UTS, 178–180
Yield, 178–180
0.1% proof, 178–180

Table of fits (shafts and holes), 83
Tangents, 36
Tangency, 54, 55
Tensile testing
Avery machine, 180
Hounsfield machine, 180
Thermoforming, 185
Thermosetting, 185
Thomas process, 200
Tin, 218
Tinplate, 218
Tolerancing, 26, 82–87
Toughness of material, 179
Transfer moulding, 197
Triangles, 37, 38, 39, 40
True lengths of lines, 124

Vickers diamond point, 180, 181

Washers
Plain, 98, 102
Shakeproof, 102
Spring, 102
Tab, 102
Whitworth, Sir Joseph, 80, 94
Wrought steel, 20

Young's modulus of elasticity, 180

Zinc
Coatings, 200
Die-casting, 200
Galvanising, 200
Plating, 200